Barron's Regents Exams and Answers

Integrated Algebra

LAWRENCE S. LEFF
Former Assistant Principal
Mathematics Supervisor
Franklin D. Roosevelt High School, Brooklyn, NY

Barron's Educational Series, Inc.

All inquiries should be addressed to:
Barron's Educational Series, Inc.
250 Wireless Boulevard
Hauppauge, NY 11788
www.barronseduc.com

ISBN 13: 978-0-7641-3870-6
ISBN 10: 0-7641-3870-7
ISSN: 1940-6134

PRINTED IN THE UNITED STATES OF AMERICA
9 8 7 6 5 4 3 2

Contents

Preface

This book is designed to prepare you for the Integrated Algebra Regents examination while strengthening your understanding and mastery of the topics on which this test is based. In addition to providing questions from previous mathematics Regents examinations, this book offers these special features:

- ***Step-by-Step Explanations of the Solutions to All Regents Questions***. Careful study of the solutions and explanations will improve your mastery of the subject. Each explanation is designed to show you how to apply the facts and concepts you have learned in class. Since the explanation for each solution has been written with emphasis on the reasoning behind each step, its value goes well beyond the application to a particular question.

- ***Unique System of Self-Analysis Charts***. Each set of solutions for a particular Integrated Algebra Regents exam ends with a Self-Analysis Chart. These charts will help you to identify weaknesses and direct your study efforts where needed. In addition, the charts classify the questions on each exam into an organized set of topic groups. This feature will enable you to locate other questions on the same topic in other Integrated Algebra Regents exams.

- ***General Test-Taking Tips***. Tips are given that will help to raise your grade on the actual Integrated Algebra Regents exam that you will take.

- ***Special Mathematics Problem-Solving Strategies***. Problem-solving strategies that you can try if you get stuck on a problem are explained and illustrated.

- ***Integrated Algebra Refresher***. A brief review of key mathematics facts and skills that are tested on the Integrated Algebra Regents exam is included for easy reference and quick study.

- ***Glossary***. Definitions of important terms related to the Integrated Algebra Regents exam are conveniently organized in a glossary.

Frequency of Topics—Integrated Algebra

Questions on the Integrated Algebra Regents exams fall into one of 28 topic categories. The "Some Key Integrated Algebra Facts and Skills" section that appears later in this book covers most of these concepts and skills. The **Frequency Chart** that follows shows which topics have been emphasized in recent years by listing the number of questions that have been asked on recent exams in each topic category.

To help guide you further in your Regents exam preparation, a **Self-Analysis Chart** is included after each Regents exam. This chart identifies exactly which questions on an exam fall within each topic category. When you score a completed Regents exam, this chart will help you identify any potential areas of weakness and then allow you to easily find additional practice questions in those same topics areas on other Regents exams.

Frequency of Topics—Integrated Algebra

	Number of Questions					
	Sample Test	Official Test Sampler	June 2008	August 2008	June 2009	August 2009
1. Sets and Numbers; Intersection and Complements of Sets; Interval Notation; Properties of Real Numbers	4	4	3	2	3	1
2. Operations on Rat'l. Numbers & Monomials	1	2	—	—	1	—
3. Laws of Exponents for Integer Exponents; Scientific Notation	2	1	1	1	1	—
4. Operations on Polynomials	1	1	2	1	2	2
5. Square Root; Operations with Radicals	1	1	1	1	1	1
6. Evaluating Formulas & Algebraic Expressions	—	—	1	—	1	1
7. Solving Linear Eqs. & Inequalities	2	4	2	4	3	3
8. Solving Literal Eqs. & Formulas for a Given Letter	1	—	—	1	1	—
9. Alg. Operations (including factoring)	1	1	2	3	2	4
10. Quadratic Equations (incl. alg. and graphical solutions; parabolas)	1	2	4	2	3	3
11. Coordinate Geometry (eq. of a line; graphs of linear eqs.; slope)	4	2	2	3	1	3
12. Systems of Linear Eqs. & Inequalities (algebraic & graphical solutions)	1	1	1	2	2	2
13. Mathematical Modeling (using: eqs.; tables; graphs)	3	3	1	2	2	—
14. Linear-Quadratic Systems	—	1	1	2	1	—
15. Perimeter, Circumference; Area of Common Figures	1	1	—	—	1	1
16. Volume and Surface Area; Area of Overlapping Figures; Relative Error in Measurement	1	2	4	2	1	2
17. Fractions and Percent	—	—	1	—	—	—

	Number of Questions					
	Sample Test	Official Test Sampler	June 2008	August 2008	June 2009	August 2009
18. Ratio & Proportion (incl. similar polygons, scale drawings, & rates)	2	—	2	2	2	—
19. Pythagorean Theorem	1	1	1	1	1	1
20. Right Triangle Trigonometry	1	1	1	2	1	1
21. Functions (def.; domain and range; vertical line test; absolute value)	2	2	1	—	1	2
22. Exponential Functions (properties; growth and decay)	2	1	1	1	1	1
23. Probability (incl. tree diagrams & sample spaces)	—	2	1	2	2	2
24. Permutations and Counting Methods (incl. Venn diagrams)	2	—	1	1	1	1
25. Statistics (mean, median, percentiles, quartiles; freq. dist., histograms; box-and-whisker plots; causality, bivariate data; qualitative vs. quantitative data; unbiased vs. biased samples; circle graphs)	3	4	4	3	3	3
26. Line of Best Fit (including linear regression, scatter plots, and linear correlation)	2	1	1	1	1	1
27. Non-routine Word Problems Requiring Arith. or Alg. Reasoning	—	1	—	—	—	4

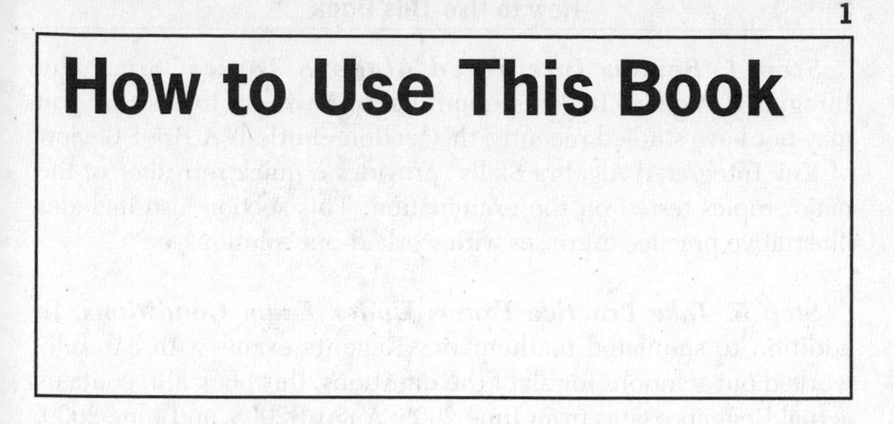

How to Use This Book

This section explains how you can make the best use of your study time.

As you work your way through this book, you will be following a carefully designed five-step study plan that will improve your understanding of the topics that the Integrated Algebra Regents examination tests while raising your exam grade.

Step 1. Know What to Expect on Test Day. Before the day of the test, you should be thoroughly familiar with the format, the scoring, and the special directions for the Integrated Algebra Regents exam. This knowledge will help you build confidence and prevent errors that may arise from misunderstanding the directions. The next section in this book, "Getting Acquainted with Integrated Algebra," provides this important information.

Step 2. Become Testwise. The section titled "Ten Test-Taking Tips" will alert you to easy things that you can do to become better prepared and to be more confident when you take the test.

Step 3. Learn Problem-Solving Strategies. The strategies explained in the section entitled "Special Problem-Solving Strategies" may help you solve problems that at first glance seem too difficult or complicated.

Step 4. Review Integrated Algebra Topics. Since the Integrated Algebra Regents exam will test you on topics that you may not have studied recently, the section entitled "A Brief Review of Key Integrated Algebra Skills" provides a quick refresher of the major topics tested on the examination. This section also includes illustrative practice exercises with worked-out solutions.

Step 5. Take Practice Exams Under Exam Conditions. In addition to simulated mathematics Regents exams with carefully worked out solutions for all of the questions, this book also contains actual Regents exams from June 2008, August 2008, and June 2009. In order to get the most benefit from the tests in this book, you should try to do the following:

- After you complete an exam, check the answer key for the entire test. Circle any omitted questions or questions that you answered incorrectly. Study the explained solutions for these questions.
- On the Self-Analysis Chart, find the topic under which each question is classified, and enter the number of points you earned if you answered the question correctly.
- Figure out your percentage on each topic by dividing your earned points by the total number of points allotted to that topic, carrying the division to two decimal places. If you are not satisfied with your percentage on any topic, reread the explained solutions for the questions you missed. Then locate related questions in other Regents examinations by using their Self-Analysis Charts to see which questions are listed for the troublesome topic. Attempting to solve these questions and then studying their solutions will provide you with additional preparation. You may also find it helpful to review the appropriate sections in "A Brief Review of Key Integrated Algebra Skills." More detailed explanations and additional practice problems with answers are available in Barron's companion book, *Let's Review: Integrated Algebra.*

Tips for Practicing Effectively and Efficiently

- When taking a practice test, do not spend too much time on any one question. If you cannot come up with a method to use or if you cannot complete the solution, put a slash through the number of the question. When you have completed as many questions as you can, return to the unanswered questions and try them again.
- After finishing a practice test, compare each of your solutions with the solutions that are given. Read the explanation provided even if you have answered the question correctly. Each solution has been carefully designed to provide additional insight that may be valuable when answering a more difficult question on the same topic.
- In the weeks before the actual test, plan to devote at least one-half hour each day to preparation. It is better to spread out your time in this way than to cram by preparing for, say, four hours the day before the exam will be given. As the test day gets closer, take at least one complete Regents exam under actual test conditions.

Getting Acquainted with Integrated Algebra

This section explains things about the Integrated Algebra Regents examination that you may not know, such as how the exam is organized, how your exam will be scored, and where you can find a complete listing of the topics tested by this exam.

WHEN DO I TAKE THE INTEGRATED ALGEBRA REGENTS EXAM?

The Integrated Algebra Regents exam is administered in January, June, and August of every school year. Most students will take this exam after successfully completing their first year of study of high school-level mathematics.

HOW IS THE INTEGRATED ALGEBRA REGENTS EXAM SET UP?

The Integrated Algebra Regents Exam is divided into four parts with a total of 39 questions. All of the questions in each of the four parts must be answered. You will be allowed a maximum of 3 hours in which to complete the test.

Part I consists of 30 standard multiple-choice questions, each with four answer choices labeled (1), (2), (3), and (4).

Parts II, III, and IV each contain three questions. The answers and the accompanying work for the questions in these three parts must be written directly in the question booklet. You must show or explain how you arrived at each answer by indicating the necessary steps, including appropriate formula substitutions, diagrams, graphs, and charts. If you use a guess-and-check strategy to arrive at an answer for a problem, you must show indicate your method and show the work for at least three guesses.

Since scrap paper is not permitted for any part of the exam, you may use the blank spaces in the question booklet as scrap paper. If you need to draw a graph, graph paper will be provided in the question booklet. All work should be done in pen, except graphs and diagrams, which should be drawn in pencil.

WHAT TYPE OF CALCULATOR DO I NEED?

Graphing calculators are *required* for the Integrated Algebra Regents examination. During the administration of the Regents exam, schools are required to make a graphing calculator available for the exclusive use of each student. You will need to use your calculator to work with trigonometric functions of angles, evaluate roots, and perform routine calculations. Knowing how to use a graphing calculator gives you an advantage when deciding how to solve a problem. Rather than solving a problem algebraically with pen and paper, it may be easier to solve the same problem using a graph or table created by a graphing calculator. A graphical or numerical solution using a calculator can also be used to help confirm an answer obtained using standard algebraic methods. You can find additional information about solving specific types of problems with a graphing calculator in Barron's companion review book, *Let's Review: Integrated Algebra.*

WHAT GETS COLLECTED AT THE END OF THE EXAMINATION?

At the end of examination, you must return:

- Any tool provided to you by your school, such as a graphing calculator.
- The question booklet. Check that you have printed your name and the name of your school in the appropriate boxes near the top of the first page.
- The Part I answer sheet and any other special answer form that your school may provide. You must sign the statement at the bottom of the Part I answer sheet indicating that you did not receive any unlawful assistance in answering any of the questions. If you fail to sign this declaration, your answer paper will not be accepted.

HOW IS THE EXAM SCORED?

Your answers to the 30 multiple-choice questions in Part I are scored as either correct or incorrect. Each correct answer receives 2 points. There is no penalty for guessing. The three questions in Part II are worth 2 points each, the three questions in Part III are worth 3 points each, and the three questions in Part IV are worth 4 points each. Solutions to questions in Parts II, III, and IV that are not completely correct may receive partial credit according to a special scoring guide provided by the New York State Education Department.

The accompanying table shows how the Integration Algebra examination breaks down.

Question Type	Number of Questions	Credit Value
Part I: Multiple choice	30	$30 \times 2 = 60$
Part II: 2-credit open ended	3	$3 \times 2 = 6$
Part III: 3-credit open ended	3	$3 \times 3 = 9$
Part IV: 4-credit open ended	3	$3 \times 4 = 12$
	Test = 39 questions	Test = 87 points

HOW IS YOUR FINAL SCORE DETERMINED?

The maximum total raw score for the Integrated Algebra Regents examination is 87 points. After the raw scores for the four parts of the test are added together, a conversion table provided by the New York State Education Department is used to convert your raw score into a final test score that falls within the usual 0 to 100 scale.

ARE ANY FORMULAS PROVIDED?

The Integrated Algebra Regents Examination test booklet will include a reference sheet containing the formulas included in the next series of boxes. This formula sheet, however, does not necessarily include all formulas that you are expected to know.

Trigonometric ratios

$$\sin A = \frac{\text{opposite}}{\text{hypotenuse}}$$

$$\cos A = \frac{\text{adjacent}}{\text{hypotenuse}}$$

$$\tan A = \frac{\text{opposite}}{\text{adjacent}}$$

Area	Trapezoid	$A = \frac{1}{2}h(b_1 + b_2)$
Volume	Cylinder	$V = \pi r^2 h$
Surface area	Rectangular prism	$SA = 2lw + 2hw + 2lh$
	Cylinder	$SA = 2\pi r^2 + 2\pi rh$
Coordinate geometry		$m = \frac{\Delta y}{\Delta x} = \frac{y_2 - y_1}{x_2 - x_1}$

WHAT IS THE *CORE CURRICULUM*?

The *Core Curriculum* is the official publication by the New York State Education Department that describes the topics and skills that are required by the Integrated Algebra Regents examination. The *Core Curriculum* for Integrated Algebra includes most of the topics previously included in Math A except for some aspects of geometry, locus, geometric constructions, and combinations. Integrated Algebra also includes an introduction to sets, functions, lines of best fit, and exponential growth and decay. If you have Internet access, you can view the *Core Curriculum* at the New York State Education Department's web site at

http://www.emsc.nysed.gov/3-8/MathCore.pdf

Ten Test-Taking Tips

1. Know What to Expect on Test Day
2. Avoid Last-Minute Studying
3. Be Well Rested and Come Prepared on Test Day
4. Know How to Use Your Calculator
5. Know When to Use Your Calculator
6. Have a Plan for Budgeting Your Time
7. Make Your Answers Easy to Read
8. Answer the Question That Is Asked
9. Take Advantage of Multiple-Choice Questions
10. Don't Omit Any Questions

These ten practical tips can help you raise your grade on the Integrated Algebra Regents examination.

TIP 1

Know What to Expect on Test Day

SUGGESTIONS

- Become familiar with the format, directions, and content of the Integrated Algebra Regents exam.

- Know where you should write your answers for the different parts of the test.
- Ask your teacher to show you an actual test booklet of a previously given mathematics Regents examination.

TIP 2

Avoid Last-Minute Studying

SUGGESTIONS

- Start your Integrated Algebra Regents examination preparation early by making a regular practice of:
  1. taking detailed notes in class and then reviewing your notes when you get home;
  2. completing all written homework assignments in a neat and organized way; and by their due date;
  3. writing down any questions you have about your homework so that you can ask your teacher about them; and
  4. saving your classroom tests for use as an additional source of practice questions.
- Get a review book early in your preparation so that additional practice examples and explanations, if needed, will be at your fingertips. The recommended review book is Barron's *Let's Review: Integrated Algebra.* This easy-to-follow book has been designed for fast and effective learning and includes numerous demonstration and practice examples with solutions as well as graphing calculator approaches.
- Build your skill and confidence by completing all of the exams in this book and studying the accompanying solutions before the day of the Integrated Algebra Regents examination. Because each exam takes up to 3 hours to complete, you should begin this process no later than several weeks before the exam is scheduled to be given.

- As the day of the actual exam nears, take the exams in this book under the timed conditions that you will encounter on the actual test. Then compare each of your answers with the explained answers given in this book.
- Use the Self-Analysis Chart at the end of each exam to help pinpoint any weaknesses.
- If you do not feel confident in a particular area, study the corresponding topic in Barron's *Let's Review: Integrated Algebra*.
- As you work your way through the exams in this book, make a list of any formulas or rules that you need to know, and learn them well before the day of the exam. The following formulas will be provided in the test booklet: sine, cosine, and tangent of an angle; area of a trapezoid; volume of a cylinder; surface areas of a rectangular prism (closed box) and cylinder; and slope of a line (see page 169).

TIP 3

Be Well Rested and Come Prepared on Test Day

SUGGESTIONS

- On the night before the Regents exam, lay out all of the things you must take with you. Check the items against the following list:

  1. Your exam room number as well as any personal identification that your school may require.
  2. Two pens.
  3. Two sharpened pencils with erasers.
  4. A ruler.
  5. A graphing calculator (with fresh batteries).
  6. A watch.

- Eat wisely and go to bed early so you will be alert and well rested when you take the exam.

- Be certain you know when your exam begins. Set your alarm clock to give you plenty of time to eat breakfast and travel to school. Also, tell your parents what time you will need to leave the house in order to get to school on time.
- Arrive at the exam room on time and with confidence that you are well prepared.

TIP 4
Know How to Use Your Calculator

SUGGESTIONS

- Bring to the Integrated Algebra Regents exam room the same calculator that you used when you completed the practice exams at home. Keep in mind that a graphing calculator is required for the Integrated Algebra Regents exam.
- If you are required to use a graphing calculator provided by your school, make sure that you practice with it in advance because not all calculators work in the same way.
- Before you begin the test, check that the calculator you are using is set to degree mode.
- Be prepared to use your graphing calculator to:
  1. find the value of the trigonometric function of an angle;
  2. find the number of degrees in an angle when the value of a trigonometric function of that angle is given;
  3. evaluate roots and powers of numbers;
  4. calculate statistical measures such as the mean and median of a set of data;
  5. enter paired data values and then calculate the equation of the line of best fit;
  6. create a table or graph from a given equation; and
  7. solve a system of equations graphically or numerically with a table.

- Avoid rounding errors. Unless otherwise directed, the π (pi) key on a calculator should be used in computations involving the constant π rather than the familiar rational approximation of 3.14 or $\frac{22}{7}$.

  When performing a sequence of calculations in which the result of one calculation is used in a second calculation, do not round off. Instead, use the full power/display of the calculator by saving intermediate results in the calculator's memory. Unless otherwise directed, rounding, if required, should be done only after the *final* answer is reached.
- Because pressing the wrong key is easy, first estimate an answer and then compare it with the answer obtained by using your calculator. If the two answers are very different, start over.

TIP 5

Know When to Use Your Calculator

SUGGESTIONS

- Do not expect to have to use your calculator on each question. Most questions will not require a calculator.
- Avoid careless errors by using your calculator to perform those routine arithmetic calculations that are not easily performed mentally.
- When a problem does not specify the solution method, be alert for opportunities to use your graphing calculator to solve a problem by creating a table or a graph.
- Whenever appropriate, check an answer obtained using standard algebraic methods by using your graphing calculator to create a suitable table or graph.

TIP 6

Have a Plan for Budgeting Your Time

SUGGESTIONS

- In the **first ninety minutes** of the 3-hour Integrated Algebra Regents exam, complete the 30 multiple-choice questions in Part I. When answering troublesome questions of this type, first rule out any choices that are impossible. If the choices are numbers, you may be able to identify the correct answer by plugging these numbers back into the original question to see which one works. If the choices contain variables, you may be able to substitute easy numbers for the variables in both the test question and in each choice. Then try to match the numerical result produced in the question to the answer choice that evaluates to the same number. This approach is explained more fully in Tip 9.
- During the **next forty minutes** of the exam, complete the three Part II questions and the three Part III questions. To maximize your credit for each question, write down clearly the steps you followed to arrive at each answer. Include any equations, formula substitutions, diagrams, tables, graphs, and so forth.
- In the **next thirty minutes** of the exam, complete the three Part IV questions. Again, be sure to show how you arrived at each answer.
- During the **last twenty minutes**, review your entire test paper for neatness, accuracy, and completeness.
    1. Check that all answers (except graphs and other drawings) are written in ink. Make sure you have answered all of the questions in each part of the exam and that all of your Part I answers have been transcribed accurately on the separate Part I answer sheet.

2. Before you submit your test materials to the proctor, check that you have written your name in the reserved spaces on the front page of the question booklet and on the Part I answer sheet. Also, do not forget to sign the declaration that appears at the bottom of the Part I answer sheet.

TIP 7

Make Your Answers Easy to Read

SUGGESTIONS

- Make sure your solutions and answers are clear, neat, and logically organized. When solving problems algebraically, define what the variables stand for, as in "Let $x = \ldots$."
- Use a pencil to draw graphs so that you can erase neatly, if necessary. Use a ruler to draw straight lines and coordinate axes.
- Label the coordinate axes. Write y at the top of the vertical axis, and write x to the right of the horizontal axis. Label each graph with its equation.
- When answering a question in Parts II, III, or IV, indicate your approach as well as your final answers. Provide enough details to enable someone who doesn't know how you think to understand why and how you moved from one step of the solution to the next. If the teacher who is grading your paper finds it difficult to figure out what you wrote, he or she may simply decide to mark your work as incorrect and to give you little, if any, partial credit.
- Draw a box around your final answer to a Part II, III, or IV question.

TIP 8

Answer the Question That Is Asked

SUGGESTIONS

- Make sure each of your answers is in the form required by the question. For example, if a question asks for an approximation, round off your answer to the required decimal position. If a question requires that the answer be expressed in lowest terms, make sure that the numerator and denominator of a fractional answer do not have any common factors other than 1 or –1. For example, instead of leaving an answer as $\frac{10x^2}{15x}$, write it as $\frac{2x}{3}$. If a question calls for the answer in simplest form, make sure you simplify a square root radical so that the radicand does not contain any perfect square factors greater than 1. For example, instead of leaving an answer as $\sqrt{18}$, write $3\sqrt{2}$.
- If a question asks only for the x-coordinate (or y-coordinate) of a point, do not give both the x-coordinate and the y-coordinate.
- After solving a word problem, check the original question to make sure your *final* answer is the quantity that the question asks you to find.
- If units of measurement are given, as in area problems, check that your answer is expressed in the correct units.
- Make sure the units used in a computation are consistent. For example, when finding the volume of a rectangular box, its three dimensions must be expressed in the same units of measurement. Suppose a rectangular box measures 2 feet in width, 6 feet in length, and 8 inches in height. To find the number of cubic feet in the volume of the box, first convert 8 inches to an equivalent

number of feet. Since 8 inches is equivalent to $\frac{8}{12} = \frac{2}{3}$ feet, the volume of the box is 2 ft. × 6 ft. × $\frac{2}{3}$ ft. = 8 ft.^{3}

- If a question requires a positive root of a quadratic equation, as in geometric problems in which the variable usually represents a physical dimension, make sure you reject the negative root.
- If the replacement set of the variable is restricted, do not include values in the solution set that are not members of the replacement set. For example, assume it is given that the replacement set for x in the inequality $x + 1 \leq 4$ is the set of positive integers. Because $x \leq 3$, the solution set is $\{1,2,3\}$.

TIP 9

Take Advantage of Multiple-Choice Questions

SUGGESTIONS

- When the answer choices of a multiple-choice question contain only numbers, try plugging each choice into the original question until you find the one that works.

EXAMPLE 1

Which ordered pair is the solution set for this system of equations?

$$x + y = 8$$
$$y = x - 3$$

(1) (2.5,5.5)
(2) (4,1)
(3) (4,4)
(4) (5.5,2.5)

Solution 1: If you do not remember how to arrive at the solution by using an algebraic method, you can get the right answer by trying each ordered pair in the system of equations until you find the pair that works for *both* equations.

Choice (1): For (2.5,5.5), let $x = 2.5$ and $y = 5.5$:

$x + y = 8 \Rightarrow 2.5 + 5.5 = 8$	It works, so test the second equation.
$y = x - 3 \Rightarrow 5.5 \neq 2.5 - 3$	It doesn't work, since $2.5 - 3 = 0.5$.

Choice (2): For (4,1), let $x = 4$ and $y = 1$:

$x + y = 8 \Rightarrow 4 + 1 \neq 8$	It doesn't work, so there is no need to test the second equation.

Choice (3): For (4,4), let $x = 4$ and $y = 4$:

$x + y = 8 \Rightarrow 4 + 4 = 8$	It works, so test the second equation.
$y = x - 3 \Rightarrow 4 \neq 4 - 3$	It doesn't work, since $4 - 3 = 1$.

Choice (4): For (5.5,2.5), let $x = 5.5$ and $y = 2.5$:

$x + y = 8 \Rightarrow 5.5 + 2.5 = 8$	It works, so test the second equation.
$y = x - 3 \Rightarrow 2.5 = 5.5 - 3$	Since it works, (5.5,2.5) is the solution.

Hence, the correct choice is **(4)**.

Solution 2: To get the solution algebraically, eliminate y in the first equation by replacing it with $x - 3$; then $x + (x - 3) = 8$ so:

$$2x - 3 = 8$$
$$2x = 11$$
$$\frac{2x}{2} = \frac{11}{2}$$
$$x = 5.5$$

The correct choice is **(4)**, the only choice for which $x = 5.5$.

- If the answer choices for a multiple-choice question contain only letters, replace the letters with easy numbers in both the question and the answer choices. Then work out the problem using these numbers.

EXAMPLE 2

Which expression is equal to $\frac{a}{x} + \frac{b}{2x}$?

(1) $\frac{2a+b}{2x}$ (2) $\frac{2a+b}{x}$ (3) $\frac{a+b}{3x}$ (4) $\frac{a+b}{2x}$

Solution 1: Let $a = b = x = 1$. Then

$$\frac{a}{x} + \frac{b}{2x} = \frac{1}{1} + \frac{1}{2(1)} = 1 + \frac{1}{2} = \frac{3}{2}$$

Next, evaluate each of the four answer choices to find the one that is equal to $\frac{3}{2}$. If more than one answer choice evaluates to $\frac{3}{2}$, start over using different values for a, b, and x.

Choice (1): $\frac{2a+b}{2x} = \frac{2(1)+1}{2(1)} = \frac{2+1}{2} = \frac{3}{2}$.

Choice (2): $\frac{2a+b}{x} = \frac{2(1)+1}{1} = \frac{3}{1} = 3$.

Choice (3): $\frac{a+b}{3x} = \frac{1+1}{3(1)} = \frac{2}{3}$.

Choice (4): $\frac{a+b}{2x} = \frac{1+1}{2(1)} = \frac{2}{2} = 1$.

The correct choice is **(1)**, the only choice that evaluates to $\frac{3}{2}$.

Solution 2: To solve algebraically, change the first fraction into an equivalent fraction that has the LCD as its denominator. Then write the sum of the numerators of the two fractions over the LCD.

$$\frac{a}{x} + \frac{b}{2x} = \left(\frac{2}{2}\right)\frac{a}{x} + \frac{b}{2x}$$

$$= \frac{2a+b}{2x}$$

Hence, the correct choice is **(1)**.

TIP 10

Don't Omit Any Questions

SUGGESTIONS

- Keep in mind that on each of the four parts of the test, you must answer all of the questions.
- If you get stuck on a multiple-choice question in Part I, try one of the problem-solving methods discussed in Tip 9. If you still cannot figure out the answer, try to eliminate any impossible answers. Then guess from the remaining choices.
- If you get stuck on a question from Parts II, III, or IV, try to maximize your partial credit by writing any formula or mathematics facts that you think might apply. If appropriate, organize and analyze the given information by making a table or a diagram. You can then try to arrive at the correct answer by guessing, checking your guess, and then revising your guess, as needed.

Special Problem-Solving Strategies

No single problem-solving strategy works for all problems. You should have a "toolbox" of strategies from which you can choose when trying to figure out how to solve an unfamiliar type of problem.

If you get stuck trying to solve a problem, use one or more of the following strategies.

STRATEGY 1: WORK BACKWARD

Reversing the steps that produced an end result can lead to the starting value that you need to find.

EXAMPLE 1

Linda paid $48 for a jacket that was on sale for 25% of the original price. What was the original price of the jacket?

(1) $60 (2) $72 (3) $96 (4) $192

Solution: Work back from the final price of $48. Since Linda paid $48 for the jacket that was on sale for 25% $\left(=\frac{1}{4}\right)$ of its original price, the original price was $4 \times \$48 = \$192$.

The correct choice is **(4)**.

EXAMPLE 2

Sara's telephone service costs $21 per month plus $0.25 for each local call; long-distance calls are extra. Last month, Sara's bill was $36.64, and it included $6.14 in long-distance charges. How many calls did she make?

Solution: Work back from the final bill of $36.64.

- The part of the bill that does not include any long-distance charges is $36.64 – $6.14 = $30.50.
- The part of the bill that does not include the long-distance charges or the basic service charge is $30.50 – $21 = $9.50.
- Since the bill for local calls is $9.50 and each local call costs $0.25, the number of local calls is $\frac{\$9.50}{\$0.25} = \mathbf{38}$.

STRATEGY 2: MAKE A TABLE OR LIST

Making a table or list can help you to organize the facts of a problem so that the answer becomes easier to find.

EXAMPLE 3

The Excel Cable Company has a monthly fee of $32.00 and an additional charge of $8.00 for each premium channel. The Best Cable Company has a monthly fee of $26.00 and an additional charge of $10.00 for each premium channel. For what number of premium channels will the total monthly subscription fee for the two cable companies be the same?

Solution: Make a table to compare the fees for each company.

Number of Premium Channels	Fee Charged by the Excel Cable Company	Fee Charged by the Best Cable Company
1	$32 + (1 × $8) = $40	$26 + (1 × $10) = $36
2	$32 + (2 × $8) = $48	$26 + (2 × $10) = $46
3	$32 + (3 × $8) = $56	$26 + (3 × $10) = $56

The fees for the two cable companies are the same for **3** premium channels.

Solution 2: Make a table using a graphing calculator. If x represents the number of premium channels ordered, then $Y1 = 32 + 8x$ is the fee charged by the Excel Cable Company and $Y2 = 26 + 10x$ is the fee charged by the Best Cable Company. By creating a table, you can compare the values of $Y1$ and $Y2$ for the same x-value. Display the table. Then use the up/down cursor arrow keys to find the row of the table on which $Y1 = Y2$, as shown in the accompanying figure.

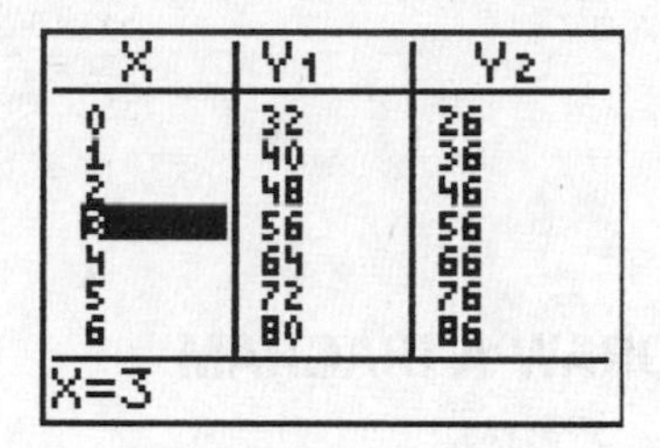

X	Y1	Y2
0	32	26
1	40	36
2	48	46
3	56	56
4	64	66
5	72	76
6	80	86

X=3

EXAMPLE 4

Solve for x: $3(2x - 1) = x + 17$.

Solution 1: Many types of equations can be solved by creating a table in which $Y1$ is equal to one side of the given equation and $Y2$ is equal to the other side. To solve $3(2x - 1) = x + 17$, set $Y1 = 3(2x - 1)$ and $Y2 = x + 17$. Display the table. Then use the up/down cursor arrow keys to find the row of the table on which $Y1 = Y2$, as was done in Example 3:

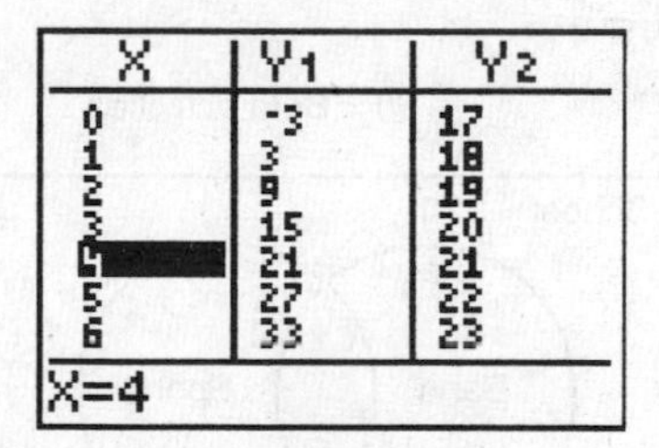

X	Y1	Y2
0	-3	17
1	3	18
2	9	19
3	15	20
4	21	21
5	27	22
6	33	23

X=4

Thus, $x = 4$.

If the original equation does not have integral roots, you may need to adjust the increment value for x in the table setup mode or use an algebraic approach.

Solution 2: Solve algebraically.

$$3(2x - 1) = x + 17$$

Remove parentheses: $6x - 3 = x + 17$

Isolate x: $6x - x = 17 + 3$

Simplify: $5x = 20$

$$\frac{5x}{5} = \frac{20}{5}$$

$$x = 4$$

STRATEGY 3: DRAW A DIAGRAM

Drawing a diagram when none is given can help you to visualize the facts of a problem that may help you arrive at a solution.

EXAMPLE 5

In a school of 320 students, 85 students are in the band, 200 students are on sports teams, and 60 students participate in both activities. How many students are *not* involved in either band or sports? Show how you arrived at your answer.

Solution: Draw a Venn diagram such as the one shown in the accompanying figure.

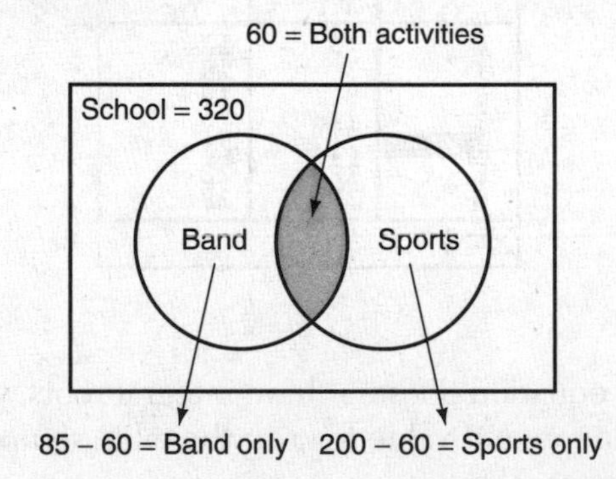

- 60 students are involved in both activities.
- 85 – 60 = 25 students are involved only in the band.
- 200 – 60 = 140 students are involved only in sports.
- Therefore, 320 – 60 – 25 – 140 = **95** students are *not* involved in either of the two activities.

STRATEGY 4: USE PARTICULAR NUMBERS

Using numbers instead of variables can greatly simplify a problem.

EXAMPLE 6

The ratio of the lengths of corresponding sides of two similar squares is 1 to 3. What is the ratio of the area of the smaller square to the area of the larger square?

(1) $1:\sqrt{3}$ (2) 1:3 (3) 1:6 (4) 1:9

Solution: Since it is given that the ratio of lengths of corresponding sides of two similar squares is 1 to 3, let the length of a side of the smaller square equal 1 and the length of the corresponding side of the larger square equal 3. Then:

- The area of the smaller square is $1 \times 1 = 1$.
- The area of the larger square is $3 \times 3 = 9$.
- The ratio of the area of the smaller square to the area of the larger square is 1:9.

Hence, the correct choice is **(4)**.

STRATEGY 5: GUESS AND CHECK

Even if you cannot solve an unfamiliar problem, you may have enough information to make a reasonable guess.

EXAMPLE 7

During a recent winter, the ratio of deer to foxes was 7 to 3 in one county of New York State. If there were 210 foxes in the county, what was the number of deer?

(1) 90 (2) 147 (3) 280 (4) 490

Solution: Make a reasonable guess, and then check it against the four answer choices.

- Because the ratio of deer to foxes was 7 to 3, the number of deer was a little more than twice the number of foxes.
- Since it is given that there were 210 foxes in the county, the number of deer must be a little more than $2 \times 210 = 420$.
- Of the four answer choices, only 490 is greater than 420.

Hence, the correct choice is **(4)**.

EXAMPLE 8

Farmer Gray has only chickens and cows in his barnyard. If these animals have a total of 60 heads and 140 legs, how many chickens and how many cows are in the barnyard?

Solution 1:

- Since there are 60 heads, the sum of the number of cows and the number of chickens is 60. A chicken has two legs and a cow has four legs.
- Guess, check, and revise while keeping track of your guesses in a list or a table.

Number of		
Chickens	**Cows**	**Total Number of Legs**
0	60	$(0 \times 2) + (60 \times 4) = 240$ ← Too high
30	30	$(30 \times 2) + (30 \times 4) = 180$ ← Too high
40	20	$(40 \times 2) + (20 \times 4) = 160$ ← Too high
50	**10**	$(50 \times 2) + (10 \times 4) = 140$ ← This is the answer!

- Check that your answer works. Two conditions must be true:

✓ The total number of heads is 60:	50 chickens + 10 cows	= 60 heads
✓ The total number of legs is 140:	50 chickens × 2 legs	= 100 legs
	10 cows × 4 legs	= 40 legs
	Total	= 140 legs

Farmer Gray has **50 chickens** and **10 cows**.

Solution 2: Solve algebraically. If x = the number of chickens, then $60 - x$ = the number of cows. Hence:

$$2x + 4(60 - x) = 140$$

$$2x + 240 - 4x = 140$$

$$-2x = 140 - 240$$

$$\frac{-2x}{-2} = \frac{-100}{-2}$$

$$x = 50$$

Farmer Gray has **50 chickens** and 60 – 50 = **10 cows**.

IMPORTANT NOTE

If you use the guess-and-check strategy to solve a problem in Part II, III, or IV of the Regents Exam, show the work for *at least three* guesses with appropriate checks. Should you reach the correct answer on the first trial, you must further illustrate your method by demonstrating that the other guesses do not work.

A General Problem-Solving Approach

1. ***Read each test question through the first time*** to get a *general idea* of the type of mathematics knowledge that is required. For example, one question may ask you to solve an algebraic equation, while for another question you may need to draw a graph or apply some geometric principle. Then ***read the question through a second time*** to pick out specific facts. Identify what is given, what you need to find, and the relationship that connects them together.
2. ***Decide how you will solve the problem***. If the solution method is not stated in the problem, you should choose an appropriate method (numerical, graphical, or algebraic) with which you are most comfortable. You may need to use one of the special problem-solving strategies discussed in this section. If you solve a word problem by using an algebraic method, decide what the variable represents and then translate the conditions of the problem into an equation or an inequality.
3. ***Carry out your plan.***
4. ***Verify that your answer is correct*** by making sure it works in the original question.

Some Key Integrated Algebra Facts and Skills

1. NUMBER SENSE AND OPERATIONS

The Set of Real Numbers and Its Subsets

1.1 INTEGERS

- **Integers** include the counting numbers (positive integers), their opposites (negative integers), and 0:

$$\ldots, -4, -3, -2, -1, 0, 1, 2, 3, 4, \ldots$$

- When two integers are multiplied together, the answer is called the **product** and the numbers being multiplied together are **factors**. Since $2 \times 3 = 6$, 2 and 3 are factors of the product, 6.
- If $a \div b$ has a 0 remainder, then a is evenly divisible by b. Thus, 6 is divisible by 3 since $6 \div 3 = 2$ with remainder 0, but 6 is not divisible by 4 since $6 \div 4 = 1$ with remainder 2.
- **Even integers** are integers that are divisible by 2:

$$\ldots, -6, -4, -2, 0, 2, 4, 6, \ldots$$

- **Odd Integers** are integers that are *not* divisible by 2:

$$\ldots, -5, -3, -1, 1, 3, 5, \ldots$$

- **Prime numbers** are integers greater than 1 that are divisible only by themselves and 1:

$$2, 3, 5, 7, 11, 13, 17, \ldots$$

1.2 GENERAL INTEGERS

- If n represents an integer, then four consecutive integers are

$$n, n+1, n+2, n+3$$

- If n represents an *even* integer, then four consecutive even integers are

$$n, n+2, n+4, n+6$$

- If n represents an *odd* integer, then four consecutive odd integers are

$$n, n+2, n+4, n+6$$

1.3 SQUARE ROOT RADICALS

The square root notation $\sqrt{n}$ means one of two equal nonnegative numbers whose product is n. For example, $\sqrt{16}=4$ since $4 \times 4 = 16$. The number underneath the radical sign $\left(\sqrt{\ }\right)$ is called the **radicand**. Whole numbers such as 1, 4, 9, 16, 25, . . . are called **perfect squares** since their square roots are whole numbers. Fractions like $\frac{4}{9}$ with numerators and denominators that are both perfect squares are also perfect squares since their square roots are fractions with whole numbers in the numerators and denominators. For example, $\sqrt{\frac{4}{9}}=\frac{2}{3}$.

1.4 REAL NUMBERS

- A **rational number** is a number that can be written as a fraction with an integer numerator and a nonzero integer denominator. The set of rational numbers includes integers, decimal numbers such as 1.25 $\left(=\frac{5}{4}\right)$, and nonending decimal numbers in which one or more nonzero digits repeat endlessly, as in 0.3333. . . $\left(=\frac{1}{3}\right)$ and 0.636363. . . $\left(=\frac{7}{11}\right)$.

- The square root of a number that is *not* a perfect square, such as $\sqrt{5}$, is an **irrational number**. The quantity π (= 3.1415926 . . .) is another example of an irrational number. An irrational number does *not* have an *exact* decimal equivalent, but can be approximated using a calculator.
- The set of **real numbers** is the union of the set of rational numbers and the set of irrational numbers.

1.5 PROPERTIES OF REAL NUMBERS

- CLOSURE PROPERTY: A set of numbers is closed for an arithmetic operation if the answer is always a number that belongs to the same set. For example, adding, subtracting, or multiplying two integers always results in another integer. Dividing two integers, however, does not necessarily result in another integer, as in $3 \div 4$. Therefore, the set of integers is closed under addition, subtraction, and multiplication, but is not closed under division. The set of real numbers is closed under addition and multiplication.
- COMMUTATIVE PROPERTIES: The order in which two real numbers are added or multiplied does not matter. Thus:

$$x + y = y + x \quad \text{and} \quad xy = yx$$

- ASSOCIATIVE PROPERTIES: The order in which three real numbers are grouped when added or multiplied does not matter. Thus:

$$x + (y + z) = (x + y) + z \quad \text{and} \quad x(yz) = (xy)z$$

- IDENTITY PROPERTIES: For addition, 0 is the identity element since, for any real number x:

$$x + 0 = 0 + x$$

For multiplication, 1 is the identity element since, for any real number x:

$$x \cdot 1 = 1 \cdot x$$

- INVERSE PROPERTIES: For addition, the inverse of a real number x is its opposite, $-x$, since

$$x + (-x) = 0$$

For multiplication, the inverse of a nonzero real number x is its reciprocal, $\frac{1}{x}$, since

$$x \cdot \left(\frac{1}{x}\right) = 1$$

- DISTRIBUTIVE PROPERTY: For three real numbers x, y, and z:

$$x\,(y + z) = xy + xz$$

1.6 COMBINING LIKE TERMS

Like terms have the same variable factors but may have different numerical coefficients, as in $2xy$ and $3xy$. To combine like terms, use the reverse of the distributive property. For example, $5x + 4x = (5 + 4)x = 9x$. Thus, for like terms, combine the numerical coefficients and keep the common variable factors, as in $7ab - 3ab = 4ab$.

1.7 OPERATIONS WITH RADICALS

- To *simplify* a radical, rewrite the radicand as the product of two whole numbers one of which is the greatest possible perfect square factor of the product. Then write the radical over each factor and simplify. For example:

$$\sqrt{75} = \sqrt{25 \cdot 3} = \sqrt{25} \cdot \sqrt{3} = 5\sqrt{3}$$

- To *combine* radicals, first rewrite the radicals so that they have the same radicand, if possible. Then treat the radicals as like terms and combine. For example:

$$\sqrt{18} + 4\sqrt{2} = \sqrt{9 \cdot 2} + 4\sqrt{2} = 3\sqrt{2} + 4\sqrt{2} = 7\sqrt{2}$$

- To *multiply* or *divide* radicals, use these rules:

$$a\sqrt{b} \times c\sqrt{d} = a \cdot c\sqrt{b \cdot d} \quad \text{and} \quad \frac{a\sqrt{b}}{c\sqrt{d}} = \frac{a}{c}\sqrt{\frac{b}{d}}$$

1.8 PERCENT

The term *percent* means "parts of 100." For example, 60% means $\frac{60}{100}$ or 0.60 or $\frac{3}{5}$.

- To find the percent of a number, multiply the number by the percent in decimal form. For example, if the rate of sales tax is 6%, the amount of sales tax on a \$40 item is 6% of \$40 or 0.06 × \$40 = \$2.40.
- To find what percent one number is of another, write the "is" number over the "of" number and multiply by 100%. For instance, if the amount of sales tax on a \$65 item is \$5.20, then the tax rate is obtained by finding what percent 5.20 is of 65. Since

$$\frac{5.20\ (\text{"is" number})}{65\ (\text{"of" number})} \times 100\% = \frac{520}{65}\%$$
$$= 8\%$$

The tax rate is 8%.

- To find a number when a percent of it is given, divide the given amount by the percent in decimal form. For example, if 30% of some number n is 12, then $n = \frac{12}{0.30} = 40$.
- To find the percent of increase or decrease when a quantity changes in value, divide the amount of the change in the quantity by its starting value, and multiply the result by 100%. To illustrate, if the price of a pair of designer jeans is reduced from \$60 to \$48, then

$$\begin{aligned}\text{percent decrease} &= \frac{\text{change in price}}{\text{original price}} \times 100\% \\ &= \frac{\$60 - \$48}{\$60} \times 100\% \\ &= \frac{12}{60} \times 100\% \\ &= 0.2 \times 100\% \\ &= 20\%\end{aligned}$$

Signed Numbers

1.9 OPERATIONS WITH SIGNED NUMBERS

Real numbers are ordered so that each number on a number line is greater than the number to its left, as shown in the accompanying graph. Thus, $2 > 1$ and $-1 > -2$.

–4 –3 –2 –1 0 1 2 3 4

The **absolute value** of a number is its distance from 0 on the number line. Since distance cannot be a negative number, the absolute value of a number is always nonnegative. The notation $|n|$ represents the absolute value of n and is equal to the number n without its sign. Thus, $|-3| = 3$.

- To *multiply or divide* signed numbers with the *same* sign, perform the operation and attach a plus sign. For example:

$$(-3) \times (-5) = +15$$

- To *multiply or divide* signed numbers with *different* signs, perform the operation and attach a minus sign. For example:

$$\frac{+15}{-3} = -5$$

- To *add* signed numbers with the *same* sign, add the numbers and attach the common sign to the sum. For example:

$$(-3) + (-4) = -7$$

- To *add* signed numbers with *different* signs, subtract the unsigned numbers and attach to the sum the sign of the number with the larger absolute value. For example:

$$(-5) + (+2) = -(5 - 2) = -3$$

- To *subtract* signed numbers, change to an equivalent addition example by taking the opposite of the number that is being subtracted. For example:

$$(-7) - (-2) = (-7) + (+2) = -5$$

Operations With Exponents

1.10 LAWS OF EXPONENTS

An **exponent** indicates how many times a number is to be used as a factor in a product. In 2^3, the exponent, 3, means that 2, called the *base*, should be used as a factor three times, as in $2 \times 2 \times 2 = 8$.

- Any quantity with a zero exponent is equal to 1. Thus, $5^0 = 1$.
- Any quantity with a negative exponent can be written with a positive exponent by inverting the base, as in

$$x^{-n} = \frac{1}{x^n} \quad (x \neq 0)$$

For example:

$$2^{-4} = \frac{1}{2^4} \quad \text{and} \quad \frac{1}{5^{-2}} = 5^2$$

- PRODUCT LAW: To **multiply** powers of the *same* base, *add* their exponents, as in

$$y^a \times y^b = y^{a+b}$$

For example:

$$2^3 \times 2^4 = 2^{3+4} = 2^7 \quad \text{and} \quad 3x^2 \cdot 2x^5 = (3 \cdot 2)(x^2 \cdot x^5) = 6x^7$$

- QUOTIENT LAW: To **divide** powers of the *same* base, *subtract* their exponents, as in

$$\frac{y^a}{y^b} = y^{a-b}$$

For example:

$$\frac{2^7}{2^3} = 2^{7-3} = 2^4 \quad \text{and} \quad \frac{8xy^3}{2y} = \frac{8x}{2} \cdot \frac{y^3}{y^1} = 4xy^2$$

- POWER LAW: To raise a power to another power, multiply the exponents, as in

$$(y^a)^b = y^{a \times b}$$

For example:

$$(2^3)^4 = 2^{3 \times 4} = 2^{12}$$

1.11 SCIENTIFIC NOTATION

A number is in **scientific notation** when it is written as the product of a number between 1 and 10 and a power of 10.

- When a number greater than 10 is written in scientific notation, the power of 10 will be the number of places the decimal point must be moved to the *left*. For example:

$$470{,}000. = 4.7 \times 10^5$$

(decimal point moved +5 places)

- When a number between 0 and 1 is written in scientific notation, the power of 10 will be the number of places the decimal point must be moved to the *right*, with a negative sign in front of the exponent. For example:

$$0.000000309 = 3.09 \times 10^{-6}$$

-6

1.12 SETS AND SET NOTATION

A **set** is a collection of objects called **elements**. The elements of a set may be listed within braces { }. If set A consists of the positive integers less than 5, then $A = \{1,2,3,4\}$. Thus, 4 is an element of set A, which is abbreviated by writing $4 \in A$.

- If set B consists of the positive odd integers less than 5, then $B = \{1,3\}$. Since each element of B is also an element of A, B is a **subset** of A, denoted by $B \subset A$.
- If A is a subset of B, then the **complement** of A, denoted by $\overline{A}$, is the set of elements that are in S but *not* in A. If $S = \{1,2,3,4,5,6,7\}$ then $\overline{A} = \{5,6,7\}$. Thus, $A \cup \overline{A} = S$. The notation A' or $\sim A$ are also used to represent the complement of a set. If E is the set of all even natural numbers, then E' (as well as $\sim E$ and $\overline{E}$) denotes the set of all odd natural numbers.
- If $C = \{0,2,4,6\}$, then sets A and C have the elements 2 and 4 in common. Thus, $A \cap C = \{2,4\}$ where $A \cap C$ represents the **intersection** of sets A and C. The **union** of sets A and C, represented by $A \cup C$, is the set comprised of each element that is a member of A, C, or both A and C. Thus, $A \cup C = \{0,1,2,3,4,6\}$.
- Since sets B and C have no elements in common, $B \cap C = \varnothing$ where $\varnothing$ represents the **empty** or **null** set which is also indicated by { }.
- The set of positive integers less than or equal to 100 can be represented using **set-builder notation** by writing

$$\{x \mid x \leq 100; \underbrace{x \text{ is a positive integer}}_{\text{replacement set for } x}\}$$

which is read, "The set of all x such that x is less than or equal to 100 such that x is a positive integer." Set-builder notation places a general rule within the braces that tells whether a particular element belongs to the set. To illustrate further, if

$$A = \{x \mid 3 < x \leq 8; x \text{ is an integer}\}$$

then $A = \{4,5,6,7,8\}$.

1.13 INTERVAL NOTATION

Interval notation is a shorthand notation that represents a set of numbers in an interval by enclosing its endpoints in parentheses or brackets, depending on whether the endpoint is included in the interval. A bracket next to an endpoint indicates that the endpoint is included in the interval. A left or right parenthesis indicates that the endpoint is not included in the interval. The accompanying table lists the possibilities where $a < b$ and x is any number in the interval:

Description	Interval Notation	Example
Interval includes a and b	$[a,b]$	$[1,5] = 1 \leq x \leq 5$
Interval includes a but not b	$[a,b)$	$[1,5) = 1 \leq x < 5$
Interval includes b but not a	$(a,b]$	$(1,5] = 1 < x \leq 5$
Interval includes neither endpoint	(a,b)	$(1,5) = 1 < x < 5$

Practice Exercises

1. Which property is illustrated by the following equation?

$$\square(\triangle + \text{O}) = \square\triangle + \square\text{O}$$

(1) distributive
(2) associative
(3) commutative
(4) transitive

2. Which equation illustrates the additive inverse property?

(1) $a + (-a) = 0$
(2) $a + 0 = a$
(3) $a \div (-a) = -1$
(4) $a \cdot \frac{1}{a} = 1$

3. In the solution of the equation $3x = 6$, which property of real numbers justifies statement **5**?

Statements	Reasons
1. $3x = 6$	1. Given
2. $\frac{1}{3}(3x) = \frac{1}{3}(6)$	2. Multiplication axiom
3. $(\frac{1}{3} \cdot 3)x - 2$	3. Associative property
4. $1 \cdot x = 2$	4. Multiplicative inverse
5. $x = 2$	5. ____________

(1) Closure
(2) Identity
(3) Commutative
(4) Inverse

4. The expression $\sqrt{300}$ is equivalent to

(1) $50\sqrt{6}$ (3) $3\sqrt{10}$
(2) $12\sqrt{5}$ (4) $10\sqrt{3}$

5. The expression $\sqrt{27} + \sqrt{12}$ is equivalent to

(1) $\sqrt{39}$ (3) $5\sqrt{6}$
(2) $13\sqrt{3}$ (4) $5\sqrt{3}$

6. The sum of $2\sqrt{3}$ and $\sqrt{12}$ is

(1) $4\sqrt{6}$ (3) $3\sqrt{15}$
(2) $8\sqrt{3}$ (4) $4\sqrt{3}$

7. The expression $\sqrt{50} + 3\sqrt{2}$ can be written in the form $x\sqrt{2}$. Find x.

8. If 0.00037 is expressed as 3.7×10^n, what is the value of n?

9. When expressed in scientific notation, the number $0.0000000364 = 3.64 \times 10^n$. The value of n is

(1) 8 (3) –10
(2) 10 (4) –8

10. The number 0.00000467 can be written in the form 4.67×10^n. Find n.

11. Which list is in order from smallest value to largest value?

(1) $\sqrt{10}, \frac{22}{7}, \pi, 3.1$ (3) $\pi, \frac{22}{7}, 3.1, \sqrt{10}$
(2) $3.1, \frac{22}{7}, \pi, \sqrt{10}$ (4) $3.1, \pi, \frac{22}{7}, \sqrt{10}$

12. The operation $*$ for the set $\{p,r,s,v\}$ is defined in the accompanying table. What is the inverse element of r under the operation $*$?

$*$	p	r	s	v
p	s	v	p	r
r	v	p	r	s
s	p	r	s	v
v	r	s	v	p

(1) r
(2) s
(3) p
(4) v

Solutions

1. If a, b, and c, are real numbers, the distributive property states that

$$a(b + c) = ab + bc$$

The given equation

$$\square(\triangle + \text{O}) = \square\triangle + \square\text{O}$$

has the form $a(b + c) = ab + bc$, where $a = \square$, $b = \triangle$, and $c = \text{O}$.

The correct choice is **(1)**.

2. The equation $a + (-a) = 0$ states that the sum of any number a and its additive inverse (opposite) is 0.

The correct choice is **(1)**.

3. The number 1 is the *identity element* for multiplication for the set of real numbers since the product of 1 and *any* real number is the same real number.

Statement 4 is $1 \cdot x = 2$. Since 1 is the identity element for multiplication, the left side of the equation in statement 4 becomes x. Then statement 5, $x = 2$, follows from statement 4 as a result of 1 being the identity element for multiplication.

The correct choice is **(2)**.

4. The given expression is: $\sqrt{300}$

Simplify $\sqrt{300}$ by finding two factors of the radicand, 300, one of which is the highest perfect square that divides into 300: $\sqrt{100(3)}$

Take the square root of the perfect square factor, 100, and place it outside the radical sign as the coefficient, 10: $10\sqrt{3}$

The correct choice is **(4)**.

5. The given expression is: $\sqrt{27} + \sqrt{12}$

You cannot add $\sqrt{27}$ and $\sqrt{12}$ in their present form because only *like radicals* may be combined. Like radicals have the same radicand (number under the radical sign) and the same index (here, understood to be 2, representing the square root). A radical can be simplified by finding two factors of the radicand, one of which is the highest possible perfect square.

Factor the radicands, using the highest possible perfect square in each: $\sqrt{9 \cdot 3} + \sqrt{4 \cdot 3}$

Simplify by taking the square root of the perfect square factor and placing it outside the radical sign as a coefficient of the radical: $3\sqrt{3} + 2\sqrt{3}$

Combine the like radicals by adding their coefficients to get the new coefficient: $5\sqrt{3}$

The given expression is equivalent to $5\sqrt{3}$.

The correct choice is **(4)**.

6. The given expression is: $2\sqrt{3} + \sqrt{12}$

Only *like radicals* can be added. Like radicals must have the same root (here both are square roots) and must have the same radicand (the number under the radical sign). Factor out any perfect square factor in the radicands: $2\sqrt{3} + \sqrt{4(3)}$

Remove the perfect square factors from under the radical sign by taking their square roots and writing them as coefficients of the radical: $2\sqrt{3} + 2\sqrt{3}$

Combine the like radicals by adding their coefficients and writing the sum as the numerical coefficient of the common radical: $4\sqrt{3}$

The correct choice is **(4)**.

7. The given expression is: $\sqrt{50}+3\sqrt{2}$

Factor out any perfect square factors in the radicand (the number under the radical sign): $\sqrt{25(2)}+3\sqrt{2}$

Remove the perfect square factors from under the radical sign by taking their square roots and writing them as coefficients of the radical: $5\sqrt{2}+3\sqrt{2}$

Combine the like radicals by combining their coefficients and using the sum as the numerical coefficient of the common radical: $8\sqrt{2}$

$8\sqrt{2}$ is in the form $x\sqrt{2}$ with $x = 8$.

$x =$ **8**.

8. To change 3.7 to 0.00037, the decimal point in 3.7 must be moved four places to the left. Each move of one place to the left is equivalent to dividing by 10. Therefore, 3.7 must be divided by 10^4 to equal 0.00037. Division by 10^4 is equivalent to multiplying by $\frac{1}{10^4}$ or by 10^{-4}. Therefore, $0.00037 = 3.7 \times 10^{-4}$, so $n = -4$.

$n =$ **–4**.

9. The given equation is $0.0000000364 = 3.64 \times 10^n$.

In order for 3.64×10^n to become 0.0000000364, the decimal point in 3.64 must be moved eight places to the left. Each move of one place to the left is equivalent to dividing 3.64 by 10, or equivalent to multiplying it by 10^{-1}. Therefore, to move the decimal point eight places to the left, 3.64 must be multiplied by 10^{-8}, that is, $n = -8$.

The correct choice is **(4)**.

10. Dividing 4.67 by 10 moves the decimal point one place to the left. Therefore, 4.67 must be divided by 10 six times to make it equal 0.00000467. Dividing by 10 six times is the same as multiplying by 10^{-6}: $0.00000467 = 4.67 \times 10^{-6}$.

$n =$ **–6**.

11. To determine which answer choice lists the given numbers from the smallest value to the largest value, locate each of the four numbers on a number line using the decimal approximations $\sqrt{10} \approx 3.162$, $\frac{22}{7} \approx 3.143$, and $\pi \approx 3.142$:

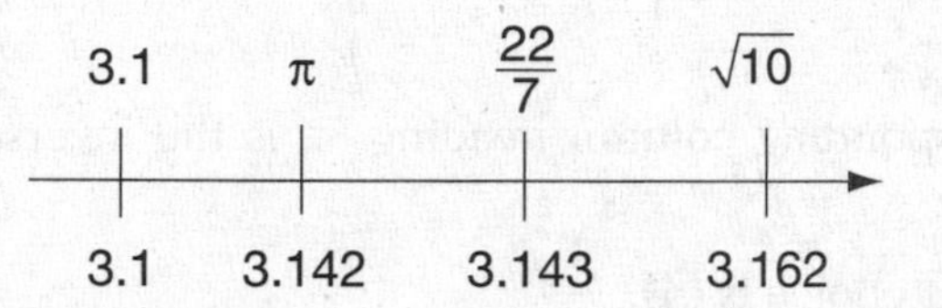

Examine each of the answer choices in turn until you find the one that lists the four numbers in the same order, from left to right, as on the number line: 3.1, π, $\frac{22}{7}$, and $\sqrt{10}$.

The correct choice is **(4)**.

12. Read the table as you would an ordinary multiplication table. For example, $p * r = v$. Before you can find the inverse of an element, you must determine the identity element for the set. To find the identity element, locate the row and column headings for which the elements in that row and column appear in the order p, r, s, v:

*	*p*	*r*	*s*	*v*
p	*s*	*v*	*p*	*r*
r	*v*	*p*	*r*	*s*
s	*p*	*r*	*s*	*v*
v	*r*	*s*	*v*	*p*

The row and column heading that satisfies this condition, s, is the identity element. You should verify that when s operates on each element of the set, the result is that same element. To find the inverse of element r, locate r on the second row and then read across until you find s, the identity element.

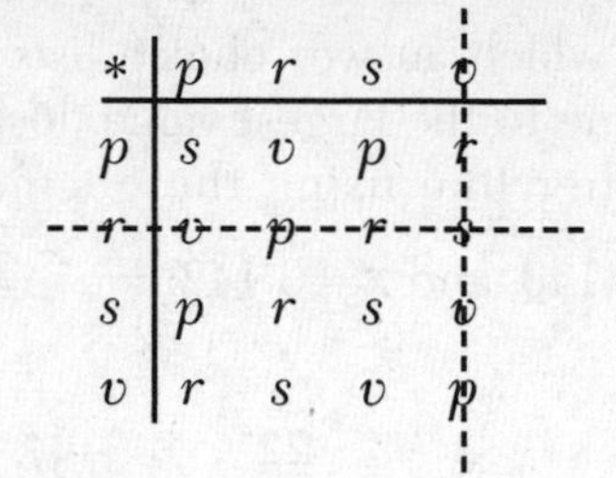

The corresponding column heading, v, is the inverse of r since $r * v = s$ and $v * r = s$.

The correct choice is **(4)**.

2. ALGEBRA

2.1 MONOMIALS

A **monomial** is a single term containing a number, a variable, or the product of numbers and variables.

- To *multiply* monomials, multiply the numerical coefficients and multiply like variable factors by adding their exponents. For example:

$$(-4a^2b)(2a^3b) = -8a^{2+3}b^{1+1} = -8a^5b^2$$

- To *divide* monomials, divide the numerical coefficients and divide like variable factors by subtracting their exponents. For example:

$$\begin{aligned}\frac{-8a^5b^2}{2a^3b} &= \left(\frac{-8}{2}\right)\left(\frac{a^5}{a^3}\right)\left(\frac{b^2}{b}\right)\\ &= -4a^{5-3}b^{2-1}\\ &= -4a^2b\end{aligned}$$

2.2 POLYNOMIALS

A **polynomial** is the sum of two or more unlike monomials.

- To *add* two polynomials, write one polynomial underneath the other with like terms aligned in the same column. Then combine like terms. For example:

$$\begin{array}{r} 3x^2 + x + 8 \\ \underline{+\,x^2 \qquad - 9} \\ 4x^2 + x - 1 \end{array}$$

- To *subtract* one polynomial from another polynomial, change to an addition example by taking the opposite of each term of the polynomial that is being subtracted. For example, to subtract $x^2 - 9$ from $4x^2 + x - 1$, write the opposite of each term of $x^2 - 9$ under the corresponding term in $4x^2 + x - 1$. Then add like terms.

$$\begin{array}{r} 4x^2 + x - 1 \\ \underline{-x^2 \qquad + 9} \\ 3x^2 + x + 8 \end{array}$$

- To *multiply* a polynomial by a monomial, use the distributive property. For example:

$$3x(-2x^3 + 5x^2 - 8) = 3x(-2x^3) + 3x(5x^2) + 3x(-8x)$$
$$= -6x^4 \quad + 15x^3 \quad -24x^2$$

- To multiply two binomials together, use "FOIL." For example:

$$(x+5)(x-3) = \overbrace{(x)(x)}^{\textit{First terms}} + \overbrace{(-3)(x)}^{\textit{Outer terms}} + \overbrace{(5)(x)}^{\textit{Inner terms}} + \overbrace{(5)(-3)}^{\textit{Last terms}}$$
$$= x^2 + 2x - 15$$

- To *divide* a polynomial by a monomial, divide each term of the polynomial by the monomial. For example:

$$\frac{12z^4 + 20z^3 - 4z^2}{-4z^2} = \frac{12z^4}{-4z^2} + \frac{20z^3}{-4z^2} + \frac{-4z^2}{-4z^2}$$
$$= -3z^2 - 5z \quad + 1$$

2.3 FACTORING

Factoring undoes multiplication.

- Factoring out the greatest common factor of the terms of a polynomial reverses the distributive property. For example:

$$x^2 + 3x = x(x + 3)$$

and

$$2y^3 - 8y^2 + 6y = 2y(y^2 - 4y + 3)$$

You can check that you have factored correctly by multiplying the two factors together and verifying that the product is the original polynomial.

- The difference of the squares of two quantities can be factored as the product of the sum and difference of the terms that are being squared. For example:

$$x^2 - 4 = (x)^2 - (2)^2 = (x + 2)(x - 2)$$

and

$$2y^3 - 18y = 2y(y^2 - 9) = 2y(y + 3)(y - 3)$$

- Quadratic trinomials that may appear on the Regents exam can be factored using the reverse of FOIL. For example:

$$x^2 + 2x - 15 = (x + a)(x + b)$$

where a and b are chosen so that $a \times b = -15$ and, at the same time, $a + b = +2$. Since $(+5) \times (-3) = -15$ and $(+5) + (-3) = +2$:

$$x^2 + 2x - 15 = (x + 5)(x - 3)$$

2.4 ALGEBRAIC FRACTIONS

If a variable is in the denominator of a fraction, you can assume that it cannot represent a number that would make the denominator evaluate to 0. For example, in the fraction $\frac{x}{x-2}$, x cannot be equal to 2 since $2 - 2 = 0$ and a fraction with a zero denominator is not defined.

- To *simplify* a fraction, factor the numerator and factor the denominator. Then divide out any factor that is contained in the numerator and in the denominator since their quotient is 1. For example:

$$\frac{2x^2y - 50y}{4x^2 + 20x} = \frac{2y(x^2 - 25)}{4x(x + 5)}$$

$$= \frac{\overset{1}{\cancel{2}}y\cancel{(x+5)}(x - 5)}{\underset{2}{\cancel{4}}x\cancel{(x+5)}}$$

$$= \frac{y(x - 5)}{2x}$$

- To *multiply* algebraic fractions, first divide out any factor that is common to a numerator and a denominator. Then write the product of the remaining factors of the numerators over the product of the remaining factors of the denominators. For example:

$$\frac{2x^3}{x^2-x-12}\cdot\frac{x^2-16}{6x}=\frac{x^2\cdot 2x}{(x+3)(x-4)}\cdot\frac{(x-4)(x+4)}{6x}$$

$$=\frac{x^2\cdot \cancel{2x}}{(x+3)\cancel{(x-4)}}\cdot\frac{\overset{1}{\cancel{(x-4)}}(x+4)}{\underset{3}{\cancel{6x}}}$$

$$=\frac{x^2(x+4)}{3(x+3)}$$

- To *divide* algebraic fractions, change to a multiplication example by inverting the second fraction.
- To *combine* fractions with *like* denominators, write the sum of their numerators over their common denominator. For example:

$$\frac{5x+1}{3}+\frac{x-13}{3}=\frac{(5x-1)+(x+13)}{3}$$

$$=\frac{6x-12}{3}$$

$$=\frac{\overset{2}{\cancel{6}}(x-2)}{\cancel{3}}$$

$$=2(x-2)$$

- To *combine* fractions with *unlike* denominators, first change each fraction into an equivalent fraction with the least common denominator (LCD) of the fractions as its denominator. Then follow the rules for combining fractions with like denominators. For example, the LCD of the fractions in the sum $\frac{x+2}{2x}+\frac{1}{3x}$ is $6x$ since that is the smallest expression into which both denominators divide evenly. Since $6x \div 2x = 3$ and $6x \div 3x = 2$, multiply the first fraction by 1 in the form of $\frac{3}{3}$ and multiply the second fraction by 1 in the form of $\frac{2}{2}$:

$$\frac{x+2}{2x}+\frac{1}{3x}=\frac{3}{3}\left(\frac{x+2}{2x}\right)+\frac{2}{2}\left(\frac{1}{3x}\right)$$
$$=\frac{3(x+2)+2(1)}{6x}$$
$$=\frac{3x+6+3}{6x}$$
$$=\frac{3x+8}{6x}$$

2.5 LINEAR EQUATIONS

An equation such as $2x - 3 = 7x$ is a **first-degree** (or **linear**) **equation** since the greatest exponent of the variable is 1.

- To solve a first-degree equation in one variable, isolate the variable by undoing operations that are connected to the variable. When undoing an operation on one side of the equation, do the same thing on the opposite side. For example, if $3x = 2.1$, undo the multiplication of x by 3 by *dividing* both sides of the equation by 3: $\frac{3x}{3}=\frac{2.1}{3}$, so $x = 0.7$. If an equation contains parentheses, remove them first. For example, to solve $3(x + 1) - x = 13$ for x:

Remove the parentheses:	$3x + 3 - x = 13$
Collect like terms on the same side of the equation:	$2x = 13 - 3$
Divide both sides of the equation by 2:	$x = \frac{10}{2} = 5$

- To solve an equation for a given variable in terms of another variable, isolate the given letter. For example, to solve $x + y = 7y - x$ for x, rearrange the terms of the equation so that like letters appear on the same side of the equation:

Add x to each side:	$2x + y = 7y$
Subtract y from each side:	$2x = 6y$
Divide each side by 2:	$x = 3y$

2.6 RATIO AND PROPORTION

- A ratio compares two quantities by division. The ratio of x to y is written as $x:y$ or as $\frac{x}{y}$. If x is three times as great as y, then $\frac{x}{y} = \frac{3}{1}$ or 3:1.
- A proportion is an equation that states that two ratios are numerically equal. To change the proportion $\frac{a}{b} = \frac{c}{d}$ into an equation without fractions, cross-multiply: $\frac{a}{b} = \frac{c}{d}$ becomes $b \times c = a \times d$. For example, if $\frac{x}{x+4} = \frac{3}{5}$, then after cross-multiplying:

$$5x = 3(x+4)$$
$$5x = 3x + 12$$
$$2x = 12$$
$$x = \frac{12}{2} = 6$$

2.7 LINEAR INEQUALITIES

The symbols for inequality comparisons are as follows:

$<$ means "is less than," as in $2 < 5$.
$>$ means "is greater than," as in $7 > 3$.
$\leq$ means "is less than *or* equal to."
$\geq$ means "is greater than *or* equal to."

Here are some examples:

1. x is at most 90 is translated as, $x \leq 90$.
2. y is *at least* 48 is translated as, $y \geq 48$.
3. "When 2 is subtracted from a number n, the result is at most 5" is translated as,

$$\underbrace{\text{When 2 is subtracted from a number } n}_{n-2}, \ \underbrace{\text{the result is } \textit{at most}}_{\leq} \ \underbrace{5}_{5}.$$

The inequality is $n - 2 \leq 5$.

4. "When twice a number x, is increased by 5, the result is at least 27" is translated as,

$$\underbrace{\text{When twice a number } x}_{2x}, \ \underbrace{\text{is increased by 5}}_{+5}, \ \underbrace{\text{the result is \textit{at least}}}_{\geq} \ \underbrace{27}_{27}.$$

   The inequality is $2x + 5 \geq 27$.

- A first-degree inequality is solved in much the same way that a first-degree equation is solved except that, when *multiplying* or *dividing* both sides of an inequality by the same *negative* number, you must *reverse* the inequality. For example, to solve $1 - 2x > 9$ for x, first subtract 1 from each side to obtain $-2x > 8$. Then divide each side by -2 and, at the same time, reverse the inequality:

$$\frac{-2x}{-2} < \frac{8}{-2}$$

$$x < -4$$

- Solve a combined inequality such as $-3 \leq 2x - 1 < 9$ by isolating the variable in the middle part of the inequality:

Add 1 to each member: $-2 \leq 2x < 10$

Divide each member by 2: $\frac{-2}{2} \leq \frac{2x}{2} < \frac{10}{2}$

Simplify: $-1 \leq x < 5$

The inequality $-1 \leq x < 5$ includes all numbers from -1 (including -1) to 5 (not including 5), as shown in the accompanying number line. A darkened circle around an endpoint indicates that the point is included in the interval, while an unshaded circle around an endpoint means that the point is *not* included.

–1 0 1 2 3 4 5

2.8 FRACTIONAL EQUATIONS

To solve an equation with fractions, clear the equation of its fractions by multiplying each member by the least common denominator of the fractional terms. For example:

The given equation is: $\frac{2x}{x} + \frac{1}{7} = \frac{4}{x}$

Multiply each term of the equation by $7x$, the lowest common multiple of the denominators: $7x\left(\frac{2}{x}\right) + 7x\left(\frac{1}{7}\right) = 7x\left(\frac{4}{x}\right)$

Subtract 14 from each side of the equation:

$$\begin{array}{rrcr} 14 & +x & = & 28 \\ -14 & & = & -14 \\ \hline & x & = & 14 \end{array}$$

The value of x is **14**.

2.9 QUADRATIC EQUATIONS

To solve a factorable quadratic equation in which 0 appears alone on one side of the equation, factor the other side. Set each factor equal to 0, and solve the two equations that result.

- If $x^2 + 3x = 0$, then $x(x + 3) = 0$, so $x = 0$ or $x + 3 = 0$, making $x = 0$ or $x = -3$.
- If $x^2 - 4x + 4 = 0$, then $(x - 2)(x - 2) = 0$, so $x - 2 = 0$, making $x = 2$.
- If $x^2 + 2x = 15$, subtract 15 from each side of the equation so that all of the nonzero terms are on the left side, as in $x^2 + 2x - 15 = 0$. Then factor: $(x + 5)(x - 3) = 0$, so $x + 5 = 0$ or $x - 3 = 0$, making $x = -5$ or $x = 3$.

Practice Exercises

1. Express $\frac{15x^2}{-3x}$ in simplest form.

2. The expression $(3x^2y^3)^2$ is equivalent to
(1) $9x^4y^6$ (2) $9x^4y^5$ (3) $3x^4y^6$ (4) $6x^4y^6$

3. The product of $(-2xy^2)(3x^2y^3)$ is
(1) $-5x^3y^5$ (2) $-6x^2y^6$ (3) $-6x^3y^5$ (4) $-6x^3y^6$

4. Subtract $4m - h$ from $4m + h$.

5. Find the sum of $2x^2 + 3x - 1$ and $3x^2 - 2x + 4$.

6. Express the product $(2x - 7)(x + 3)$ as a trinomial.

7. When $3x^3 + 3x$ is divided by $3x$, the quotient is
(1) x^2 (2) $x^2 + 1$ (3) $x^2 + 3x$ (4) $3x^3$

8. Solve for y: $6(y + 3) = 2y - 2$.

9. Solve for y: $\frac{y}{3} + 2 = 5$.

10. Factor: $x^2 - 49$.

11. Factor: $x^2 + x - 30$.

12. Factor: $x^2 + 5x - 14$.

13. The solution set of $x^2 - x - 6 = 0$ is

(1) $\{1,-6\}$ (2) $\{-3,2\}$ (3) $\{3,-2\}$ (4) $\{5,1\}$

14. When expressed in factored form, the binomial $4a^2 - 9b^2$ is equivalent to

(1) $(2a - 3b)(2a - 3b)$ (3) $(4a - 3b)(a + 3b)$
(2) $(2a + 3b)(2a - 3b)$ (4) $(2a - 9b)(2a + b)$

15. What is the value of x in the equation $\frac{3}{4}x + 2 = \frac{5}{4}x - 6$?

(1) -16 (2) 16 (3) -4 (4) 4

16. What is the solution set of the equation $3x^2 = 48$?

(1) $\{-2,-8\}$ (2) $\{2,8\}$ (3) $\{4,-4\}$ (4) $\{4\}$

17. One of the roots of the equation $x^2 + 3x - 18 = 0$ is 3. What is the other root?

(1) 15 (2) 6 (3) -6 (4) -21

18. If $bx - 2 = K$, then x equals

(1) $\frac{K}{b} + 2$ (2) $\frac{K-2}{b}$ (3) $\frac{2-K}{b}$ (4) $\frac{K+2}{b}$

19. Seth has one less than twice the number of compact discs (CDs) that Jason has. Raoul has 53 more CDs than Jason. If Seth gives Jason 25 CDs, Seth and Jason will have the same number of CDs. How many CDs did each of the three boys have originally?

20. The Eye Surgery Institute just purchased a new laser machine for $500,000 to use during eye surgery. The institute must pay the inventor $550 each time the machine is used. If the institute charges $2,000 for each laser surgery, what is the *minimum* number of surgeries that must be performed in order for the institute to make a profit?

21. Solve for x: $\frac{x}{x+3} = \frac{5}{x+7}$.

22. A rectangular park is three blocks longer than it is wide. The area of the park is 40 square blocks. If w represents the width, write an equation in terms of w for the area of the park. Find the length and the width of the park.

Solutions

1. The given expression is: $\frac{15x^2}{-3x}$

To divide two monomials, first divide their numerical coefficients to find the numerical coefficient of the quotient: $(15) \div (-3) = -5$

Divide the literal factors to find the literal factor of the quotient. Remember that powers of the same base are divided by subtracting their exponents: $(x^2) \div (x^1) = x^1$

Combine the two results: $\frac{15x^2}{-3x} = -5x$

The fraction in simplest form is $\mathbf{-5x}$.

2. The given expression, $(3x^2y^3)^2$, means $(3x^2y^3)(3x^2y^3)$.

The numerical coefficient of the product is the product of the two numerical coefficients: $(3)(3) = 9$

In multiplying powers of the same base, add the exponents to obtain the exponent for that base in the product: $(x^2y^3)(x^2y^3) = x^4y^6$

Hence: $(3x^2y^3)^2 = 9x^4y^6$

The correct choice is **(1)**.

3. The given expression is: $(-2xy^2)(3x^2y^3)$

To multiply two monomials, first find the numerical coefficient of the product by multiplying the two numerical coefficients together: $(-2)(3) = -6$

Find the literal factor of the product by multiplying the literal factors together. The product of two powers of the same base is found by adding the exponents of that base: $(x^1y^2)(x^2y^3) = x^3y^5$

Combine the two results: $(-2xy^2)(3x^2y^3) = -6x^3y^5$

The correct choice is **(3)**.

4. Subtraction is the inverse operation of addition. Thus, to subtract, add the additive inverse of the subtrahend (expression to be subtracted) to the minuend (expression subtracted from). Therefore, to subtract $4m - h$ from $4m + h$, add $-4m + h$ to $4m + h$:

Subtract: $\begin{array}{r} 4m + h \\ \underline{4m - h} \end{array}$ means add: $\begin{array}{r} 4m + h \\ \underline{-4m + h} \\ 2h \end{array}$

The difference is $\mathbf{2h}$.

5. Write the second trinomial under the first, placing like terms in the same column:

$$\begin{array}{r} 2x^2 + 3x - 1 \\ \underline{3x^2 - 2x + 4} \end{array}$$

Add each column by adding the numerical coefficients algebraically and bringing down the literal factor:

$$5x^2 + x + 3$$

The sum is $\mathbf{5x^2 + x^2 + 3}$.

6. The product of two binomials may be found by multiplying each term of one by each term of the other, using a procedure analogous to that used in arithmetic to multiply multidigit numbers:

$$\begin{array}{rrr} & 2x & -\ 7 \\ & \underline{x} & \underline{+\ 3} \\ 2x^2 & -\ 7x & \\ & \underline{6x} & \underline{-\ 21} \end{array}$$

Combine like terms:

$$2x^2 - x - 21$$

The trinomial is $\mathbf{2x^2 - x - 21}$.

7. Indicate the division in fractional form:

$$\frac{3x^3 + 3x}{3x}$$

Apply the distributive law by dividing each term of $3x^3 + 3x$ in turn by $3x$. Remember that, in dividing powers of the same base, the exponents are subtracted. Also note that $3x \div 3x = 1$:

$$x^2 + 1$$

The correct choice is **(2)**.

8. The given expression is: $6(y + 3) = 2y - 2$

Remove parentheses by applying the distributive law of multiplication over addition: $6y + 18 = 2y - 2$

Add –18 (the additive inverse of 18) and also add $-2y$ (the additive inverse of $2y$) to both sides of the equation:

$$\begin{array}{r} -2y - 18 = -2y - 18 \\ \hline 4y \quad = \quad -20 \end{array}$$

Divide both sides by 4: $\dfrac{4y}{4} = \dfrac{-20}{4}$

The solution is $y = \mathbf{-5}$. $y = -5$

9. The given expression is: $\dfrac{y}{3} + 2 = 5$

To clear fractions in the equation, multiply each term by the least common denominator, in this case, by 3: $3\left(\dfrac{y}{3}\right) + 3(2) = 3(5)$

Simplify: $y + 6 = 15$

Add –6 (the additive inverse of 6) to both sides:

$$\begin{array}{r} -6 = -6 \\ \hline y \quad = 9 \end{array}$$

The solution is $y = \mathbf{9}$.

10. The given binomial, $x^2 - 49$ represents the difference between two perfect squares, x^2 and 49. To factor such an expression, take the square root of each perfect square:

$$\sqrt{x^2} = x \quad \text{and} \quad \sqrt{49} = 7$$

One factor will be the sum of the respective square roots, and the other factor will be the difference of the square roots:

$$x^2 - 49 = (x + 7)(x - 7)$$

The factored form is $\mathbf{(x + 7)(x - 7)}$.

11. The given expression is a quadratic trinomial: $x^2 + x - 30$

The factors of a quadratic trinomial are two binomials. The factors of the first term, x^2, are x and x, and they become the first terms of the binomials: $(x \quad)(x \quad)$

The factors of the last term, –30, become the second terms of the binomials, but they must be chosen in such a way that the sum of the product of the inner terms and the product of the outer terms is equal to the middle term, $+x$, of the original trinomial. Try +6 and –5 as the factors of –30:

$+6x$ = inner product

$(x + 6)(x - 5)$

$-5x$ = outer product

Since $+6x$ and $-5x$ add up to $+x$, these are the correct factors: $(x + 6)(x - 5)$

The factored form is $\mathbf{(x + 6)(x - 5)}$.

12. The given expression is a quadratic trinomial: $x^2 + 5x - 14$

The factors of a quadratic trinomial are two binomials. The first terms of the binomial are the factors, x and x, of the first term, x^2, of the trinomial: $(x \quad)(x \quad)$

The second terms of the binomials are the factors of the last term, –14, of the trinomial. These factors must be chosen in such a way that the sum of the product of the inner terms and the product of the outer terms is equal to the middle term, $+5x$, of the original trinomial. The term

−14 has factors of 14 and −1, −14 and 1, +7 and −2, and −7 and +2. Try +7 and −2:

$+7x$ = inner product

$(x + 7)(x - 2)$

$-2x$ = outer product

Since $(+7x) + (-2x) = +5x$, these are the correct factors:

$(x + 7)(x - 2)$

The factored form is $\mathbf{(x + 7)(x - 2)}$.

13. The given expression is a quadratic equation:

$x^2 - x - 6 = 0$

The solution set is found by solving the equation by factoring. The left side is a *quadratic trinomial*, which can be factored into two binomials.

The factors of the first term, x^2, are x and x, and they constitute the first terms of each binomial factor:

$(x \quad)(x \quad) = 0$

The last term, −6, must be factored into two factors that will be the second terms of the binomials. The factors must be chosen in such a way that the sum of the product of the two inner terms and the product of the two outer terms is equal to the middle term of the original trinomial. Try $-6 = (-3)(+2)$:

$-3x$ = inner product

$(x - 3)(x + 2) = 0$

$+2x$ = outer product

Thc sum of the inner and outer products is $-3x + 2x$ or $-x$, which is equal to the middle term of the original trinomial. Thus, the correct factors have been chosen, and the equation can be written as:

$(x - 3)(x + 2) = 0$

Since the product of two factors is 0, either factor may be equal to 0: $x - 3 = 0$ *or* $x + 2 = 0$

Add the appropriate additive inverse to each side, +3 in the case of the left equation and –2 in the case of the right one:

$$\frac{\begin{array}{r}3 = 3\end{array}}{x = 3} \qquad \frac{\begin{array}{r}-2 = -2\end{array}}{x = -2}$$

Thus, the solution set is {3,–2}.

The correct choice is **(3)**.

14. Factor $4a^2 - 9b^2$ as the sum and difference of two squares:

$$\begin{aligned} 4a^2 - 9b^2 &= (2a)^2 - (3b)^2 \\ &= (2a + 3b)(2a - 3b) \end{aligned}$$

The correct choice is **(2)**.

15. The given equation is: $\frac{3}{4}x + 2 = \frac{5}{4}x - 6$

Remove the fractions by multiplying both sides of the equation by 4:

$$4\left(\frac{3}{4}x + 2\right) = 4\left(\frac{5}{4}x - 6\right)$$

$$4\left(\frac{3}{4}x\right) + 4(2) = 4\left(\frac{5}{4}x\right) + 4(-6)$$

$$3x + 8 = 5x - 24$$

Isolate x:

$$3x - 5x = -24 - 8$$

$$-2x = -32$$

$$\frac{-2x}{-2} = \frac{-32}{-2}$$

$$x = 16$$

The correct choice is **(2)**.

16. The given equation is: $3x^2 = 48$

Divide both sides by 3: $\frac{3x^2}{3} = \frac{48}{3}$

$x^2 = 16$

Solve for x: $x = \pm\sqrt{16}$

$= \pm 4$

The solution set is $\{4,-4\}$.

The correct choice is **(3)**.

17. It is given that one of the roots of the equation $x^2 + 3x - 18 = 0$ is 3. To find the other root of the given equation, solve the quadratic equation by factoring the quadratic trinomial as the product of two binomials:

$$x^2 + 3x - 18 = 0$$
$$(x + ?)(x + ?) = 0$$

Since 3 is a root, $x - 3$ is a factor of $x^2 + 3x - 18$: $(x + ?)(x - 3) = 0$

The product of the constant terms of the two binomial factors is -18, the constant term of $x^2 + 3x - 18$.

Since $(-3)(+6) = -18$, the missing term in the binomial factor is $+6$: $(x + 6)(x - 3) = 0$

If the product of two numbers is 0, then at least one of the factors is equal to 0: $x + 6 = 0$ or $x - 3 = 0$

$x = -6$ $\quad$ $x = 3$

The other root of the equation is -6.

The correct choice is **(3)**.

18. The given equation is: $bx - 2 = K$

Add 2 to both sides: $bx = K + 2$

Divide both sides by b: $\frac{bx}{b} = \frac{K+2}{b}$

$$x = \frac{K+2}{b}$$

The correct choice is **(4)**.

19. It is given that Seth has one less than twice the number of compact discs (CDs) that Jason has and that Raoul has 53 more CDs than Jason.

- If x represents the number of CDs that Jason has, then $2x - 1$ represents the number of CDs that Seth has, and Raoul has $x + 53$ CDs.
- After Seth gives Jason 25 CDs, Seth has $(2x - 1) - 25 = 2x - 26$ CDs remaining and Jason has $x + 25$ CDs.
- Since it is given that Seth and Jason now have the same number of CDs:

$$2x - 26 = x + 25$$
$$2x = x + 25 + 26$$
$$2x = x + 51$$
$$2x - x = 51$$
$$x = 51$$

To begin with,

Jason had $x = 51$ CDs,
Seth had $2x - 1 = 2(51) - 1 = 102 - 1 = 101$ CDs,
and Raoul had $x + 53 = 51 + 53 = 104$ CDs.

20. Let x represent the number of surgeries that are performed. Since profit is the difference between revenue and cost, we must find the smallest positive integer value of x for which

$$\overbrace{2{,}000x}^{\text{Revenue}} - \overbrace{(550x + 500{,}000)}^{\text{Cost}} > 0$$

$$2{,}000x - 550x - 500{,}000 > 0$$

$$\frac{1{,}450x}{1{,}450} > \frac{500{,}000}{1{,}450}$$

$$x > 344.8275862$$

Since x must be a whole number, the minimum number of surgeries that must be performed to make a profit is **345**.

21. The given equation is: $\frac{x}{x+3} = \frac{5}{x+7}$

Set the cross products equal to each other: $x(x + 7) = 5(x + 3)$

Remove the parentheses by multiplying each term inside the parentheses by the term in front of the parentheses: $x^2 + 7x = 5x + 15$

Collect all of the nonzero terms on left side of the equation so that 0 is on the right side of the equation: $x^2 + 2x - 15 = 0$

Factor the left side of the equation as the product of two binomials: $(x + ?)(x - ?) = 0$

The missing terms of the binomial factors are the two numbers whose product is –15, the last term of $x^2 + 2x - 15$ and whose sum is +2, the coefficient of the x-term of $x^2 + 2x - 15$. Because $(+5)(-3) = -15$ and $(+5) + (-3) = +2$, the missing terms are –3 and +5: $(x - 3)(x + 5) = 0$

If the product of two numbers is 0, then at least one of these numbers is 0: $x - 3 = 0$ or $x + 5 = 0$

$x = 3$ or $x = -5$

The values of x that satisfy the equation are **3** and **–5**.

22. It is given that a rectangular park is three blocks longer than it is wide. The area of the park is 40 square blocks. If w represents the width of the park, then $w + 3$ represents the length of the park. Since the area of a rectangle is the product of its width and length:

$$w(w + 3) = 40$$

To find the dimensions of the park, solve the equation algebraically or use guess and check.

Method I: Solve algebraically.

If $w(w + 3) = 40$, then $w^2 + 3w = 40$ and $w^2 + 3w - 40 = 0$.

Factor the left side of the quadratic equation as the product of two binomials: $(w + ?)(w - ?) = 0$

The missing terms of the binomial factors are the two integers whose sum is +3, the coefficient of the w-term of $w^2 + 3w = 40$, and whose product is –40, the last term of $w^2 + 3w = 40$. Since $(+8) + (-5) = +3$ and $(+8)(-5) = -40$, the missing terms of the binomial factors are +8 and –5: $(w + 8)(w - 5) = 0$

If the product of two terms is 0, then at least one of these terms is 0:

$$w + 8 = 0 \quad \text{or} \quad w - 5 = 0$$

reject since width must be positive ⟶ $[w = -8]$ or $w = 5$

$$w + 3 = 5 + 3 = 8$$

The length of the park is **8** blocks, and the width of the park is **5** blocks.

Method II: Solve by guess and check.

Since $w(w + 3) = 40$ and $w > 0$, you need to find two positive integers whose product is 40 such that the larger of these two integers is 3 more than the smaller integer.

- Guess 1: Try 4 and 7: $4 \times 7 = 28$. The product is too small.
- Guess 2: Try 7 and 10: $7 \times 10 = 70$. The product is too large.
- Guess 3: Try 5 and 8: $5 \times 8 = 40$. ✔

3. GEOMETRY

3.1 PERIMETER AND AREA FORMULAS

The distance around a circle is called its **perimeter**. The **area** of a figure is the number of 1×1 squares that it can enclose. Perimeter and area formulas for six types of figures are given in the accompanying table.

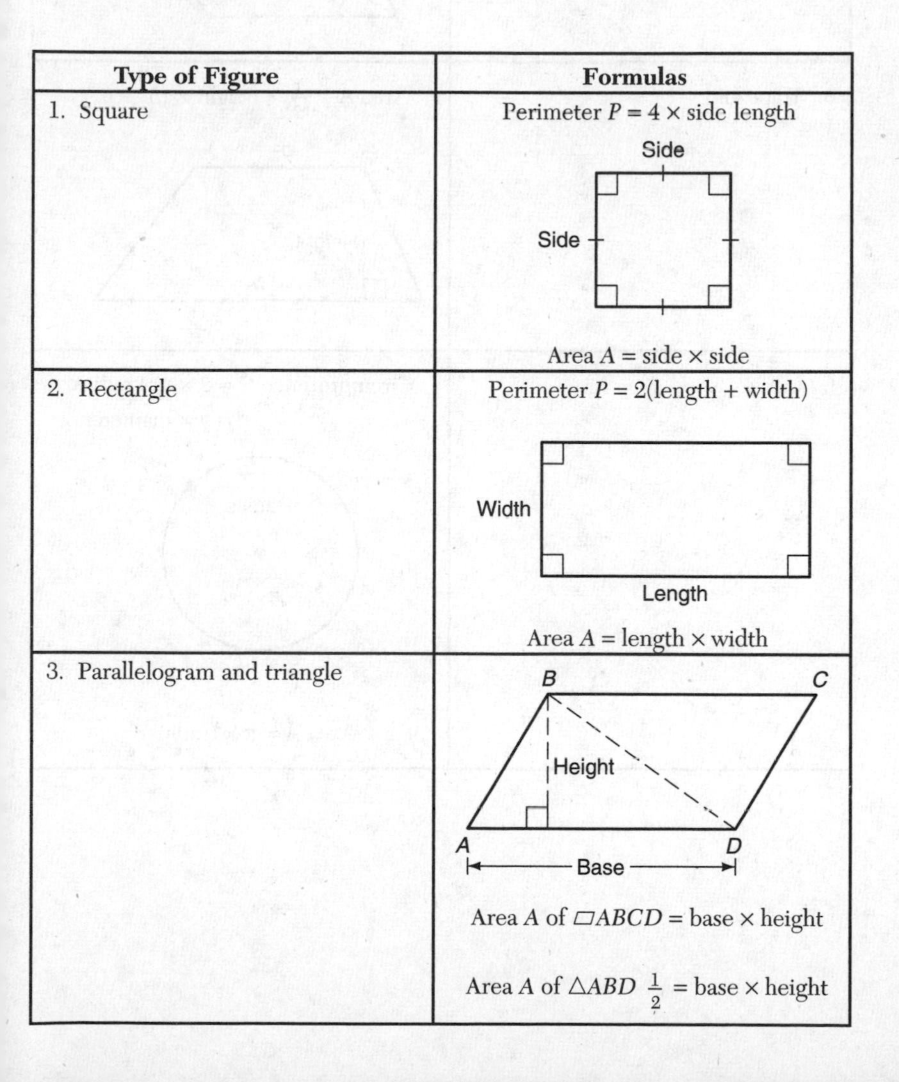

Type of Figure	Formulas
1. Square	Perimeter $P = 4 \times$ side length Area $A =$ side $\times$ side
2. Rectangle	Perimeter $P = 2$(length + width) Area $A =$ length $\times$ width
3. Parallelogram and triangle	Area A of $\square ABCD =$ base $\times$ height Area A of $\triangle ABD$ $\frac{1}{2}$ = base $\times$ height

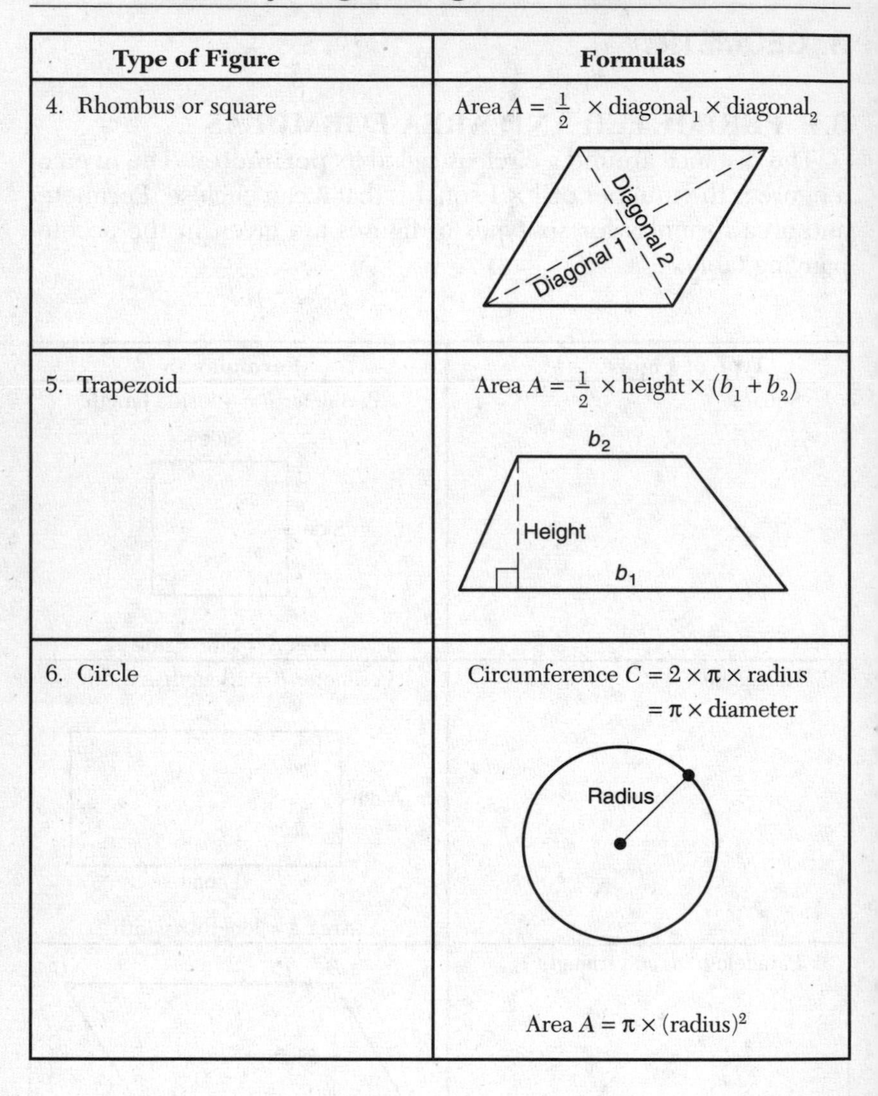

Type of Figure	Formulas
4. Rhombus or square	Area $A = \frac{1}{2} \times \text{diagonal}_1 \times \text{diagonal}_2$
5. Trapezoid	Area $A = \frac{1}{2} \times \text{height} \times (b_1 + b_2)$
6. Circle	Circumference $C = 2 \times \pi \times \text{radius}$ $= \pi \times \text{diameter}$ Area $A = \pi \times (\text{radius})^2$

3.2 FINDING AREAS INDIRECTLY

To find the area of the region in which two figures overlap, we usually need to subtract the areas of the two figures.

- Concentric circles are circles with the same center. To find the area, A, of the region between two concentric circles, subtract the areas of the two circles. If in the accompanying diagram, $R = 5$ and $r = 3$, then the area of the ring between the two concentric circles is $(\pi \times 5^2) - (\pi \times 3^2) = 16\pi$.

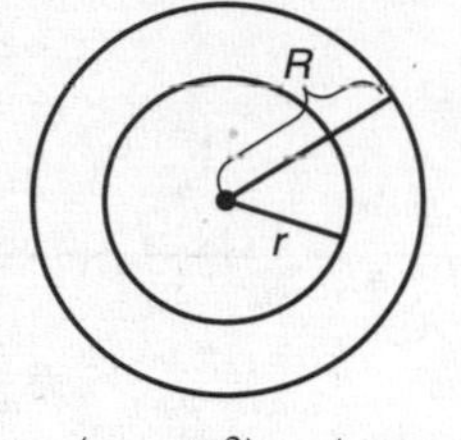

$$A = (\pi \times R^2) - (\pi \times r^2)$$

- To find the area of the shaded region in the accompanying figure, subtract the sum of the areas of the two circles from the area of the rectangle. If the radius of each circle is 5, then the sum of the areas of the two circles is $(\pi \times 5^2) + (\pi \times 5^2) = 50\pi$.

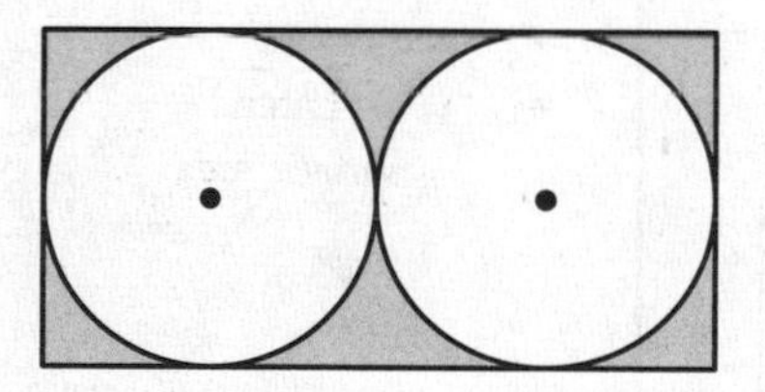

The width of the rectangle is one diameter or 10, and the length of the rectangle is two diameters or 20. Hence, the area of the rectangle is $20 \times 10 = 200$. The area of the shaded region in terms of π is $200 - 50\pi$.

3.3 VOLUME AND SURFACE AREA

The accompanying table shows how to find the volume and the surface area of several three-dimensional solids.

Figure	Surface Area	Volume
1. Rectangular Solid 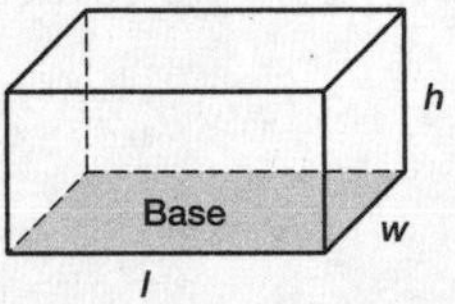	$2(\ell \times w) + 2(\ell \times h) + 2(w \times h)$	$\ell \times w \times h$
2. Cube 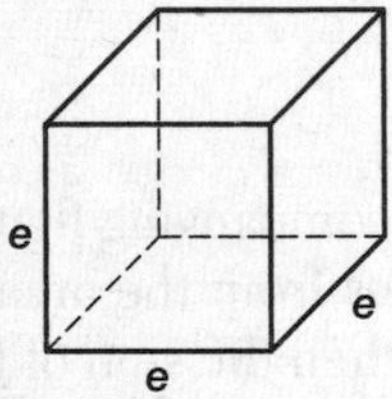	$6e^2$	e^3
3. Cylinder 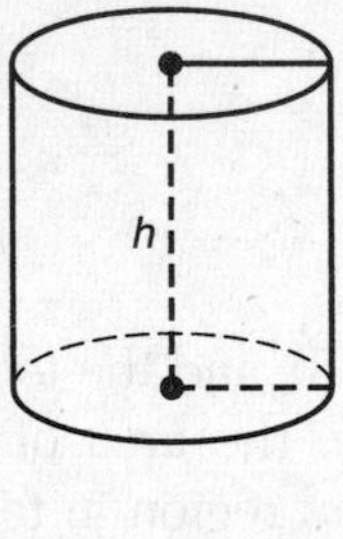	$2\pi r^2 + \underbrace{2\pi rh}_{\text{Lateral surface area}}$	$\pi r^2 h$

Coordinate Geometry

3.4 COORDINATE PLANE

When a horizontal number line (called the **x-axis**) intersects a vertical number line (the **y-axis**) at their common zero point, called the **origin**, a rectangular coordinate system is created. The two axes divide the plane into four quadrants that are numbered, using roman numerals, in the counterclockwise direction, as shown in the accompanying figure. Points are located using an ordered pair of numbers, called **coordinates**, of the form (x,y).

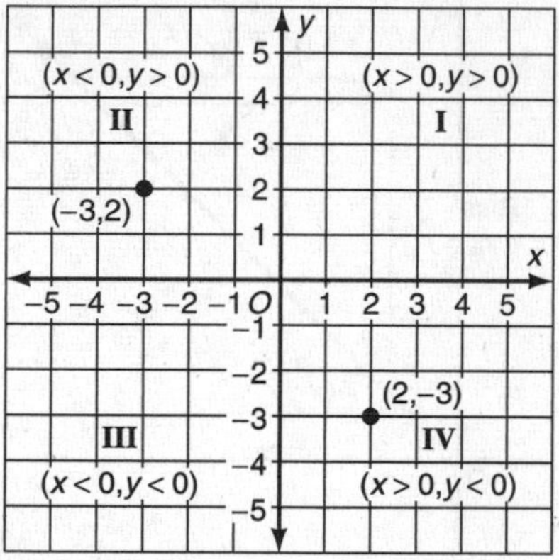

- The x-coordinate gives the horizontal distance and direction of the point from the origin. When x is positive, the point lies to the right of the origin; when x is negative, the point lies to the left of the origin. For example, point $(-3,2)$ is located in Quadrant II since $x = -3$ and $y = 2$.
- The y-coordinate gives the vertical distance and direction of the point from the origin. When y is positive, the point lies above the origin; when y is negative, the point lies below the origin. For example, point $(2,-3)$ is located in Quadrant IV since $x = 2$ and $y = -3$.

3.5 SLOPE FORMULA

The slope of a line is a number that represents the steepness of a line. The slope of the nonvertical line that passes through $A(x_A,y_A)$ and $B(x_B,y_B)$ can be determined using the formula

$$\text{slope} = \frac{\text{rise (difference in } y)}{\text{run (difference in } x)}$$

$$= \frac{y_B - y_A}{x_B - x_A}$$

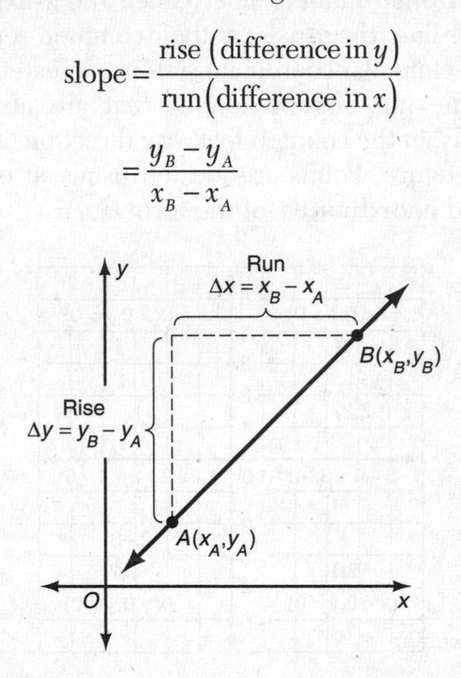

It is customary to represent slope using the letter *m*, to represent $y_B - y_A$ by Δy (read "delta *y*"), and to represent $x_B - x_A$ by Δx (read "delta *x*"). Thus,

$$\text{slope} = m = \frac{\Delta y}{\Delta x}$$

For example, to find the slope, *m*, of the line that passes through $A(-1,2)$ and $B(1,8)$, use the slope formula:

$$m = \frac{\Delta y}{\Delta x} = \frac{8-2}{1-(-1)} = \frac{6}{2} = 3\,.$$

3.6 SOME FACTS ABOUT SLOPE

The slope of a line represents the rate at which y changes with respect to x. The slope of a line may be positive, negative, 0, or undefined.

- A line that rises as x increases has a positive slope. A line that falls as x increases has a negative slope.

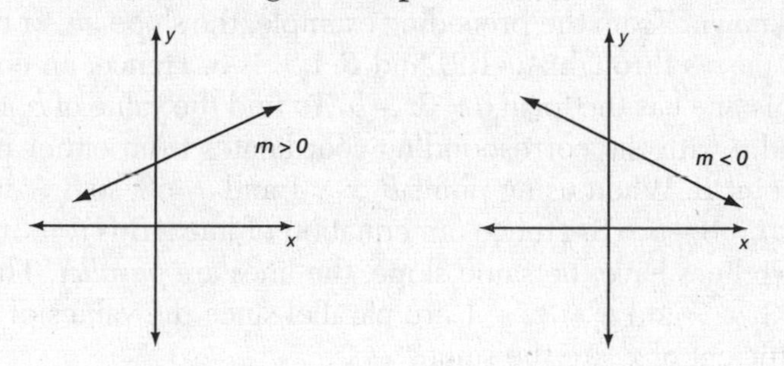

- The slope of a horizontal line is 0.

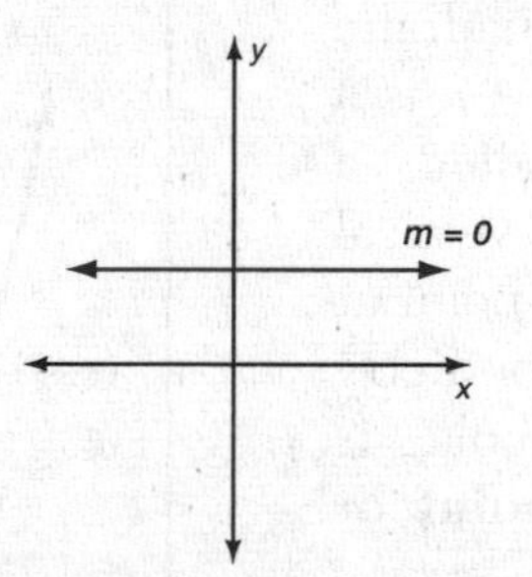

- The slope of a vertical line is undefined.

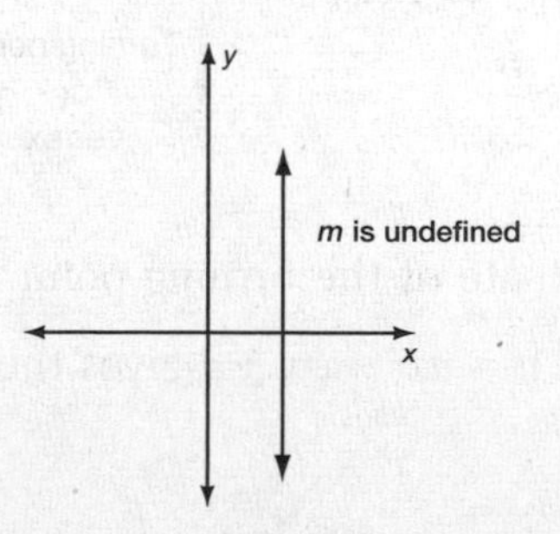

3.7 SOME FACTS ABOUT SLOPE

The set of points (x,y) on a nonvertical line satisfy the equation $y = mx + b$ where m represents the slope of the line and b is the y-coordinate of the point at which the line crosses the y-axis. This point is called the ***y*-intercept**.

- An equation of a line can be written when its slope and y-intercept are known. From the preceding example, the slope, m, of the line that passes through $A(-1,2)$ and $B(1,8)$ is 3. Hence, an equation of this line has the form $y = 3x + b$. To find the value of b, replace x and y with the corresponding coordinates from either point A or point B. When using point B, $x = 1$ and $y = 8$, so $8 = 3(1) + b$ and $b = 8 - 3 = 5$. Hence, an equation of line AB is $y = 3x + 5$.
- If two lines have the same slope, the lines are *parallel*. The lines $y = 3x - 7$ and $y = 3x + 1$ are parallel since the values of m, the coefficient of x, are the same.

3.8 PARABOLAS

The graph of the quadratic equation $y = ax^2 + bx + c$ is a U-shaped curve, called a **parabola**, that has a vertical axis of symmetry. The point at which the axis of symmetry intersects the parabola is called the **turning point** or **vertex**.

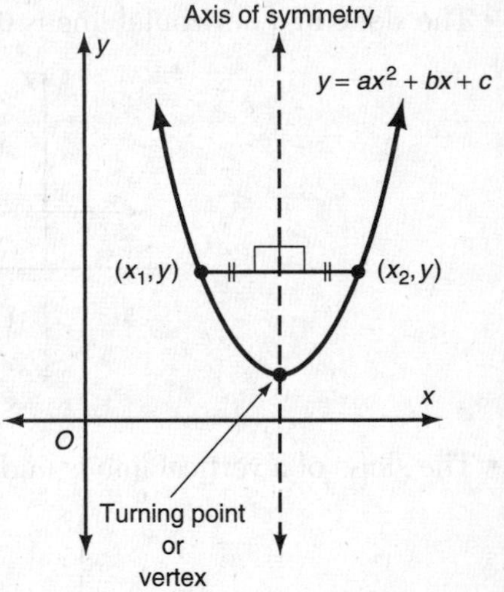

- An equation of the axis of symmetry is $x = -\dfrac{b}{2a}$.

- Since the x-coordinate of the turning point is $-\dfrac{b}{2a}$, substituting this value for x in $y = ax^2 + bx + c$ gives the y-coordinate of the turning point.

- The sign of a, the coefficient of the x^2-term, determines whether the turning point of a parabola is a minimum or a maximum point on the graph. If $a > 0$, the turning point is the *lowest* point on the parabola. If $a < 0$, the turning point is the *highest* point on the parabola.

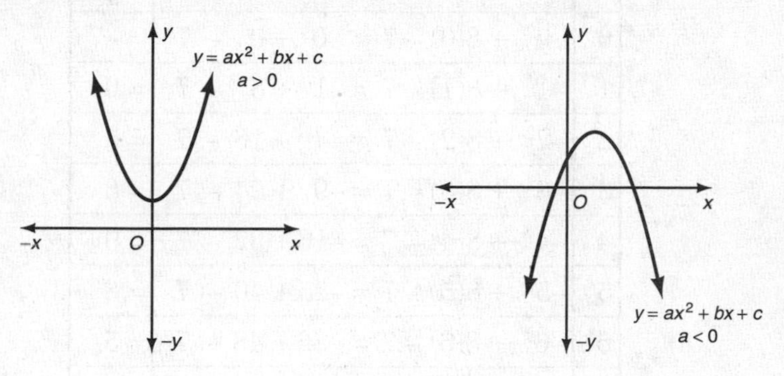

3.9 GRAPHING A PARABOLA: AN EXAMPLE

a. Write an equation of the axis of symmetry of the graph of $y = -x^2 + 8x - 7$.

b. Draw the graph of the equation $y = -x^2 + 8x - 7$, including all integral values of x such that $0 \leq x \leq 8$.

c. From the graph drawn in part ***b***, find the roots of the equation $-x^2 + 8x - 7 = 0$.

a. If an equation of a parabola is in the form $y = ax^2 + bx + c$, an equation of its axis of symmetry is in the form $x = \dfrac{-b}{2a}$.

The given equation, $y = -x^2 + 8x - 7$, is in the form $y = ax^2 + bx + c$, with $a = -1$, $b = 8$, and $c = -7$. An equation of its axis of symmetry is

$$x = \frac{-8}{2(-1)}$$

$$= \frac{-8}{-2}$$

$$= 4$$

An equation of the axis of symmetry is $\boldsymbol{x = 4}$.

b. To graph $y = -x^2 + 8x - 7$ for $0 \leq x \leq 8$, prepare a table of values for x and y by substituting each integral value of x, from 0 to 8 inclusive, in the equation to determine the corresponding value of y.

x	$-x^2 + 8x - 7$	$= y$
0	$-0^2 + 8(0) - 7 = 0 + 0 - 7$	$= -7$
1	$-1^2 + 8(1) - 7 = -1 + 8 - 7$	$= 0$
2	$-2^2 + 8(2) - 7 = -4 + 16 - 7$	$= 5$
3	$-3^2 + 8(3) - 7 = -9 + 24 - 7$	$= 8$
4	$-4^2 + 8(4) - 7 = -16 + 32 - 7$	$= 9$
5	$-5^2 + 8(5) - 7 = -25 + 40 - 7$	$= 8$
6	$-6^2 + 8(6) - 7 = -36 + 48 - 7$	$= 5$
7	$-7^2 + 8(7) - 7 = -49 + 56 - 7$	$= 0$
8	$-8^2 + 8(8) - 7 = -64 + 64 - 7$	$= -7$

Plot points (0,–7) (1,0), (2,5), (3,8), (4,9), (5,8), (6,5), (7,0), and (8,–7), and draw a smooth curve through them. This curve is the graph of $y = -x^2 + 8x - 7$ for $0 \leq x \leq 8$.

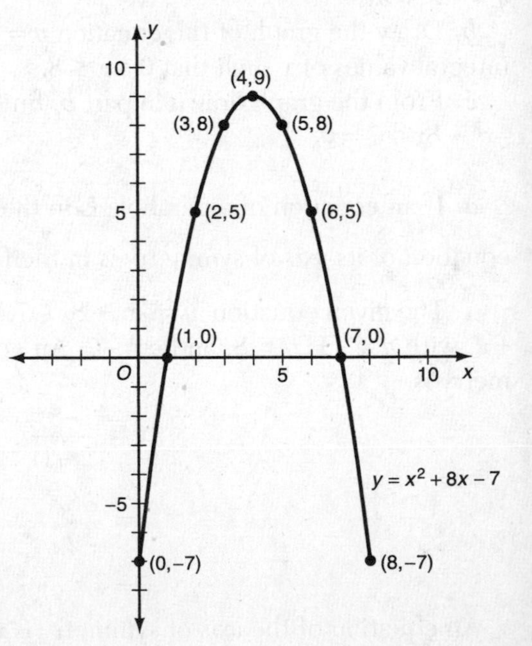

c. The roots of $-x^2 + 8x - 7 = 0$ are the values of x on the graph at the points where $y = 0$, which is the x-axis. The graph crosses the x-axis at the two points where $x = 1$ and $x = 7$.

The roots are **1** and **7**.

3.10 GRAPHING A PARABOLA USING TECHNOLOGY

A graphing calculator can be used to graph a parabola or create the corresponding table of values quickly.

- To graph $y = -x^2 + 8x - 7$, open the $Y =$ editor and set $Y_1 = -X \wedge 2 + 8X - 7$ by pressing

  [(–)] [x, T, θ, n] [^] [2] [+] [8] [x, T, θ, n] [–] [7]

- To create a table of values for $y = -x^2 + 8x - 7$:

  1. Press [2nd] [TBLSET] .
  2. Change the **TblStart** value to 0 since 0 is 4 units less than the x-coordinate of the vertex. If necessary, set ΔTbl = 1 so that x increases in steps of 1 unit.
  3. Press [2nd] [TABLE] to view the table. If you need to look at table entries that are not currently in view, use a cursor key to scroll up or down.

Relations and Functions

3.11 DEFINITIONS

A relation is any set of ordered pairs of numbers of the form (x,y). A **function** is a relation in which no two ordered pairs have the same first member but different second members. The relation $f = \{(1,1), (2,4), (3,9)\}$ is a function while the relation $g = \{(9,3), (0,1), (9,-3)\}$ is not a function since (9,3) and (9,–3) have the same first member but different second members.

The set of all first members or x-values of a function is called the **domain**, and the corresponding set of all second members or y-values is called the **range**. For the function $f = \{(1,1), (2,4), (3,9)\}$, the domain is $\{1,2,3\}$ and the range is $\{1,4,9\}$. Unless otherwise specified, the domain of a function is assumed to be the largest possible set of real numbers.

A function may be represented in any one of the following ways:

- A set of ordered pairs. The ordered pairs may be displayed as a pairing of the elements of two sets, as shown in the accompanying figure.

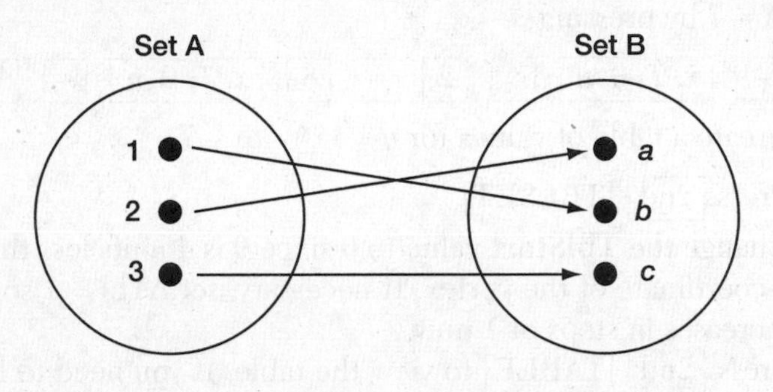

The set of ordered pairs that belong to the function is

$$\{(1,b), (2,a), (3,c)\}.$$

- An algebraic equation in two variables such as $y = 2x + 1$. Not all equations represent functions. The equation $y = x^2$ is a function, but the equation $x = y^2$ is not a function since it is possible to produce ordered pairs such as (4,2) and (4,–2) that have the same first member but different second members.
- A graph. Not all graphs represent functions. A graph represents a function only when no vertical line intersects it in more than one point, as illustrated in the accompanying figures.

The graph represents a function since it passes the vertical line test.

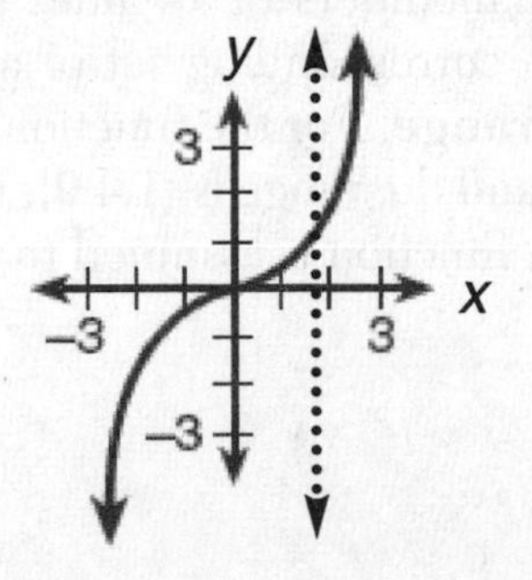

The graph does not represent a function since it fails the vertical line test.

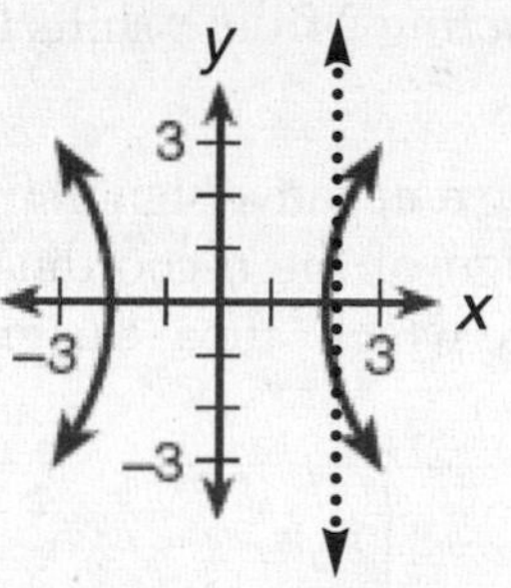

3.12 ABSOLUTE VALUE FUNCTION

The absolute value function $y = |x|$ assigns to y the absolute value of x so that y is always non-negative for each value of x in its domain. The graph of the absolute value function is "**V**-shaped," contains the origin, is symmetric with respect to the y-axis, and consists of those parts of the graphs of $y = x$ and $y = -x$ that are above the x-axis, as shown in the accompanying figure.

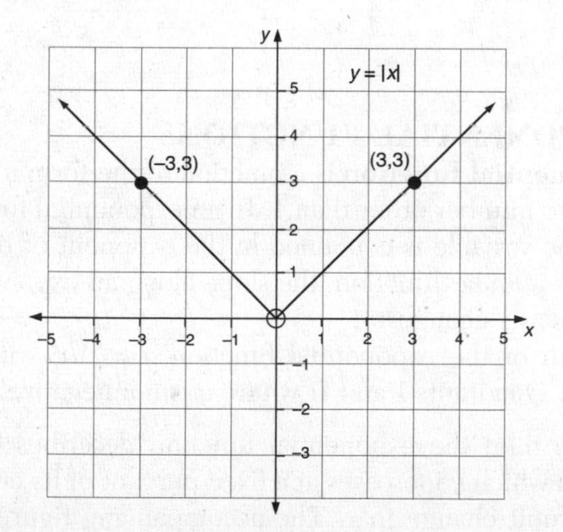

The graph of $y = -|x - 2|$ is shown in the accompanying figure.

- The effect of *subtracting* 2 from x shifts the graph of $y = |x|$ to the *right* 2 units.
- The effect of placing a negative sign in front of the absolute value sign changes the sign of the y-coordinate of each point on the graph to its opposite which "flips" the graph over the x–axis.

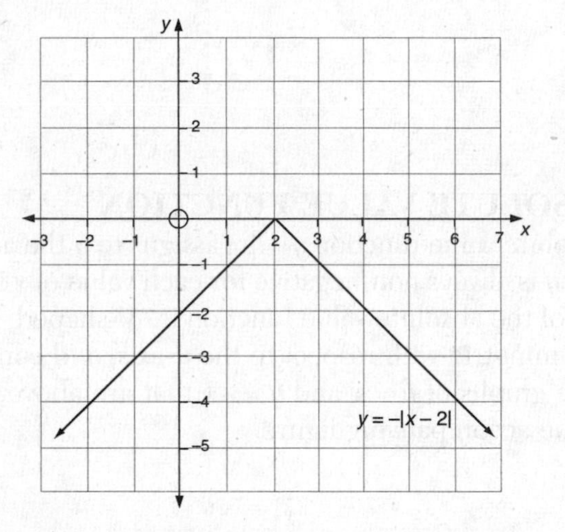

3.13 EXPONENTIAL FUNCTIONS

An **exponential function** is a function of the form $y = b^x$ where b is a positive number other than 1. In an exponential function such as $y = 2^x$, the variable is contained in the exponent of the constant base. Unlike a linear function, the slope along an exponential function changes as x changes.

The graph of the exponential function $y = Ab^x$ with $A > 0$ is restricted to Quadrants I and II where y is nonnegative.

- If $b > 1$, then the exponential function describes exponential growth in which y increases at a fixed percent of its current value for each unit change in x. The accompanying figure shows the graph of $y = 2^x$, which is restricted to Quadrants I and II, where y is always positive. The graph passes the vertical line test, rises as x increases, has $y = 1$ as its y-intercept, and never touches the x-axis.

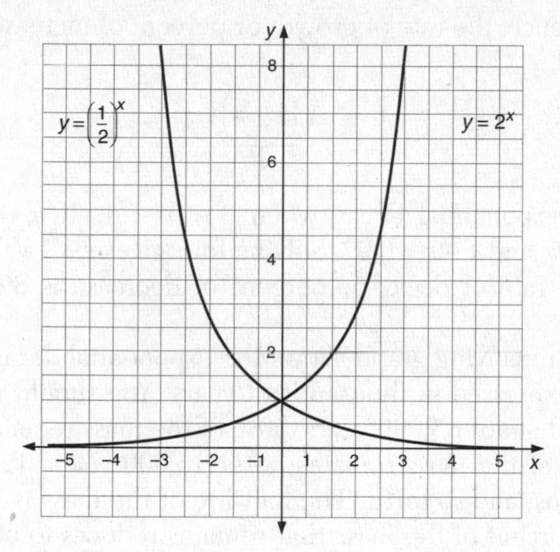

- If $0 < b < 1$, then the exponential function describes exponential decay in which y decreases at a fixed percent of its current value for each unit change in x. The accompanying figure shows the graph of $y = \left(\frac{1}{2}\right)^x$, which is restricted to Quadrants I and II, where y is always positive. The graph passes the vertical line test, falls as x increases, has $y = 1$ as its y-intercept, and never touches the x-axis.

3.14 EXPONENTIAL GROWTH AND DECAY

In a linear function of the form $y = mx + b$, y changes at a constant rate. In an exponential function of the form $y = Ab^x$, y changes as a percent of its current value. If A = the initial amount of a substance, r = the rate of growth or decay, and x = the number of consecutive time periods over which the growth or decay process is being studied, then

$$y = A\underbrace{(1+r)}_{\text{“}b\text{”}}^{x}$$

represents exponential growth when $r > 0$. If $y = 320(1.07)^5$, then $A = 320$, $x = 5$, and $1 + r = 1.07$. Solving $1 + r = 1.07$ for r gives $r = 0.07$. Hence, the rate of growth or percent of increase is 7% per time period.

$$y = A\underbrace{\left(1-r\right)}_{\text{“}b\text{”}}{}^{x}$$

represents exponential decay when $0 < r < 1$. If $y = 12(0.92)^5$, $A = 12$, $x = 5$, and $1 - r = 0.92$. Solving for r gives $r = 1 - 0.92 = 0.08$. Hence, the rate of decay or percent of decrease is 8% per time period.

The accompanying graph shows the exponential decay of a mass over time expressed in thousands of years. You should verify from that graph that after 5,000 years, 50% of the mass remains; after 10 years, 25% of the mass remains; after 15,000 years, 12.5% of the mass remains; and so forth. The half-life of the mass is 5,000 years since the percent of the mass that remains reduces to 50% or one-half of its previous value every 5,000 years.

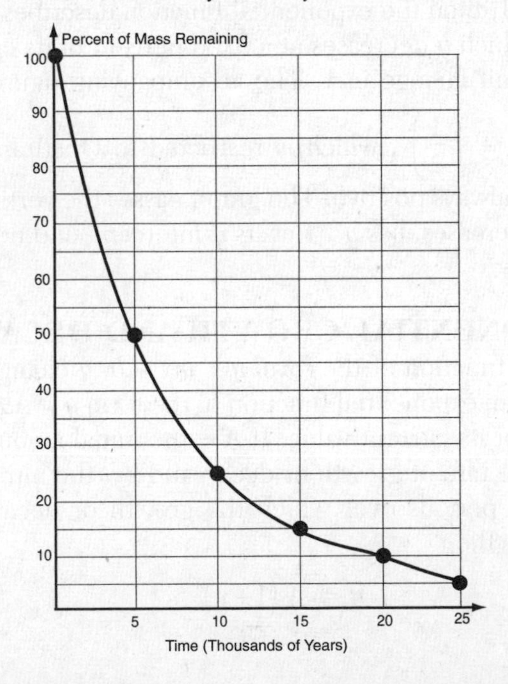

EXAMPLE

On January 1 of a certain year, the price of gasoline was $2.79 per gallon. Throughout the year, the price of gasoline increased by 1.5% per month. What was the cost of one gallon of gasoline, to the nearest cent, one year later on January 1?

Solution

If y represents the cost of one gallon of gasoline 12 months (one year) later on January 1, then $y = A(1 + r)^x$ where $A = 2.79$, $r = 1.5\% = 0.015$, and $x = 12$. Thus:

$$\begin{aligned} y &= A(1 + r)^x \\ &= 2.79(1 + 0.015)^{12} \\ &= 2.79(1.015)^{12} \\ &= 3.34 \end{aligned}$$

The cost of one gallon of gasoline, to the nearest cent, was **$3.34**.

Practice Exercises

1. What is the area of a parallelogram if the coordinates of its vertices are $A(0,-2)$, $B(3,2)$, $C(8,2)$, and $D(5,-2)$?

2. Find the area of $\triangle ABC$ whose vertices are $A(2,1)$, $B(10,7)$, and $C(4,10)$.

3. The coordinates of the turning point of the graph of $y = 2x^2 - 4x + 1$ are

(1) $(1,-1)$ (3) $(-1,5)$
(2) $(1,1)$ (4) $(2,1)$

4. Find the distance between points $(-1,5)$ and $(-7,3)$.

5. What is the slope of the line that passes through the points whose coordinates are $(-1,4)$ and $(1,5)$?

6. A point on the graph of $x + 3y = 13$ is

(1) $(4,4)$ (2) $(-2,3)$ (3) $(-5,6)$ (4) $(4,-3)$

7. What is the slope of the line whose equation is $4y = 3x + 16$?

8. The graph of $y = 3x - 4$ is parallel to the graph of

(1) $y = 4x - 3$ (3) $y = -3x + 4$
(2) $y = 3x + 4$ (4) $y = 3$

9. The equation of a line whose slope is 2 and whose y-intercept is –2 is

(1) $2y = x - 2$ (3) $y = -2x + 2$
(2) $y = -2$ (4) $y = 2x - 2$

10. If the domain of $y = 2x + 1$ is $\{x \mid -2 \leq x \leq 3\}$, which integer is *not* in the range?

(1) –4 (2) –2 (3) 0 (4) 7

11. Which graph represents a function?

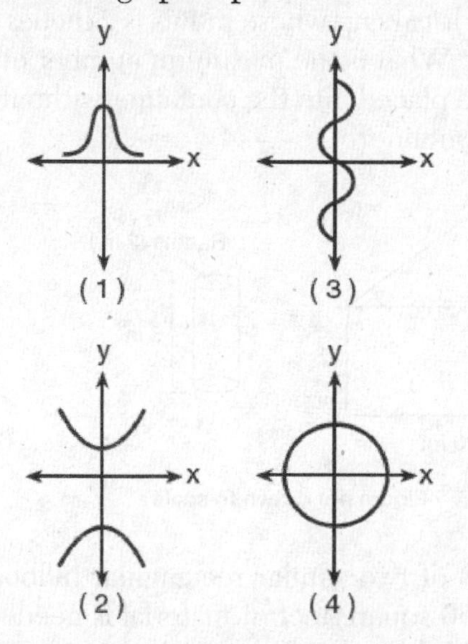

12. Graph and label the following lines:

$$y = 5$$
$$x = -4$$
$$y = \frac{5}{4}x + 5$$

Calculate the area, in square units, of the triangle formed by the three points of intersection.

13. A closed cardboard box has length $x - 2$, width $x + 1$, and height $2x$. Express the surface area of the box as a polynomial in x written in simplest form.

14. The volume of a rectangular pool is 1,080 cubic meters. Its length, width, and depth are in the ratio of 10:4:1. Find the number of meters in each of the dimensions of the pool.

15. In the accompanying diagram, a rectangular container with the dimensions 10 inches by 15 inches by 20 inches is to be filled with water, using a cylindrical cup whose radius is 2 inches and whose height is 5 inches. What is the maximum number of full cups of water that can be placed into the container without the water overflowing the container?

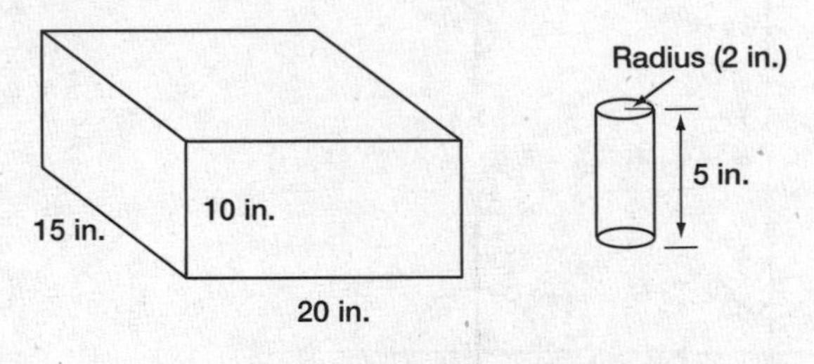

Figure not drawn to scale

16. The lengths of the sides of two similar rectangular billboards are in the ratio 5:4. If 250 square feet of material is needed to cover the larger billboard, how much material, in square feet, is needed to cover the smaller billboard?

Solutions

1. Drop altitude $\overline{BE}$ perpendicular to $\overline{AD}$. The coordinates of E are (3,–2).

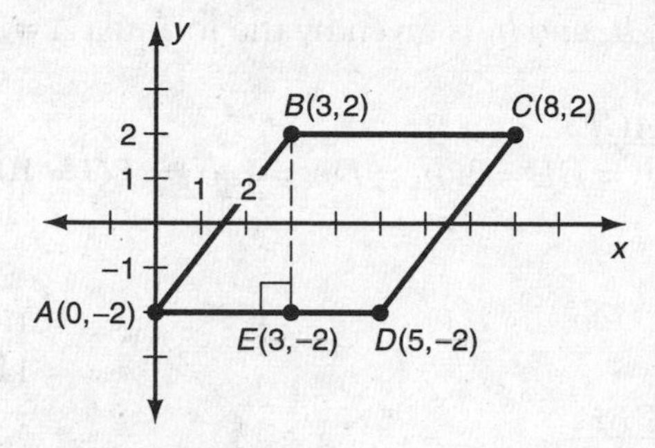

The area of a parallelogram equals the product of the length of its base and its altitude: Area of $\square ABCD = AD \times BE$

Since $AD = 5 - 0 = 5$ and $BE = 2 - (-2) = 2 + 2 = 4$: $= 5 \times 4$

$= 20$

The area is **20** square units.

2. Drop perpendiculars $\overline{AD}$, $\overline{CE}$, and $\overline{BF}$ to the x-axis.

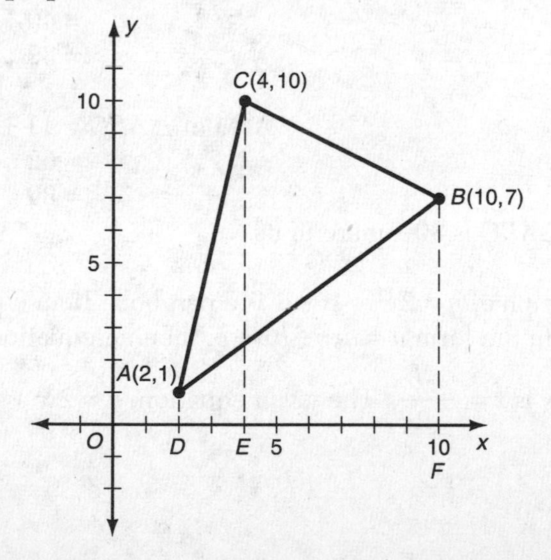

Area of $\triangle ABC$ = area of trapezoid $DACE$ + area of trapezoid $ECBF$ – area of trapezoid $DABF$.

The area, A, of a trapezoid whose altitude is h and the length of whose bases are b_1 and b_2 is given by the formula $A = \frac{1}{2}h(b_1 + b_2)$.

Trapezoid $DACE$:

$$h = DE = 2,\ b_1 = DA = 1,\ b_2 = EC = 10:$$

$$\begin{aligned} A &= \frac{1}{2}(2)(1 + 10) \\ &= 1(11) \\ &= 11 \end{aligned}$$

Trapezoid $ECBF$:

$$h = EF = 6,\ b_1 = FB = 7,\ b_2 = EC = 10:$$

$$\begin{aligned} A &= \frac{1}{2}(6)(7 + 10) \\ &= 3(17) \\ &= 51 \end{aligned}$$

Trapezoid $DABF$:

$$h = DF = 8,\ b_1 = DA = 1,\ b_2 = FB = 7:$$

$$\begin{aligned} A &= \frac{1}{2}(8)(1 + 7) \\ &= 4(8) \\ &= 32 \end{aligned}$$

$$\begin{aligned} \text{Area of } \triangle ABC &= 11 + 51 - 32 \\ &= 62 \quad - 32 \\ &= 30 \end{aligned}$$

Area of $\triangle ABC$ is **30** square units.

3. The graph of $y = 2x^2 - 4x + 1$ is a parabola. If an equation of a parabola is in the form $y = ax^2 + bx + c$, then an equation of its axis of symmetry is $x = -\frac{b}{2a}$. The given equation, $y = 2x^2 - 4x + 1$, is in the form $y = ax^2 + bx + c$, with $a = 2$, $b = -4$, and $c = 1$. An equation of its axis of symmetry is

$$x = -\frac{-4}{2(2)}$$
$$= -\frac{-4}{4}$$
$$= 1.$$

The axis of symmetry of a parabola passes through its turning point. Therefore the x-coordinate of the turning point in this case is 1. Since the turning point is on the parabola, its coordinates must satisfy the equation of the parabola; substitute 1 for x in the equation to determine the y-coordinate of the turning point:

$$y = 2(1)^2 - 4(1) + 1$$
$$= 2(1) - 4 + 1$$
$$= 2 - 4 + 1$$
$$= -1$$

The coordinates of the turning point are (1,–1).

The correct choice is **(1)**.

4. To find the distance, d, between (–1,5) and (–7,3) use the distance formula where $\Delta x = -7 - (-1) = -7 + 1 = -6$ and $\Delta y = 3 - 5 = -2$.

$$d = \sqrt{(\Delta x)^2 + (\Delta y)^2}$$
$$= \sqrt{(-6)^2 + (-2)^2}$$
$$= \sqrt{36 + 4}$$
$$= \mathbf{\sqrt{40}}$$

5.

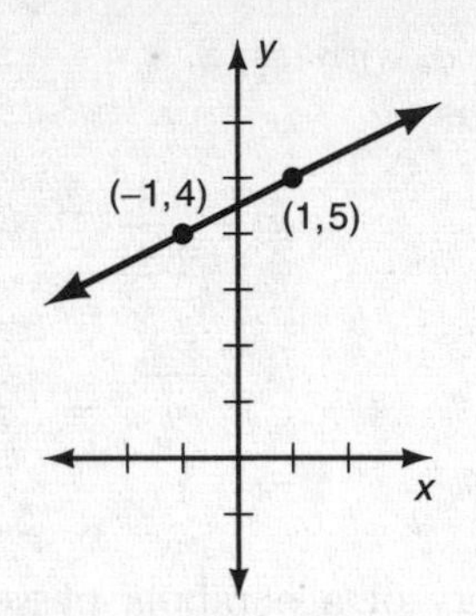

Use the formula for the slope, m, of a line joining points (x_1,y_1) and (x_2,y_2):

$$m = \frac{y_2 - y_1}{x_2 - x_1}$$

Since $x_1 = -1$, $y_1 = 4$, and $x_2 = 1$, $y_2 = 5$:

$$= \frac{5-4}{1-(-1)}$$

$$= \frac{1}{1+1}$$

$$= \frac{1}{2}$$

The slope = $\mathbf{\frac{1}{2}}$.

6. If a point is on the graph of $x + 3y = 13$, its coordinates must satisfy this equation.

Try each choice in turn by substituting the given coordinates in the equation $x + 3y = 13$:

(1) (4,4): $4 + 3(4) \stackrel{?}{=} 13$

$4 + 12 \stackrel{?}{=} 13$

$16 \neq 13$ (4,4) is *not* on the graph.

(2) (–2,3): $-2 + 3(3) \stackrel{?}{=} 13$

$-2 + 9 \stackrel{?}{=} 13$

$7 \neq 13$ (–2,3) is *not* on the graph.

(3) (–5,6): $-5 + 3(6) \stackrel{?}{=} 13$
$-5 + 18 \stackrel{?}{=} 13$
$13 = 13$✓ (–5,6) *is* on the graph.

(4) (4,–3): $4 + 3(-3) \stackrel{?}{=} 13$
$4 - 9 \stackrel{?}{=} 13$
$-5 \neq 13$ (4,–3) is *not* on the graph.

The correct choice is **(3)**.

7. If an equation of a straight line is in the form $y = mx + b$, then m represents the slope.

To get the given equation, $4y = 3x + 16$, into the $y = mx + b$ form, divide each term by 4:

$$y = \frac{3}{4}x + 4$$

Here, $m = \frac{3}{4}$.

The slope is $\mathbf{\frac{3}{4}}$.

8. If two lines are parallel, they must have the same slope.

If an equation of a line is in the form $y = mx + b$, then m represents the slope. The given equation, $y = 3x - 4$, is in the $y = mx + b$ form with $m = 3$, so its slope is 3.

Examine each choice in turn to see which one also has a slope of 3:

(1) $y = 4x - 3$ is in the $y = mx + b$ form with $m = 4$.
Therefore, it is *not* parallel to $y = 3x - 4$.

(2) $y = 3x + 4$ is in the $y = mx + b$ form with $m = 3$.
Therefore, it *is* parallel to $y = 3x - 4$.

(3) $y = -3x + 4$ is in the $y = mx + b$ form with $m = -3$.
Therefore, it is *not* parallel to $y = 3x - 4$.

(4) $y = 3$ can be written in the $y = mx + b$ form as $y = 0x + 3$. In this form, $m = 0$. Therefore, $y = 3$ is *not* parallel to $y = 3x - 4$.

The correct choice is **(2)**.

9. An equation of a line whose slope is m and whose y-intercept is b can be written in the form $y = mx + b$.

In this case, $m = 2$ and $b = -2$. Hence, an equation of the line is $y = 2x = -2$.

The correct choice is **(4)**.

10. When $x = -2.5$, $y = 2(-2.5) + 1 = -4$. Since -2.5 is not in the domain, -4 is not in the range.

The correct choice is **(1)**.

11. The only graph that passes the vertical line test is the graph in choice (1).

The correct choice is **(1)**.

12. To graph the three given lines: $y = 5$, $x = -4$, and $y = \frac{5}{4}x + 5$, on the same set of axes, proceed as follows:

- To graph $y = 5$, draw a horizontal line through (0,5).
- To graph $x = -4$, draw a vertical line through (–4,0).
- To graph $y = \frac{5}{4}x + 5$, first pick two convenient values of x that are divisible by 4, and find the corresponding values of y. If $x = 0$, $y = 5$; and if $x = -8$, $y = \frac{5}{4}(-8) + 5 = -10 + 5 = -5$. Then draw a line through points (0,5) and (–8,–5), as shown in the accompanying diagram.

Label each line.

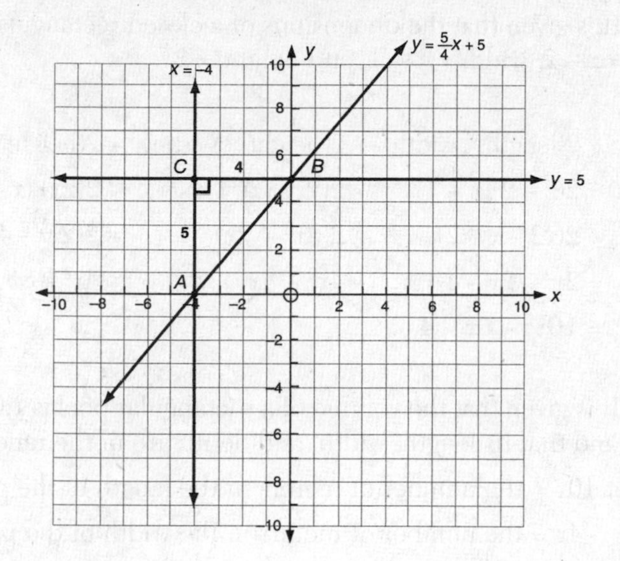

The three points of intersection are the vertices of a right triangle with vertical leg $AC = 5$ and horizontal leg $BC = 4$, as indicated on the diagram. Thus:

$$\begin{aligned}\text{Area of right triangle } ABC &= \frac{1}{2}(AC)(BC)\\ &= \frac{1}{2}(5)(4)\\ &= 10 \text{ square units}\end{aligned}$$

The area, in square units, of the triangle formed by the three points of intersection is **10**.

13. It is given that the dimensions of a closed rectangular box are length = $x - 2$, width = $x + 1$, and height = $2x$.

Surface area			
	= 2(length × width)	+ 2(height × width)	+ 2(height × length)
	$= 2(x-2)(x+1)$	$+ 2(2x)(x+1)$	$+ 2(2x)(x-2)$
	$= 2(x2 - x - 2)$	$+ 2(2x^2 + 2x)$	$+ 2(2x^2 - 4x)$
	$= 2x^2 - 2x - 4$	$+ 4x^2 + 4x$	$+ 4x^2 - 8x$
	$= \mathbf{10x^2 - 6x - 4}$		

14. It is given that the volume of a rectangular pool is 1,080 cubic meters and that its length, width, and depth are in the ratio 10:4:1.

Let $10x$ = the number of meters in the length of the pool;
$4x$ = the number of meters in the width of the pool; and
x = the number of meters in the depth of the pool.

Since the volume of a rectangular pool is the product of its length, width, and depth:

$$(10x)(4x)(x) = 1{,}080 \text{ m}^3$$
$$40x^3 = 1{,}080 \text{ m}^3$$
$$x^3 = \frac{1{,}080}{40} \text{m}^3$$
$$x^3 = 27 \text{ m}^3$$
$$x = \sqrt[3]{27} \text{ m} = 3 \text{ m since } 3 \times 3 \times 3 = 27.$$

Hence, $10x = 10(3) = 30$ meters and $4x = 4(3) = 12$ meters.

The depth of the pool is **3** meters, the width of the pool is **12** meters, and the length of the pool is **30** meters.

15. It is given that a rectangular container with the dimensions 10 inches by 15 inches by 20 inches is to be filled with water, using a cylindrical cup whose radius is 2 inches and whose height is 5 inches. To find the number of full cups of water that can be poured into the container without the water overflowing, divide the volume of the rectangular container by the volume of the cylindrical cup, and round the quotient down to the nearest whole number.

The volume of the rectangular container is the product of its length, width, and height:

$$\text{Volume} = 10 \times 15 \times 20$$
$$= 3000 \text{ cubic inches}$$

The volume of a cylinder is the area of its circular base times its height. Thus:

$$\text{Volume} = (\pi \times 2^2) \times 5$$
$$= 20\pi$$
$$\approx 62.8$$

Thus, $\frac{3000}{62.8} \approx 47.77 \approx 47$.

The maximum number of *full* cups of water than can be poured into the container without the water overflowing the container is **47**.

16. If two polygons are similar, then the ratio of their areas is equal to the *square* of the ratio of the lengths of any pair of corresponding sides.

It is given that the lengths of the (corresponding) sides of two similar rectangular billboards are in the ratio 5:4 and the area of the larger billboard is 250 square feet.

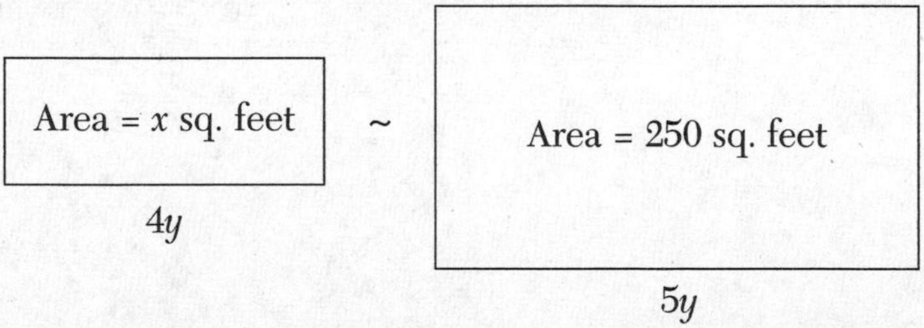

Area = x sq. feet ~ Area = 250 sq. feet

$4y$ $5y$

If x represents the number of square feet needed to cover the smaller billboard, then:

$$\frac{\text{larger billboard}}{\text{smaller billboard}} = \left(\frac{5y}{4y}\right)^2 = \frac{250}{x}$$

$$\frac{25}{16} = \frac{250}{x}$$

Set the cross products equal: $25x = (16)(250)$

Solve for x: $x = \dfrac{(16)(250)}{25}$

$$= \frac{(16)(\overset{10}{\cancel{250}})}{\cancel{25}}$$

$$= (16)(10)$$

$$= 160$$

160 square feet of material is needed to cover the smaller billboard.

Systems of Equations and Inequalities

3.15 SOLVING A SYSTEM OF LINEAR EQUATIONS GRAPHICALLY: AN EXAMPLE

Solve graphically each of the following systems of equations:

$$y = x + 3$$
$$2x + y = 3$$

Draw the graphs of both equations on the same set of axes. The coordinates of the point of intersection of the two graphs represent the solution to the system.

Step 1. Draw the graph of $y = x + 3$. Set up a table of values by choosing three convenient values for x and substituting them in the equation to find the corresponding values of y:

x	$x + 3$	$= y$
0	$0 + 3$	$= 3$
3	$3 + 3$	$= 6$
-3	$-3 + 3$	$= 0$

Plot points (0,3), (3,6), and (–3,0). They should lie on a straight line. Draw this line; it is the graph of $y = x + 3$.

Step 2. Draw the graph of $2x + y = 3$. To do this, it is advisable to first rearrange the equation so that it is in a form in which it is solved for y in terms of x.

The given equation is: $2x + y = 3$

Add $-2x$ (the additive inverse of $2x$) to both sides: $-2x \quad = \; -2x$

$$y = 3 - 2x$$

Set up a table of values by choosing three convenient values for x and substituting them in the equation to find the corresponding values of y:

x	$3-2x$		$= y$
0	$3-2(0)$	$= 3-0$	$= 3$
3	$3-2(3)$	$= 3-6$	$= -3$
-3	$3-2(-3)$	$= 3+6$	$= 9$

Plot points (0,3), (3,–3), and (–3,9), as shown in the accompanying figure. They should lie in a straight line. Draw this line; it is the graph of $2x + y = 3$.

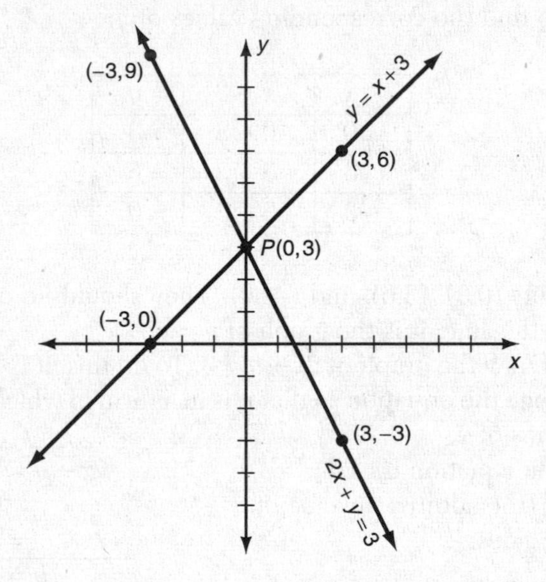

Step 3. Locate P, the point of intersection of the two graphs, which represents the common solution. The coordinates of P are (0,3) or $x = 0$ and $y = 3$.

The solution to the system is $\boldsymbol{x = 0}$, $\boldsymbol{y = 3}$ or **{(0,3)}**.

CHECK: The solution is checked by substituting 0 for x and 3 for y in *both* of the two *original* equations to see whether they are satisfied:

$$\begin{aligned} y &= x + 3 \\ 3 &\stackrel{?}{=} 0 + 3 \\ 3 &\stackrel{\checkmark}{=} 3 \end{aligned} \qquad\qquad \begin{aligned} 2x + y &= 3 \\ 2(0) + 3 &\stackrel{?}{=} 3 \\ 0 + 3 &\stackrel{?}{=} 3 \\ 3 &\stackrel{\checkmark}{=} 3 \end{aligned}$$

3.16 SOLVING A SYSTEM OF LINEAR EQUATIONS ALGEBRAICALLY: TWO EXAMPLES

Solve algebraically each of the following systems of equations that follow.

1. $x + y = 7$
$2x - y = 2$

Add the two equations together to eliminate y:

$$\begin{aligned} x + y &= 7 \\ 2x - y &= 2 \\ \hline 3x \qquad &= 9 \end{aligned}$$

Multiply both sides by $\frac{1}{3}$ (the multiplicative inverse of 3):

$$\frac{1}{3}(3x) = \frac{1}{3}(9)$$
$$x = 3$$

The solution for x is **3**.

2. $2x + y = 6$
$x = 3y + 10$

Rearrange the second equation by adding $-3y$ (the additive inverse of $3y$) to both sides:

$$\begin{aligned} x &= 3y + 10 \\ -3y &= -3y \\ \hline x - 3y &= 10 \end{aligned}$$

Multiply each term in the first equation by 3: $6x + 3y = 18$

Add the new form of the second equation, thus eliminating y: $x - 3y = 10$

$$7x = 28$$

Multiply both sides by $\frac{1}{7}$ (the multiplicative inverse of 7): $\frac{1}{7}(7x) = \frac{1}{7}(28)$

Substitute 4 for x in the first equation:

$$x = 4$$
$$2(4) + y = 6$$
$$8 + y = 6$$

Add –8 (the additive inverse of 8) to both sides: $-8 = -8$

$$y = -2$$

The solution is $\boldsymbol{x = 4, y = -2}$.

Systems of Linear Inequalities

3.16 SOLVING A SYSTEM OF LINEAR INEQUALITIES GRAPHICALLY: AN EXAMPLE

Solve graphically.

a. $y > x + 4$
$x + y \leq 2$

b. Determine which point lies in the solution set:

(1) (2,3)
(2) (–5,2)
(3) (0,6)
(4) (–1,0)

a. Step 1. Graph the solution set of the inequality $y > x + 4$. The graph of the inequality $y > x + 4$ is represented by all the points on the coordinate plane for which y, the ordinate, is greater than $x + 4$. Hence, first draw the graph of the line for which $y = x + 4$; having this graph will enable you to locate the region for which $y > x + 4$.

Select any three convenient values for x and substitute them in the equation $y = x + 4$ to find the corresponding values of y:

x	$x + 4$	$= y$
0	$0 + 4$	$= 4$
3	$3 + 4$	$= 7$
-4	$-4 + 4$	$= 0$

Plot points (0,4), (3,7), and (–4,0). Draw a *dotted line* through these three points to get the graph of $y = x + 4$. The dotted line is used to signify that points on it are *not* part of the solution set of the inequality $y > x + 4$.

To find the *region* or *half-plane* on one side of the line $y = x + 4$ that represents $y > x + 4$, select a test point, say (1,8), one one side of the line. Substituting in the inequality $y > x + 4$ gives $8 > 1 + 4$, or $8 > 5$, which is true. Thus, the side of the line on which (1,8) lies (above and to the left) is the region representing $y > x + 4$. Shade it with cross-hatching extending up and to the left, as shown in the accompanying figure.

Step 2. Graph the solution set of $x + y \leq 2$. This graph is represented by all the points on the coordinate plane for which $x + y < 2$ in addition to the points on the line for which $x + y = 2$. Hence, the line $x + y = 2$ is first graphed. To make it convenient to find points on the line, solve for y in terms of x:

The given equation is: $x + y = 2$

Add $-x$ (the additive inverse of x) to both sides of the equation: $-x \quad = \quad -x$

$$y = 2 - x$$

Set up a table by selecting any three convenient values for x and substituting them in the equation $y = 2 - x$ to find the corresponding values of y:

x	$2 - x$	$= y$
0	$2 - 0$	$= 2$
3	$2 - 3$	$= -1$
5	$2 - 5$	$= 3$

Plot points (0,2), (3,–1), and (5,–3). Draw a *solid line* through these three points to get the graph of $x + y = 2$. The solid line is used to signify that points on it are part of the solution set of $x + y \leq 2$.

To find the *region* or *half-plane* on one side of the line $x + y = 2$ for which $x + y < 2$, select a test point, say (–2,–1), on one side of the line. Substituting (–2,–1) in the inequality $x + y < 2$ results in $-2 - 1 < 2$, or $-3 < 2$, which is true. Thus, the side of the line where (–2,–1) is located (below and to the left) is the region representing $x + y < 2$. Shade this region with cross-hatching extending down and to the left, as shown in the accompanying figure.

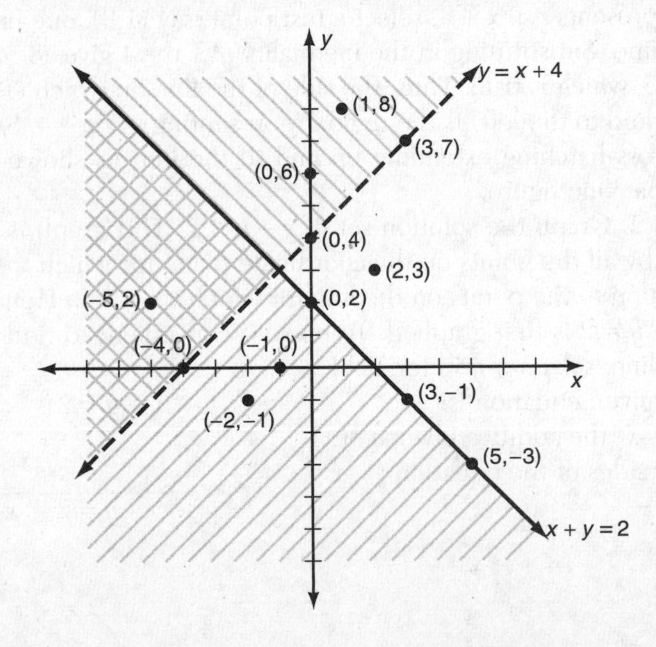

b. If a point is in the solution set, it will lie in the region in which the graph shows *both* sets of cross-hatching; points in this region satisfy *both* inequalities. Test each point in turn by locating it on the graph:

(1) (2,3) does *not* lie in the cross-hatched area.

(2) (–5,2) lies in the region with both sets of cross-hatching. Therefore, it is in the solution set of the graph.

(3) (0,6) lies in a region cross-hatched only for $y > x + 4$; it satisfies this inequality but *not* the other.

(4) (–1,0) lies in a region cross-hatched only for $x + y \leq 2$; it satisfies this inequality but *not* the other.

The only point in the solution set of the graph is **(–5,2)**.

Linear-Quadratic Systems

3.17 SOLVING A LINEAR-QUADRATIC SYSTEM GRAPHICALLY: AN EXAMPLE

Solve graphically the following systems of equations:

$$y = x^2 - 6x + 5$$
$$y + 7 = 2x$$

Step 1. Graph the second-degree equation. The graph of $y = x^2 - 6x + 5$ is a parabola. The x-coordinate of the turning point of the parabola is

$$x = -\frac{b}{2a} = -\frac{-6}{2(1)} = 3$$

Set up a table of values by picking three consecutive integer values of x on either side of $x = 3$ and substituting them in the equation to find their y-coordinates:

x	$x^2 - 6x + 5$	$= y$
0	$0^2 - 6(0) + 5$	$= 5$
1	$1^2 - 6(1) + 5 = 1 - 6 + 5$	$= 0$
2	$2^2 - 6(2) + 5 = 4 - 12 + 5$	$= -3$
3	$3^2 - 6(3) + 5 = 9 - 18 + 5$	$= -4$
4	$4^2 - 6(4) + 5 = 16 - 24 + 5$	$= -3$
5	$5^2 - 6(5) + 5 = 25 - 30 + 5$	$= 0$
6	$6^2 - 6(6) + 5 = 36 - 36 + 5$	$= 5$

TIP: Use a graphing calculator to create this table of values.

Plot points (0,5), (1,0), (2,–3), (3,–4), (4,–3), (5,0), and (6,5) on a coordinate plane, and connect them with a smooth curve that has the shape of a parabola. The graph is labeled $y = x^2 - 6x + 5$ on the accompanying figure.

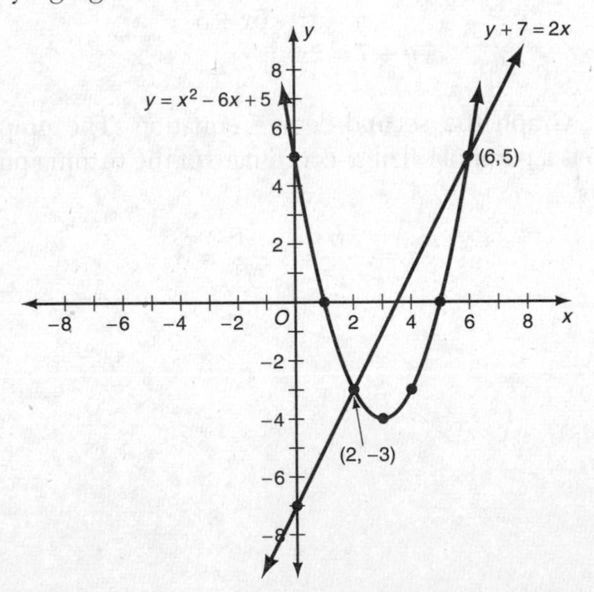

Step 2. Graph the first-degree equation using the same set of axes. The graph of $y + 7 = 2x$ is a line. Since $y = 2x - 7$, the line crosses the y-axis at (0,–7). To find a second point on the line, pick any convenient value of x. For example, if $x = 5$, then $y = 2x - 7 = 2(5) - 7 = 3$. Plot (0,–7) and (5,3). Then connect the two points with a straight line. The graph is labeled $y + 7 = 2x$ on the accompanying figure.

Step 3. Find the coordinates of the points(s) at which the graphs intersect.

The solutions are **(2,–3)** and **(6,5)**.

3.18 SOLVING A LINEAR-QUADRATIC SYSTEM ALGEBRAICALLY: AN EXAMPLE

Solve algebraically the following system of equations:

$$y = x^2 - 6x + 5$$
$$y + 7 = 2x$$

Solve the first-degree equation for y by subtracting 7 from each side of the equation:

$$y + 7 = 2x$$
$$y = 2x - 7$$

Substitute $2x - 7$ for y in the first equation, thereby obtaining a quadratic equation only in x:

$$2x - 7 = x^2 - 6x + 5$$

Put the quadratic equation in standard form with all terms on one side equal to 0:

$$0 = x^2 - 8x + 12$$

Factor the right side of the equation as the product of two binomials:

$$0 = (x + ?)(x + ?)$$

Find the two missing numbers whose sum is –8 and whose product is +12. The two numbers are –2 and –6:

$$0 = (x - 2)(x - 6)$$

If the product of two factors is 0, either factor may equal 0:

$$x - 2 = 0 \quad \text{or} \quad x - 6 = 0$$
$$x = 2 \quad \text{or} \quad x = 6$$

To find the corresponding values of y, substitute each of the solutions for x in the original first-degree equation:

Let $x = 2$	Let $x = 6$
$y + 7 = 2x$	$y + 7 = 2x$
$= 2(2)$	$= 2(6)$
$y = 4 - 7$	$y = 12 - 7$
$= -3$	$= 5$

The solutions are **(2,–3)** and **(6,5)**.

CHECK: Substitute each pair of values of x and y in *both* of the *original* equations to verify that both equations are satisfied.

	$y = x^2 - 6x + 5$	$y + 7 = 2x$
Let $x = 2$ and $y = -3$:	$-3 \overset{?}{=} 2^2 - 6(2) + 5$	$-3 + 7 \overset{?}{=} 2(2)$
	$-3 \overset{?}{=} 4 - 12 + 5$	$4 \overset{\checkmark}{=} 4$
	$-3 \overset{?}{=} -8 + 5$	
	$-3 \overset{\checkmark}{=} -3$	
Let $x = 6$ and $y = 5$:	$5 \overset{?}{=} 6^2 - 6(6) + 5$	$5 + 7 \overset{?}{=} 2(6)$
	$5 \overset{?}{=} 36 - 36 + 5$	$12 \overset{\checkmark}{=} 12$
	$5 \overset{?}{=} 0 + 5$	
	$5 \overset{\checkmark}{=} 5$	

Practice Exercises

1. What is a solution for the system of equations $x - y = 2$ and $y = 2x - 4$?
(1) (0,2) (2) (2,0) (3) (3,2) (4) (4,2)

2. Solve the following system of equations:

$$y = x^2 + 4x + 1$$

$$y = 5x + 3$$

3. Tanisha and Rachel had lunch at the mall. Tanisha ordered three slices of pizza and two colas. Rachel ordered two slices of pizza and three colas. Tanisha's bill was $6.00, and Rachel's bill was $5.25. What were the prices of one slice of pizza and one cola?

4. A rocket is launched from the ground and follows a parabolic path represented by the equation $y = -x^2 + 10x$. At the same time, a flare is launched from a height of 10 feet and follows a straight path represented by the equation $y = -x + 10$. Find the coordinates of the point or points where the paths intersect.

Solutions

1. The given system of equations is:

$$x - y = 2$$
$$y = 2x - 4$$

Eliminate y in the first equation by substituting its equal, $2x - 4$:

$$x - (2x - 4) = 2$$

Remove the parentheses by taking the opposite of each term that is inside the parentheses:

$$x - 2x + 4 = 2$$

Solve for x:

$$-x + 4 = 2$$
$$x = 2$$

Find the corresponding value of y by substituting 2 for x in the second equation:

$$y = 2(2) - 4 = 0$$

The solution is (2,0).

The correct choice is **(2)**.

2. Method I: Solve algebraically.

- Eliminate y in the quadratic equation by replacing it with $5x + 3$:

$$\overbrace{5x + 3}^{y} = x^2 + 4x + 1$$
$$0 = x^2 + 4x - 5x + 1 - 3$$
$$0 = x^2 - x - 2$$
$$\textit{or} \quad x^2 - x - 2 = 0$$

- Solve the quadratic equation by factoring the quadratic trinomial as the product of two binomials:

$$x^2 - x - 2 = 0$$
$$(x + 1)(x - 2) = 0$$
$$x + 1 = 0 \quad \textit{or} \quad x - 2 = 0$$
$$x = -1 \quad \textit{or} \quad x = 2$$

- Find the corresponding values of y by substituting the solutions for x into the linear equation:

  When $x = -1$, $y = 5(-1) + 3 = -5 + 3 = -2$, so $(-1,-2)$ is a solution.

  When $x = 2$, $y = 5(2) + 3 = 13$, so $(2,13)$ is a solution.

The solutions are **(–1,–2)** and **(2,13)**.

Method II: Solve graphically.

To solve the given system of equations graphically, graph each equation on the same set of axes. Then determine the coordinates of the points of intersection.

- To graph $y = 5x + 3$, plot at least two convenient points that lie on the line. For example, find the corresponding values of y when $x = -2$, $x = 0$, and $x = 1$, where the third x-value serves as a check point. The calculations are summarized in the accompanying table.

x	$y = 5x + 3$	(x,y)
-2	$y = 5(-2) + 3 = -10 + 3 = -7$	$(-2,-7)$
0	$y = 5(0) + 3 = 3$	$(0,3)$
1	$y = 5(1) + 3 = 8$	$(1,8)$

TIP: Use a graphing calculator to create this table of values.

  Plot $(-2,-7)$, $(0,3)$, and $(1,8)$. Then connect these points with a straight line as shown on the next page.

- The graph of a quadratic equation of the form $y = ax^2 + bx + c$ is a parabola with an axis of symmetry at $x = -\dfrac{b}{2a}$. For the given quadratic equation, $a = 1$ and $b = 4$, so $x = -\dfrac{4}{2(1)} = -2$. To graph $y = x^2 + 4x + 1$, plot at least three points on either side of $x = -2$. Find the coordinates of these points as shown in the accompanying table:

x	$y = x^2 + 4x + 1$	(x,y)
–5	$y = (-5)^2 + 4(-5) + 1 = 25 - 20 + 1 = 6$	(–5,6)
–4	$y = (-4)^2 + 4(-4) + 1 = 16 - 16 + 1 = 1$	(–4,1)
–3	$y = (-3)^2 + 4(-3) + 1 = 9 - 12 + 1 = -2$	(–3,–2)
–2	$y = (-2)^2 + 4(-2) + 1 = 4 - 8 + 1 = -3$	(–2,–3)
–1	$y = (-1)^2 + 4(-1) + 1 = 1 - 4 + 1 = -2$	(–1,–2)
0	$y = (0)^2 + 4(0) + 1 = 1$	(0,1)
1	$y = (1)^2 + 4(1) + 1 = 1 + 4 + 1 = 6$	(1,6)

TIP: Use a graphing calculator to create this table of values.

Plot (–5,6), (–4,1), (–3,–2), (–2,–3), (–1,–2), (0,1), and (1,6). Then connect these points with a smooth U-shaped curve as shown in the accompanying diagram.

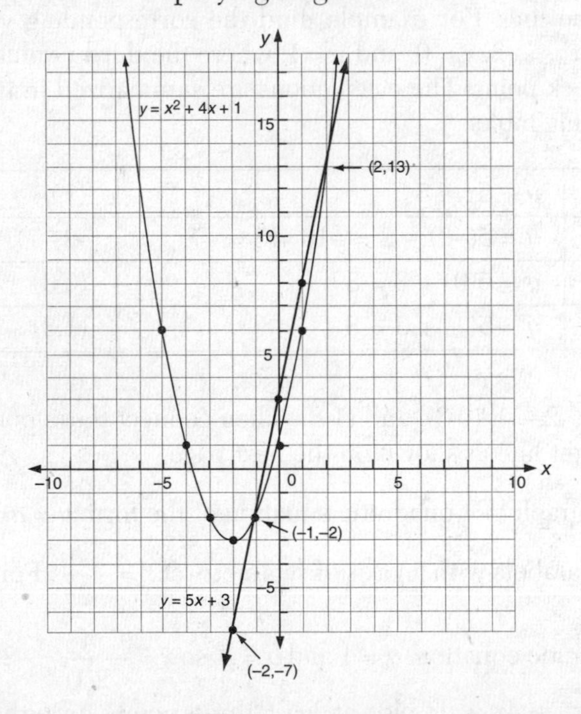

- Determine the coordinates of the points of intersection of the two graphs. The graphs intersect at (–1,–2) and (2,13).

The solutions of the given system of equations are **(–1,–2)** and **(2,13)**.

3. Tanisha and Rachel had lunch at the mall. Let x represent the cost of one slice of pizza and y represent the cost of one cola.

It is given that Tanisha ordered three slices of pizza and two colas and that her bill was \$6.00. Hence, $3x + 2y = \$6.00$

It is also given that Rachel ordered two slices of pizza and three colas and that her bill was \$5.25. Hence, $2x + 3y = \$5.25$.

Solve the system of two equations algebraically:

$$\begin{aligned} 3x + 2y &= \$6.00 \\ 2x + 3y &= \$5.25 \end{aligned}$$

Eliminate one of the variables by adding corresponding sides of the two equations together. To eliminate y, multiply each member of the first equation by 3 and then multiply each member of the second equation by –2 so that the coefficients of y in the resulting equations will be additive inverses (exact opposites):

$$\begin{aligned} 3(3x + 2y) &= 3(\$6.00) \\ -2(2x + 3y) &= -2(\$5.25) \end{aligned} \quad \Longrightarrow \quad \begin{aligned} 9x + 6y &= \$18.00 \\ -4x - 6y &= -\$10.50 \\ 5x \quad\quad &= \$7.50 \\ x &= \frac{\$7.50}{5} \\ &= \$1.50 \end{aligned}$$

Find y by substituting \$1.50 for x in either of the two equations. Substituting into the first equation gives:

$$\begin{aligned} 3(\$1.50) + 2y &= \$6.00 \\ \$4.50 + 2y &= \$6.00 \\ 2y &= \$6.00 - \$4.50 \\ y &= \frac{\$1.50}{2} \\ &= \$0.75 \end{aligned}$$

The price of one slice of pizza was **$1.50**, and the price of one cola was **$0.75**.

4. It is given that the path of a rocket is represented by the equation $y = -x^2 + 10x$ and the path of a flare is represented by the equation $y = -x + 10$.

Method I: Solve algebraically.

$$y = -x + 10 = -x^2 + 10x$$

$$x^2 - 11x + 10 = 0$$

$$(x - 1)(x - 10) = 0$$

$$x - 1 = 0 \quad \text{or} \quad x - 10 = 0$$

$$x = 1 \qquad\qquad x = 10$$

Substitute each value of x into the linear equation to find the corresponding value of y:

- If $x = 1$, then $y = -x + 10 = -1 + 10 = 9$.
- If $x = 10$, then $y = -x + 10 = -10 + 10 = 0$.

The coordinates of the points of intersection of the paths are **(1,9)** and **(10,0)**.

Method II: Solve using a graphing calculator.
Set $Y_1 = -x^2 + 10x$ and $Y_2 = -x + 10$.

Use the table feature of the graphing calculator to find those values of x for which $Y_1 = Y_2$.

X	Y1	Y2
0	0	10
1	9	9
2	16	8
3	21	7
4	24	6
5	25	5
6	24	4
X=1		

X	Y1	Y2
4	24	6
5	25	5
6	24	4
7	21	3
8	16	2
9	9	1
10	0	0
X=10		

Alternatively, graph Y_1 and Y_2 in an appropriate viewing window. Then use the intersect feature of the calculator to find the coordinates of the points of intersection of the two graphs.

4. MEASUREMENT

4.1 SIMILAR TRIANGLES

When two polygons are similar, you may conclude that:

- The lengths of their corresponding sides are in proportion.
- Their perimeters have the same ratio as the lengths of any pair of corresponding sides.

Two triangles are similar when two angles of one triangle have the same degree measures as the corresponding angles of a second triangle.

4.2 PYTHAGOREAN THEOREM

In a right triangle, the longest side is opposite the right (90°) angle and is called the **hypotenuse**. The other two sides are called the **legs**. The lengths of the sides of a right triangle are related by the Pythagorean theorem:

$$(\text{leg } 1)^2 + (\text{leg } 2)^2 = (\text{hypotenuse})^2$$

Referring to the figure below:

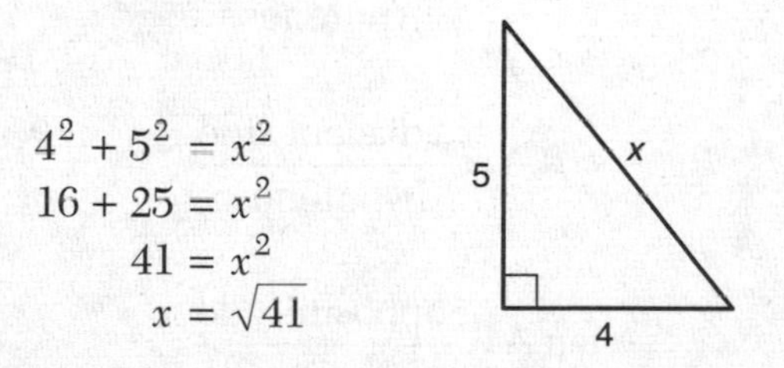

$$\begin{aligned} 4^2 + 5^2 &= x^2 \\ 16 + 25 &= x^2 \\ 41 &= x^2 \\ x &= \sqrt{41} \end{aligned}$$

It is not necessary to use the Pythagorean theorem if the lengths of two sides of a right triangle fit one of the following patterns, where n is a positive integer:

- leg 1 : leg 2 : hypotenuse = $3n{:}4n{:}5n$
- leg 1 : leg 2 : hypotenuse = $5n{:}12n{:}13n$

A set of three positive integers that satisfy the Pythagorean relationship is called a **Pythagorean triple**. For example, if the length and width of a rectangle are 8 and 6, respectively, you do not need the Pythagorean theorem to find the diagonal (hypotenuse). Since 3:4:5 = 6:8:10 (where $n = 2$), the length of a diagonal is 10.

4.3 RIGHT-TRIANGLE TRIGONOMETRY

The Pythagorean theorem relates the lengths of the three sides of a right triangle, while trigonometric ratios relate the measures of two sides and an acute angle of a right triangle:

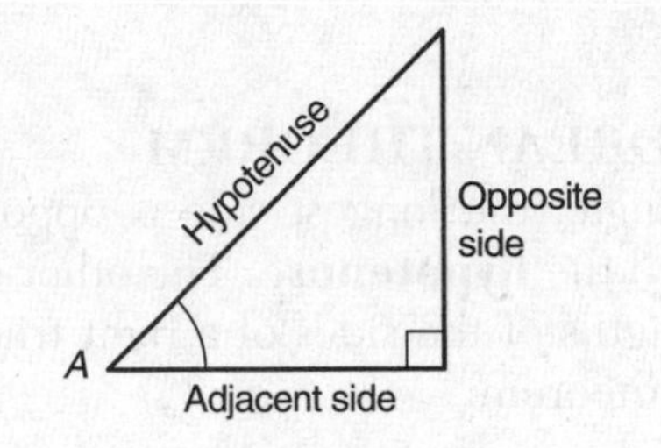

$$\sin A = \frac{\text{opposite side}}{\text{hypotenuse}}$$

$$\cos A = \frac{\text{adjacent side}}{\text{hypotenuse}}$$

$$\tan A = \frac{\text{opposite side}}{\text{adjacent side}}$$

These three definitions are included in the Regents formula sheet.

4.4 RELATIVE ERROR IN MEASUREMENT

Sometimes a measurement made with tools such as a ruler or tape measure differs from what is known to be the actual measurement. The relative error in measurement can be calculated using the following relationship:

$$\text{Relative error in measurement} = \frac{\begin{pmatrix}\text{Measurement}\\ \text{with Error}\end{pmatrix} - \begin{pmatrix}\text{Actual}\\ \text{Measurement}\end{pmatrix}}{\text{Actual Measurment}}$$

For example, according to a floor plan, a living room floor measures 12 feet by 20 feet. Tom the carpet installer measured the same floor to be 12.1 feet by 20.3 feet. The relative error in measurement of the area of the living room floor is:

$$\text{Relative error in measurement} = \frac{(12.1)(20.3) - (12)(20)}{(12)(20)}$$

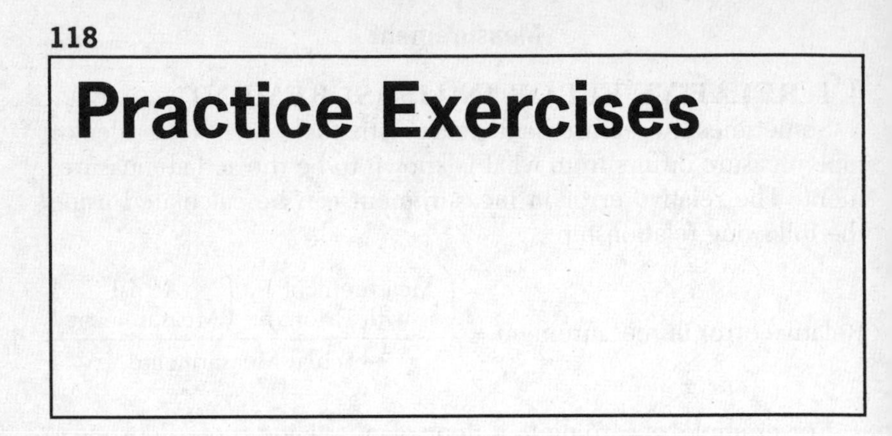

Practice Exercises

1. The angle of elevation from a point 25 feet from the base of a tree on level ground to the top of the tree is 30°. Which equation can be used to find the height of the tree?

(1) $\tan 30° = \frac{x}{25}$ (3) $\sin 30° = \frac{x}{25}$

(2) $\cos 30° = \frac{x}{25}$ (4) $30^2 + 25^2 = x^2$

2. Delroy's sailboat has two sails that are similar triangles. The larger sail has sides of 10 feet, 24 feet, and 26 feet. If the shortest side of the smaller sail measures 6 feet, what is the perimeter of the *smaller* sail?

3. If the length of a rectangular television screen is 20 inches and its height is 15 inches, what is the length of its diagonal, in inches?

4. Ron and Francine are building a ramp for performing skateboard stunts, as shown in the accompanying diagram. The ramp is 7 feet long and 3 feet high. What is the measure of the angle, x, that the ramp makes with the ground, to the *nearest tenth of a degree*?

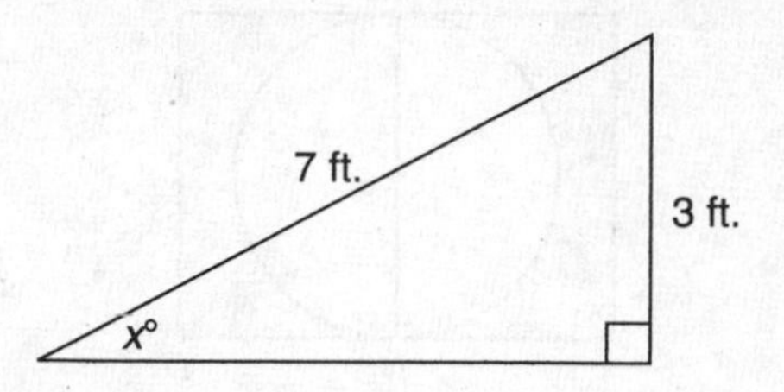

5. In the accompanying diagram of right triangles ABD and DBC, $AB = 5$, $AD = 4$, and $CD = 1$. Find the length of $\overline{BC}$, to the *nearest tenth*.

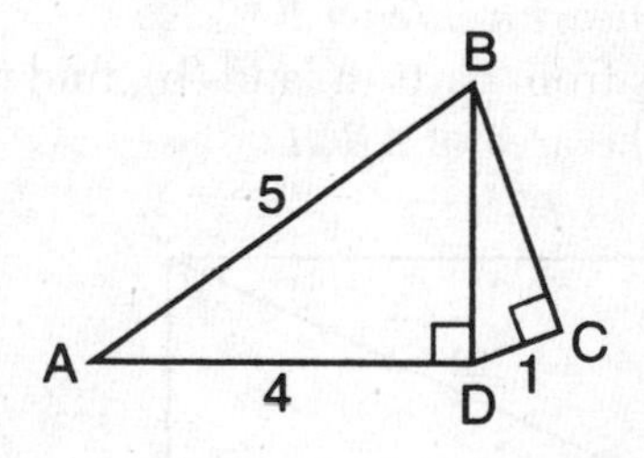

6. In the accompanying diagram, $\triangle IHJ \sim \triangle LKJ$. If $IH = 5$, $HJ = 2$, and $LK = 7$, find KJ.

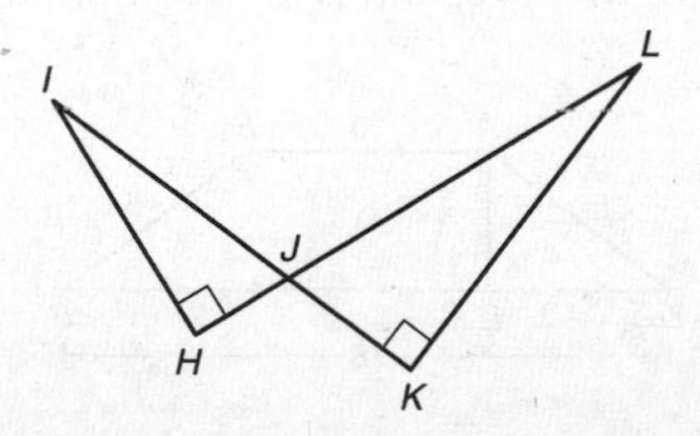

7. In the accompanying diagram, $ABCD$ is a rectangle. Diameter $\overline{MN}$ of circle O is perpendicular to $\overline{BC}$ at M and to $\overline{AD}$ at N, $AD = 8$, and $CD = 6$. Find in terms of π the area of the shaded region.

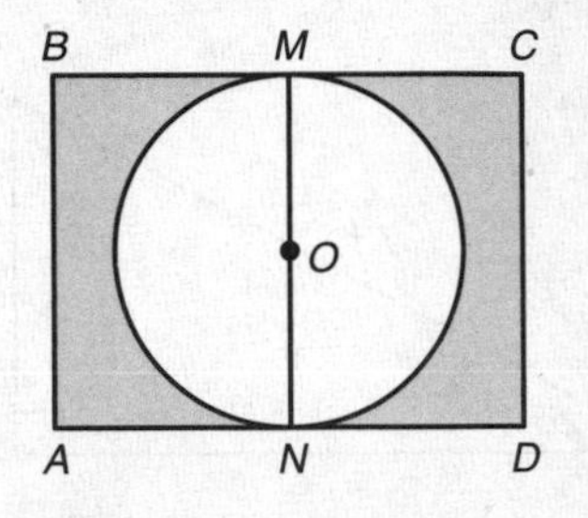

8. In the accompanying diagram of rectangle $ABCD$, $AC = 22$ and $m\angle CAB = 24$.

a. Find AB to the *nearest integer*.

b. Find BC to the *nearest integer*.

c. Using the results from parts (a) and (b), find the number of square units in the area of $ABCD$.

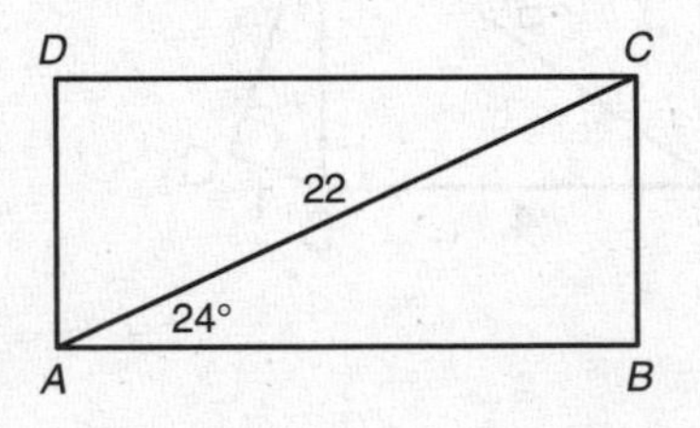

9. In the accompanying diagram, $ABCD$ is an isosceles trapezoid, $AD = BC = 5$, $AB = 10$, and $DC = 18$. Find the area of trapezoid $ABCD$.

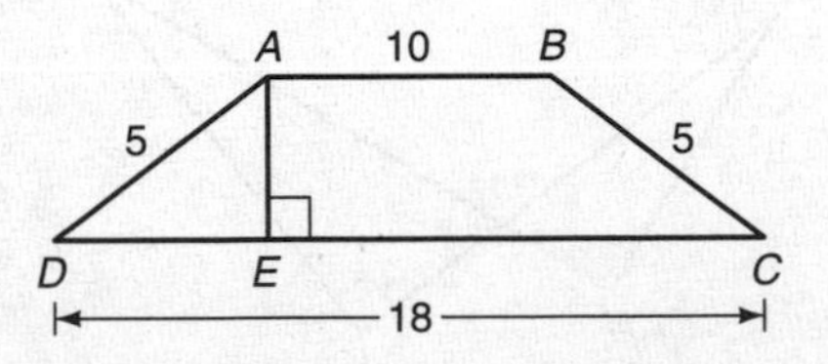

10. A person measures the angle of depression from the top of a wall to a point on the ground. The point is located on level ground 62 feet from the base of the wall and the angle of depression is 52°. How high is the wall, to the *nearest tenth of a foot*?

11. Two hikers started at the same location. One traveled 2 miles east and then 1 mile north. The other traveled 1 mile west and then 3 miles south. At the end of their hikes, how many miles apart are the two hikers?

12. In the accompanying diagram, right triangle ABC is inscribed in circle O, diameter $AB = 26$, and $CB = 10$. Find, to the *nearest square unit*, the area of the shaded region.

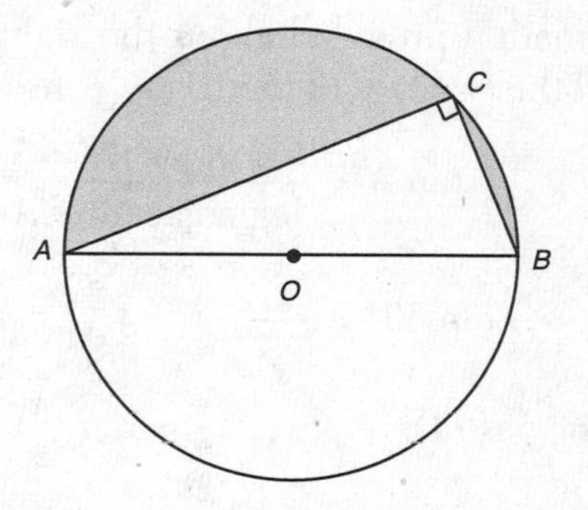

Solutions

1. It is given that the angle of elevation from a point 25 feet from the base of a tree on level ground to the top of the tree is 30°. Draw a right triangle. Label the vertical leg x to indicate that it represents the height of the tree.

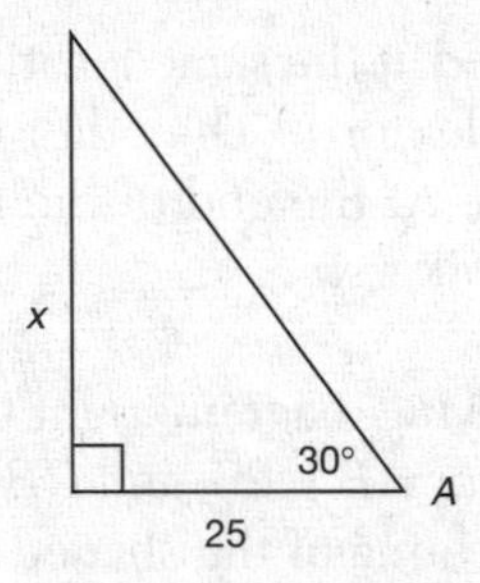

Since the information provided about the right triangle involves an acute angle and the two legs of the triangle, use the tangent ratio:

$$\tan A = \frac{\text{leg opposite } \angle A}{\text{leg adjacent } \angle A}$$

$$\tan 30° = \frac{x}{25}$$

The correct choice is **(1)**.

2. If two triangles are similar, then their perimeters have the same ratio as the lengths of any pair of corresponding sides.

- The perimeter of the larger sail is $10 + 24 + 26 = 60$.
- The ratio of corresponding sides is $\frac{6}{10}$ or, equivalently, $\frac{3}{5}$.
- If x represents the perimeter of the smaller sail, then

$$\frac{x}{60} = \frac{3}{5}$$

$$5x = 180$$

$$x = \frac{180}{5} = 36 \text{ ft.}$$

3. The diagonal, x, of the screen is the hypotenuse of a right triangle whose legs measure 20 inches and 15 inches. Use the Pythagorean theorem to find x:

$$\begin{aligned} x^2 &= 20^2 + 15^2 \\ &= 400 + 225 \\ &= 625 \\ x &= \sqrt{625} \\ &= 25 \end{aligned}$$

The length of the diagonal is **25 inches**.

4. Use the sine ratio:

$$\sin x° = \frac{3}{7}$$

$$x° = \underbrace{\sin^{-1}\left(\frac{3}{7}\right)}_{\text{"the angle whose sine is } \frac{3}{7}\text{"}}$$

$$\approx 25.37693353°$$

To the *nearest tenth of a degree*, angle x measures **25.4°**.

5. It is given that, in the accompanying diagram of right triangles ABD and DBC, $AB = 5$, $AD = 4$, and $CD = 1$.

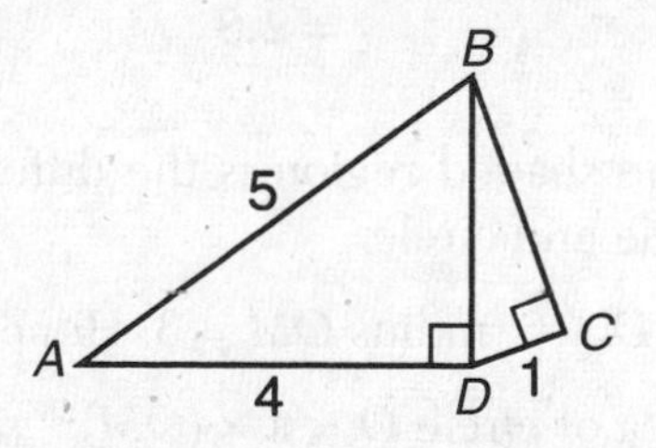

You are required to find the length of $\overline{BC}$, to the *nearest tenth*.

- The lengths of right triangle ABD form a 3-4-5 Pythagorean triple in which $\overline{AB}$ is the hypotenuse, so leg $BD = 3$.
- In right triangle BCD, $\overline{BD}$ is the hypotenuse. Use the Pythagorean theorem to find the length of leg $\overline{BC}$:

$$\begin{aligned}(BC)^2 + (CD)^2 &= (BD)^2\\(BC)^2 + (1)^2 &= (3)^2\\(BC)^2 + 1 &= 9\\(BC)^2 &= 9 - 1\\BC &= \sqrt{8}\\BC &\approx 2.83\end{aligned}$$

The length of $\overline{BC}$ is **2.8**, to the *nearest tenth*.

6. Since it is given that $\triangle IHJ \sim \triangle LKJ$:

$$\begin{aligned}\frac{IH}{JH} &= \frac{LK}{KJ}\\\frac{5}{2} &= \frac{7}{KJ}\\5(KJ) &= 14\\KJ &= \frac{14}{5}\\&= \mathbf{2.8}\end{aligned}$$

7. The area of the shaded region is the difference between the areas of the rectangle and circle.

- Since $MN = CD = 6$, radius $OM = 3$. Hence,

$$\begin{aligned}\text{area of circle } O &= \pi \times (OM)^2\\&= \pi \times (3)^2\\&= 9\pi.\end{aligned}$$

- Area of rectangle $ABCD = 8 \times 6 = 48$.
- Area of shaded region $= 48 - 9\pi$.

In terms of π, the area of the shaded region is $\mathbf{48 - 9\pi}$.

8. a. $\cos 24^\circ = \dfrac{AB}{BC}$

$$\cos 24^\circ = \frac{AB}{22}$$

$$AB = 22 \times \cos 24^\circ$$

$$AB \approx \mathbf{20}$$

b. $\sin 24^\circ = \dfrac{BC}{AC}$

$$\sin 24^\circ = \frac{BC}{22}$$

$$BC = 22 \times \sin 24^\circ$$

$$BC \approx \mathbf{9}$$

b. area $ABCD = AB \times BC$

$$= 20 \times 9$$

$$= \mathbf{180}$$

9. Draw the altitude from B to $\overline{CD}$, intersecting $\overline{CD}$ at F:

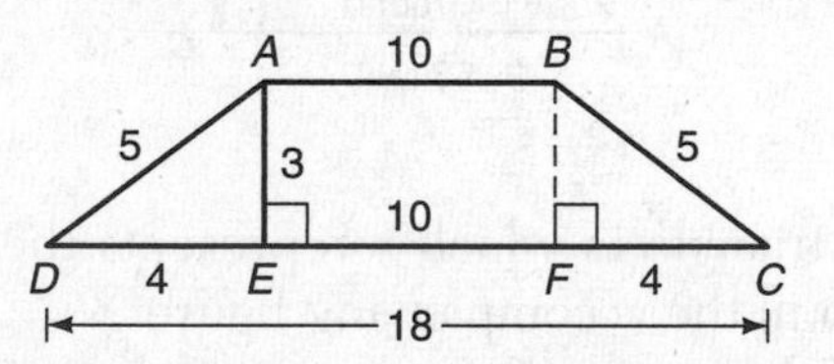

- Since $AEFB$ is a rectangle, $EF = 10$, so $DE + FC = 18 - 10 = 8$. As $DE = FC$, $DE = \dfrac{1}{2} \times 8 = 4$.

- Right triangle AED is a 3-4-5 right triangle in which $AE = 3$.
- Use the area of a trapezoid formula:

$$\text{area of trapezoid } ABCD = \frac{1}{2}(AE)(AB + CD)$$
$$= \frac{1}{2}(3)(10 + 18)$$
$$= \frac{1}{2}(3)(28)$$
$$= \frac{1}{2}(84)$$
$$= \mathbf{42}$$

10. It is given that a person measures 52° as the angle of depression from the top of a wall to a point located on level ground 62 feet from the base of the wall. You are required to find the height of the wall, to the *nearest tenth of a foot*.

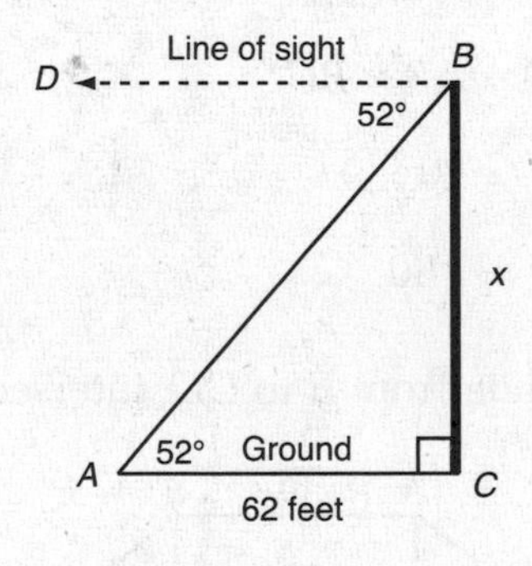

- Draw a right triangle in which x represents the height of the wall, as shown in the accompanying figure.
- The angle of depression is represented by $\angle DBA$. Because $\overrightarrow{BD} \parallel \overrightarrow{AC}$, $\angle A = \angle DBA = 52°$.

- In right $\triangle ABC$, use the tangent ratio to find x:

$$\tan\angle A = \frac{\text{length of leg opposite } \angle A}{\text{length of leg adjacent to } \angle A}$$

$$\tan 52^\circ = \frac{x}{62}$$

$x = 62 \times \tan 52^\circ$ ← Use your calculator and the stored value of $\tan 52^\circ$.

≈ 79.3563812 ← Round off as the last step.

The height of the wall, correct to the *nearest tenth of a foot*, is **79.4** feet.

11. Method I: Use the Pythagorean theorem.

- The first hiker travels 2 miles east from A to C and then 1 mile north from C to D. The second hiker travels 1 mile west from A to B and then 3 miles south from B to E. See the accompanying figure.

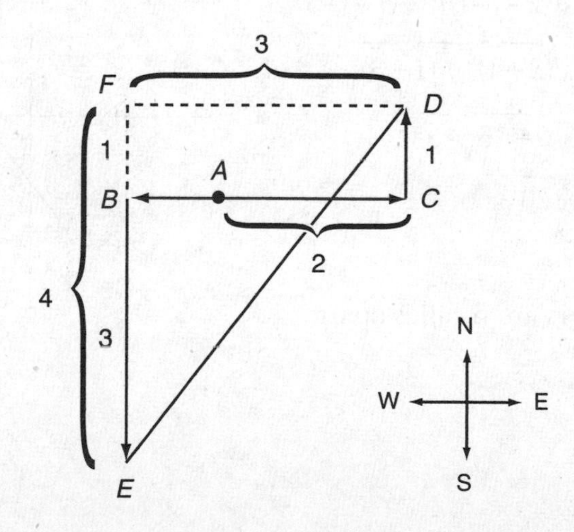

- Draw $\overline{DE}$, which represents the distance the hikers are from each other.
- Form a right triangle in which $\overline{DE}$ is the hypotenuse by completing rectangle *BCDF*.
- In rectangle *BCDF*, $BF = CD = 1$ and $DF = AB + AC = 1 + 2 = 3$.
- Because $EF = 3 + 1 = 4$, the lengths of the sides of right $\triangle EFD$ form a 3-4-5 Pythagorean triple in which the length of hypotenuse $\overline{DE}$ is 5.

The hikers are **5** miles apart.

Method II: Plot the distances traveled on graph paper.

- Assume that the common starting point is the origin. After the first hiker travels 2 miles east ($x = +2$) and then 1 mile north ($y = +1$), the hiker is at $D(2,1)$.
- After the second hiker travels 1 mile west ($x = -1$) and then 3 miles south ($y = -3$), the hiker is at $E(-1, -3)$.
- Find *DE* by using the distance formula:

$$\begin{aligned} DE &= \sqrt{(\Delta x)^2 + (\Delta y)^2} \\ &= \sqrt{(2-(-1))^2 + (1-(-3))^2} \\ &= \sqrt{(2+1)^2 + (1+3)^2} \\ &= \sqrt{9+16} \\ &= \sqrt{25} \\ &= 5 \end{aligned}$$

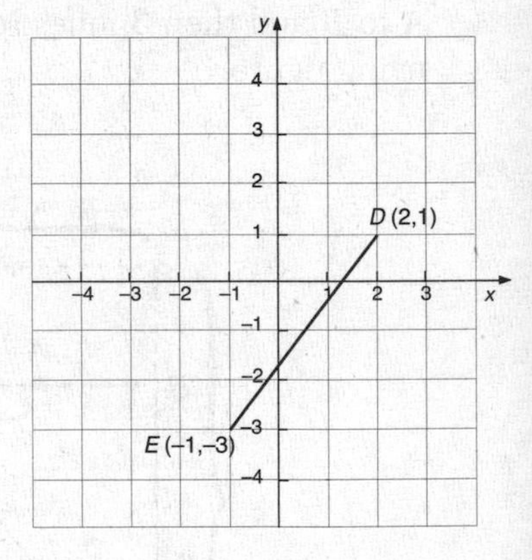

The hikers are **5** miles apart.

12. To find the area of the shaded region, subtract the area of $\triangle ABC$ from the area of semicircle ACB.

- Find the area of semicircle ACB:

$$\text{radius } OA = \frac{1}{2} \times AB = \frac{1}{2} \times 26 = 13$$

$$\text{area of circle } O = \pi \times (OA)^2 = 169\pi$$

$$\text{area of semicircle } ACB = \frac{1}{2} \times 169\pi.$$

- Use the Pythagorean theorem to find AC:

$$(AC)^2 + (CB)^2 = (AB)^2$$

$$(AC)^2 + (10)^2 = (26)^2$$

$$(AC)^2 = 676 - 100$$

$$AC = \sqrt{576} = 24$$

- Find the area of right triangle ABC:

$$\text{area} = \frac{1}{2} \times BC \times AC$$

$$= \frac{1}{2} \times 10 \times 24$$

$$= 120$$

- Find the area of the shaded region:

$$\text{area of shaded region} = \left(\frac{1}{2} \times 169\pi\right) - 120$$

$$\approx 265.46 \quad - 120$$

$$\approx 145$$

To the *nearest square unit*, the area of the shaded region is **145**.

5. PROBABILITY, COUNTING, AND STATISTICS

5.1 KEY STATISTICS

Each of the following statistics is a single number that helps describe how the individual data values in a list are distributed.

- The **mode** is the number in a list of data values that occurs the most often. The mode of {2, 2, 3, 3, 3, 4} is 3. A list of values may have one mode, more than one mode, or no modes.
- The **arithmetic mean** or **average** of a set of data values is the sum of the data values divided by the number of values. The average or mean of {1, 2, 3, 4, 5} is $\frac{1+2+3+4+5}{5} = \frac{15}{5} = 3$.
- The **median** is the number in the middle position of a list of data values that is arranged in size order. The median of {1, 2, 3, 4, 5} is 3. If the list contains an even number of data values, the median is the average of the two middle values. The median of {2, 5, 8, 14, 15, 19} is $\frac{8+14}{2} = \frac{22}{2} = 11$.
- The **range** is the difference obtained by subtracting the lowest value from the highest value in the list.

5.2 LINEAR TRANSFORMATIONS OF DATA

When the same arithmetic operation is performed on each data value, the mean, median, and mode of the transformed set of data values can be obtained by performing the same operation on their original values. For example, if each data value is doubled, then the mean of the transformed set of data values is two times the mean of the original set of data values.

5.3 FITTING A REGRESSION LINE TO DATA

The line that best fits a paired set of data values can be obtained using the regression feature of a graphing calculator.

Minutes Studied (x)	15	40	45	60	70	75	90
Test Grade (y)	60	67	75	75	73	89	93

Using the statistics editor of your graphing calculator, enter the x-values in the accompanying table in list L1 and the y-values in list L2. The **regression line** or **line of best fit** can be calculated by pressing STAT ▷ 4 to choose the **LinReg**($ax + b$) option. Press ENTER to show a summary of the regression statistics where a is the slope of the regression line, b is the y-intercept, and r is the coefficient of linear correlation.

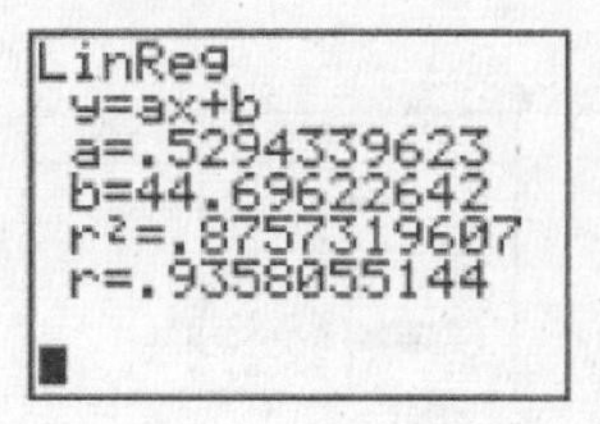

An equation of the line of best fit with the regression coefficients rounded off to the nearest hundredth is $y = 0.53x + 44.70$. Since the coefficient of linear correlation, r, is close to 1, the line is a good fit to the data.

You can store the regression equation as Y_1 if immediately after selecting the linear regression option, you press:

VARS ▷ ENTER ENTER

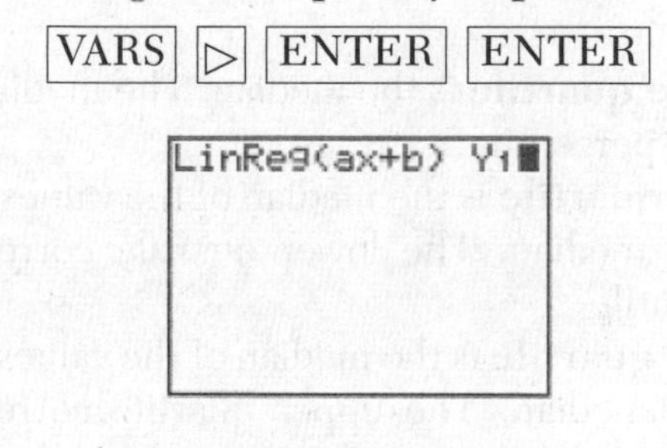

Press $Y =$ to see the regression equation.

- To predict the value of y when $x = 80$ minutes, substitute 80 for x in the regression equation:

$$\begin{aligned} y &= 0.53x + 44.70 \\ &= 0.53 \times 80 + 44.7 \\ &\approx 87 \end{aligned}$$

If the regression equation has been stored as Y_1, you can also obtain the predicted y-value by using the table feature of your calculator:

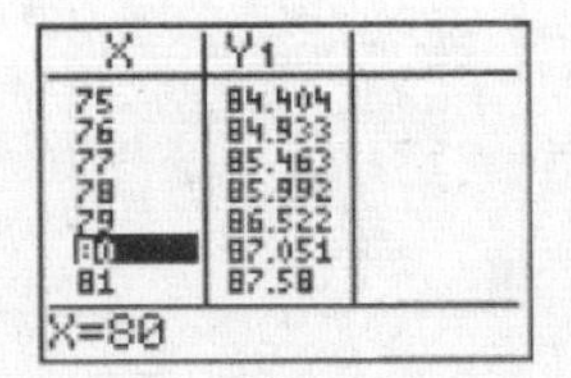

5.4 PERCENTILES AND QUARTILES

The ***p*th percentile** is the data point at or below which $p\%$ of the data values fall. The median is the 50th percentile. Quartiles divide a an ordered set of data values into four groups containing the same number of values.

- The **middle quartile** is the median. The median corresponds to the 50th percentile.
- The **lower quartile** is the median of the values that fall below the overall median. The lower quartile corresponds to the 25th percentile.
- The **upper quartile** is the median of the values that fall above the overall median. The upper quartile corresponds to the 75th percentile

5.5 BOX-AND-WHISKER PLOT

A **box-and-whisker plot** visually summarizes a set of values using five key values: the lowest value (L), the lower quartile (Q_1), the median (Q_2), the upper quartile (Q_3), and the highest value (H), as shown in the accompanying diagram.

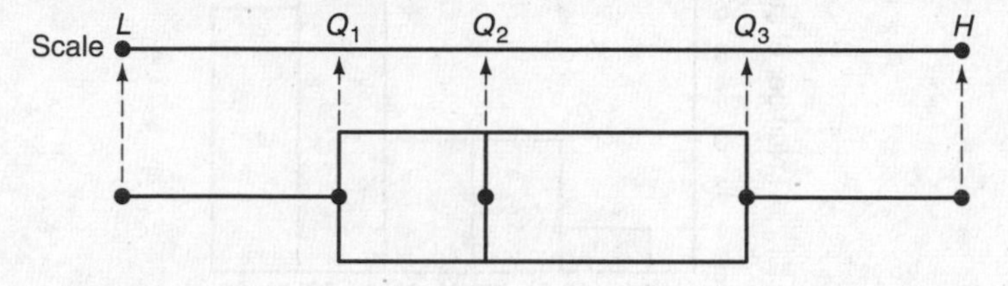

5.6 FREQUENCY TABLES AND HISTOGRAMS: EXAMPLE

The table below gives the distribution of test scores for a class of 20 students.

Test Score Interval	Number of Students (frequency)
91–100	1
81–90	3
71–80	3
61–70	7
51–60	6

a. Draw a *frequency histogram* for the given data.
b. Which interval contains the median?
c. Which interval contains the lower quartile?

Solution

a.

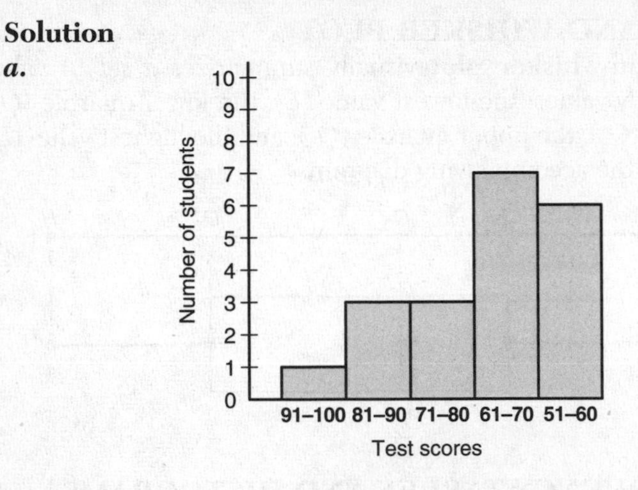

b. Find the total number of students by adding all the frequencies:

$$1 + 3 + 3 + 7 + 6 = 20.$$

- If the scores are arranged in order of size, the *median* is the middle score if the number of scores is odd and is the score midway between the two middle scores if, as here, there is an even number of scores. Since there are 20 scores, the median score will lie midway between the 10th and 11th scores.
- When counting up from the bottom, there are six scores in the 51–60 interval. To count up to the 10th score, four more are needed from the next (the 61–70) interval; five more are needed from the same interval to reach the 11th score. Both the 10th and 11th scores are in the 61–70 interval. The score midway between them (the median) must also lie in this interval.

The **61–70** interval contains the median.

c. The *lower quartile* is the score separating the lowest one-quarter of all scores from the remaining three-quarters.

- Since there are 20 scores in all and $\frac{20}{4} = 5$, the lowest five scores fall below the lower quartile. Hence, the lower quartile is the score midway between the 5th and 6th scores.

Counting up from the bottom, the 5th and 6th scores both lie in the lowest (the 51–60) interval.

The **51–60** interval contains the lower quartile.

5.7 CUMULATIVE FREQUENCY HISTOGRAMS: EXAMPLE

The accompanying frequency table on the left shows the distribution of weight, in pounds, of 32 students.

Interval	Frequency
169–179	9
140–159	8
120–139	6
100–119	2
80–99	7

Interval	Cumulative Frequency
80–179	
80–159	
80–139	
80–119	
80–99	

a. *On another sheet of paper*, copy and complete the accompanying cumulative frequency table on the right, using the data given in the frequency table.

b. Construct a cumulative frequency histogram using the table completed in part ***a***.

Solutions

a.

Interval	Cumulative Frequency
80–179	32
80–159	23
80–139	15
80–119	9
80–99	7

b.

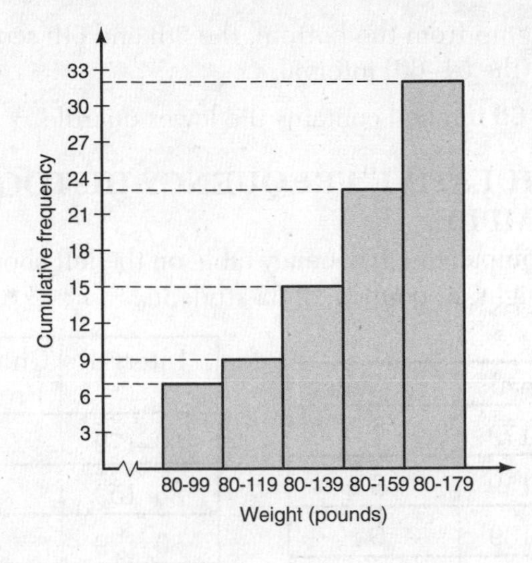

Counting Methods

5.8 MULTIPLICATION PRINCIPLE OF COUNTING

The **multiplication principle of counting** states that, if an event can happen in n ways and a second event can happen next in m ways, the first event followed by the second event can happen in $n \times m$ ways. This principle applies to a sequence of two or more events. For example, if John has five shirts, four ties, and six pairs of pants, then the number of different outfits, each consisting of a shirt, a tie, and a pair of pants, that John can put together is $5 \times 4 \times 6 = 120$.

5.9 COUNTING ORDERED ARRANGEMENTS OF OBJECTS

- The number of ways in which n objects can be arranged in a line is $n!$ (n factorial), where $n!$ represents the product of consecutive whole numbers from n down to 1:

  $$n! = n \times (n-1) \times (n-2) \times \cdots \times 1$$

  For example, the number of ways in which five students can be arranged in a line is $5! = 5 \times 4 \times 3 \times 2 \times 1 = 120$.

- The number of ways in which n objects can be arranged in r available positions, where $r \leq n$, is the product of the r greatest factors of n, denoted by ${}_nP_r$. For example, the number of ways in which five students ($n = 5$) can be seated in a row of three chairs ($r = 3$) is

$$ {}_5P_3 = 5 \times 4 \times 3 = 60 $$

- The number of ways in which a set of n objects can be arranged when x objects in the set are identical, y objects are identical, and z objects are identical is

$$ \frac{n!}{x! \cdot y! \cdot z!} $$

For example, the number of different ways in which 3 blue flags, 2 red flags, and 1 white flag can be arranged on a vertical flag pole is

$$ \begin{aligned} \frac{(3+2+1)!}{3! \cdot 2! \cdot 1!} &= \frac{6!}{3! \cdot 2! \cdot 1!} \\ &= \frac{6 \times 5 \times 4 \times 3 \times 2 \times 1}{(3 \times 2 \times 1) \cdot (2 \times 1) \cdot 1} \\ &= \frac{720}{12} \\ &= 60 \end{aligned} $$

5.10 TREE DIAGRAMS

The set of all possible outcomes for one event, or a series of events, is called a **sample space**. The multiplication principle of counting tells how many outcomes are in a sample space but does not describe the individual outcomes. When you need to know the identity of each of the outcomes in a sample space, you can draw a **tree diagram** or make an organized list.

Suppose a jar contains a red (R) marble, a blue (B) marble, and a white (W) marble. After a marble is selected at random and its color is noted, it is put back into the jar. Then another marble is picked at random and its color is noted. Using the multiplication principle of counting, you know that 3×3 or 9 different outcomes are possible. These outcomes can be described as shown in the accompanying tree diagram:

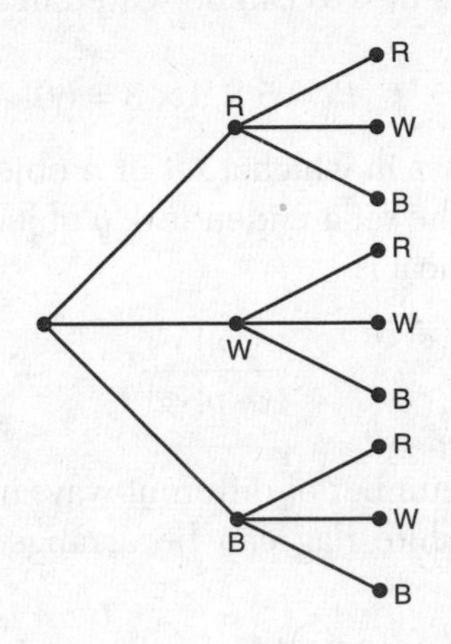

The set of all possible outcomes can also be described by making a list of ordered pairs of the form (color of first marble, color of second marble):

(R, R)	(W, R)	(B, R)
(R, W)	(W, W)	(B, W)
(R, B)	(W, B)	(B, B)

5.11 VENN DIAGRAMS

A Venn diagram uses overlapping circles to represent sets that have members in common.

EXAMPLE

In Clark Middle School, 60 students are in seventh grade. If 25 of these students take art only, 18 take music only, and 9 do not take either art or music, how many take both art and music?

Solution

It is given that, of the 60 students in seventh grade in Clark Middle School, 25 take art only, 18 take music only, and 9 do not take either art or music. To find how many of the 60 students take both art and music, organize the facts in a Venn diagram in which x represents the number of students who take both art and music.

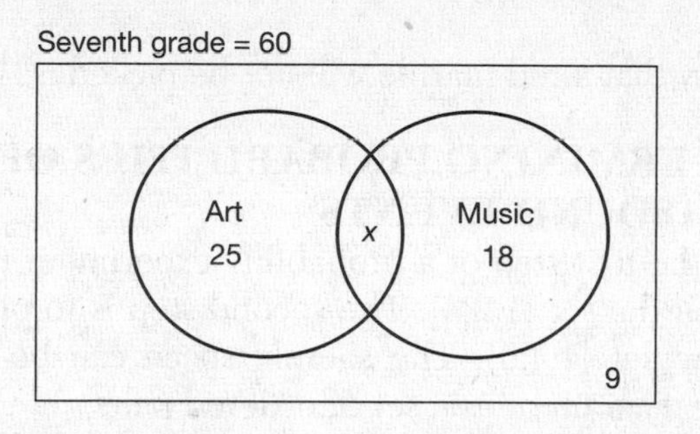

The sum of the different, nonoverlapping groups of students must add up to 60. Thus:

$$\begin{aligned} 25 + x + 18 + 9 &= 60 \\ x + 52 &= 60 \\ x &= 60 - 52 \\ &= 8 \end{aligned}$$

In the seventh grade, **8** students take both art and music.

5.12 MEANING OF PROBABILITY

A **probability** value expresses the likelihood that a future event will happen. If an event is certain to happen, its probability is 1. If an event is impossible, its probability is 0. All other events have probability values between 0 and 1.

- If an event E can happen in r of n equally likely ways, the probability that it will happen, denoted by $P(E)$, is defined as follows:

$$P(E) = \frac{\text{number of favorable outcomes}}{\text{total number of possible outcomes}} = \frac{r}{n}.$$

- If the probability that an event will happen is p, the probability that the event will *not* happen is $1 - p$. For example, the probability of picking a red marble from a jar that contains only three red marbles and four green marbles is $\frac{3}{3+4}$ or $\frac{3}{7}$. The probability that a red marble will *not* be picked is $1-\frac{3}{7}=\frac{4}{7}$.

5.13 DETERMINING PROBABILITIES OF COMPOUND EVENTS

Suppose the first step of a probability experiment is to pick one number from the set {1,2,3}. The second step is to pick one number from the set {1,4,9}. The sample space can be displayed as either a tree diagram or as a set of ordered pairs.

The tree diagram contains three branches leading from "Start" since either 1, 2, or 3 may be picked on the first step. For each of the first-step branches, three second-step branches lead to 1, 4, or 9 as the possible second number to be picked, as shown in the accompanying figure.

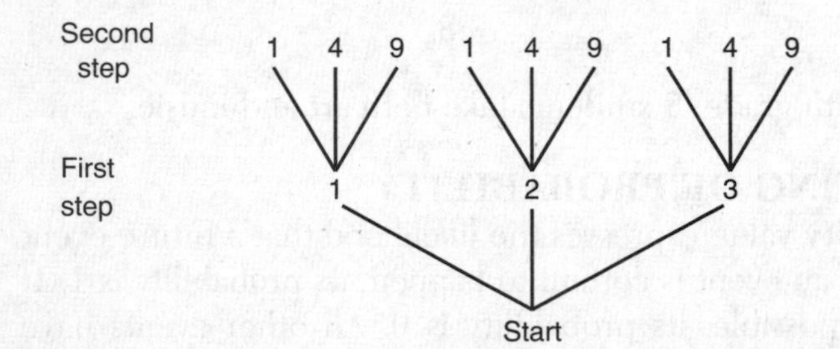

The accompanying table representing the sample space contains two columns, one for each of the two steps in the experiment. The column for the first step shows the possible numbers, 1, 2, and 3, for that step. In the second column, each number in the first column is paired with 1, 4, and 9, the possible numbers that may be picked in the second step.

First Step	Second Step	Ordered Pair
1	1	(1,1)
1	4	(1,4)
1	9	(1,9)
2	1	(2,1)
2	4	(2,4)
2	9	(2,9)
3	1	(3,1)
3	4	(3,4)
3	9	(3,9)

Once the sample space is described, the probabilities of various events can be determined.

- What is the probability that both numbers selected are the same? The only way in which both numbers can be the same is if 1 is chosen both times. The tree diagram shows that only one path leads to two selections of 1—the leftmost path. There are nine possible paths in all. If the sample space table is used, there is only one line on which both columns contain 1—the first row. There are nine rows in all. Therefore, the probability that both numbers are the same is $\frac{1}{9}$.
- What is the probability that the second number is the square of the first? If the second number is the square of the first, the possible selections are 1 and 1, 2 and 4, and 3 and 9. On the tree diagram, one path leads to 1 and 1, a second path leads to 2 and 4, and a third path leads to 3 and 9. Thus, three paths of the possible total of nine are favorable cases. If the sample space table is used to obtain the information, the combinations of 1 and 1, 2 and 4, and 3 and 9 are shown on one row each. A total of three rows are favorable cases out of the nine possible rows. Therefore, the probability that the second number is the square of the first is $\frac{3}{9}$ or $\frac{1}{3}$.

- What is the probability that both numbers selected are odd? The tree diagram shows that the paths leading to both odd numbers are 1–1, 1–9, 3–1, and 3–9. Thus, four possible paths of the total of nine represent the successful outcomes. In the table for the sample space, the rows containing 1 and 1, 1 and 9, 3 and 1, and 3 and 9 represent the successful outcomes of rows containing both odd numbers. There are nine rows in all. Therefore, the probability that both numbers are odd is $\frac{4}{9}$.

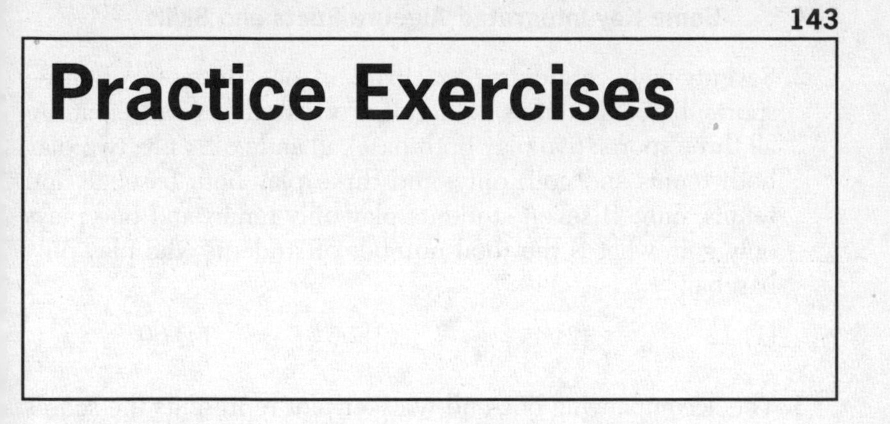

Practice Exercises

1. The accompanying graph shows the high temperatures in Elmira, New York, for a 5-day period in January.

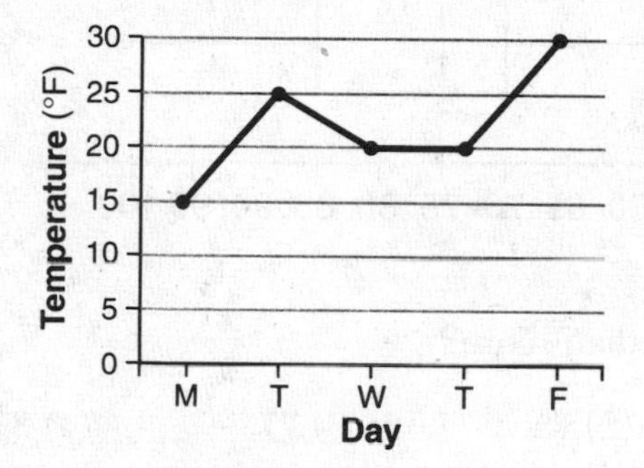

Which statement describes the data?

(1) median = mode (3) mean < mode
(2) median = mean (4) mean = mode

2. Seventy-eight students participate in one or more of three sports: baseball, tennis, and golf. Four students participate in all three sports; five play both baseball and golf, only; two play both tennis and golf, only; and three play both baseball and tennis, only. If seven students play only tennis and one plays only golf, what is the total number of students who play only baseball?

(1) 12 (2) 44 (3) 56 (4) 60

3. The accompanying box-and-whisker plot represents the scores earned on a science test.

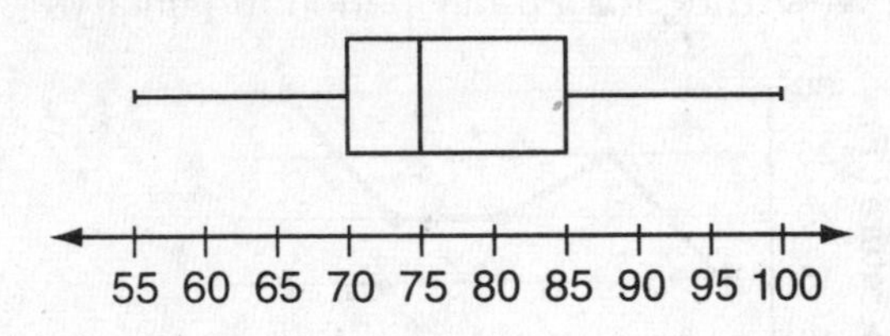

What is the median score?

(1) 70 (2) 75 (3) 77 (4) 85

4. If a boy has five shirts and three pairs of pants, how many possible outfits, each consisting of one shirt and one pair of pants, can he choose?

5. How many different arrangements of seven letters can be made using the letters in the name "ULYSSES"?

6. Six members of a school's varsity tennis team will march in a parade. How many different ways can the players be lined up if Angela, the team captain, is always at the front of the line?

7. The student scores on Mrs. Frederick's mathematics test are shown on the stem-and-leaf plot below.

Stem	Leaves
4	3
6	0 5 5 7 9
7	2 5 6 8 9 9 9
9	0 1 2 5 9

Key: 4 | 3 = 43 points

Find the median of these scores.

8. The accompanying cumulative frequency histogram shows the scores that 24 students received on an English test. How many students had scores between 71 and 80?

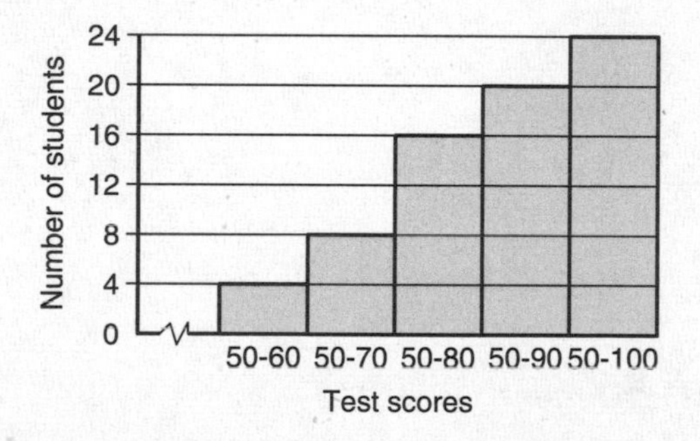

9. José surveyed 20 of his friends to find out what equipment they use to play recorded movies. He found that 12 of his friends have only DVD players, 5 have both DVD players and VCRs, and 2 have neither type of player. The rest of his friends have only VCRs. What is the total number of his friends that have VCRs?

10. All seven-digit telephone numbers in a town begin with 245. How many telephone numbers may be assigned in the town if the last four digits do *not* begin or end in a zero?

11. Selena and Tracey play on a softball team. Selena has 8 hits out of 20 times at bat, and Tracey has 6 hits out of 16 times at bat. Based on their past performance, what is the probability that both girls will get a hit next time at bat?

12. Tamika could not remember her scores from five mathematics tests. She did remember that the mean (average) was exactly 80, the median was 81, and the mode was 88. If all her scores were integers with 100 the highest score possible and 0 the lowest score possible, what was the lowest score she could have received on any one test?

Solutions

1. According to this graph, the temperatures were 15, 25, 20, 20, and 30.

- The median is the middle value when a set of numbers is arranged in size order. The median of the set of 5 temperatures is 20.
- The mode is the value that occurs the greatest number of times in a set of numbers. The mode of the set of 5 temperatures is 20.
- The mean of a set of numbers is the sum of the numbers divided by the number of values in the set. The mean of the set of 5 temperatures is

$$\text{mean} = \frac{15 + 25 + 20 + 20 + 30}{5}$$

$$= \frac{110}{5}$$

$$= 22$$

Compare the values of the median (= 20), mode (= 20), and mean (= 22). The median is equal to the mode.

The correct choice is **(1)**.

2. It is given that seventy-eight students participate in one or more of three sports: baseball, tennis, and golf, as follows:

- four participate in all three sports;
- five play both baseball and golf, only;
- two play both tennis and golf, only;
- three play both baseball and tennis, only;
- seven play only tennis and one plays only golf.

Summarize the given information in a Venn diagram in which the three overlapping circles represent the different sports activities:

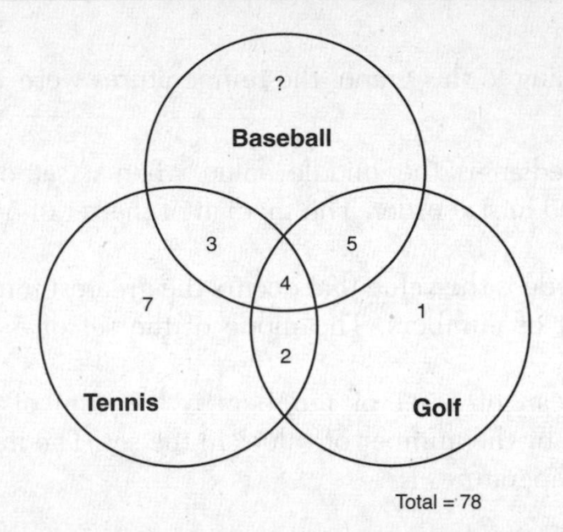

The only information missing from the diagram is the number of students who play only baseball. Since there is a total of 78 students represented by the Venn diagram, the number of students who play only baseball is

$$78 - (4 + 5 + 2 + 3 + 7 + 1) = 78 - 22 = 56$$

The correct choice is **(3)**.

3. As shown in the accompanying figure, a box-and-whisker plot uses five key data values: lowest score (L), highest score (H), first quartile ($Q1$), second quartile or median ($Q2$), and third quartile ($Q3$).

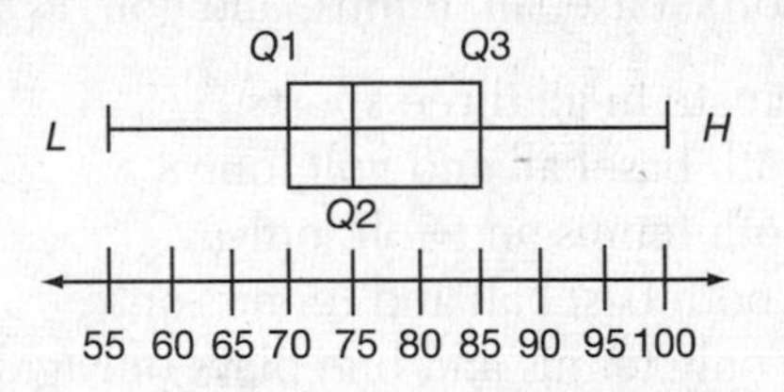

Since $Q2$ is aligned with 75 on the number line, the median score is 75.

The correct choice is **(2)**.

4. The boy has five choices for shirts. For each of these five choices, he has three choices for pairs of pants. The total number of possible outfits is therefore $5 \times 3 = 15$.

The number of possible outfits is **15**.

5. The number of different arrangements of p things where r are alike is $\frac{p!}{r!}$. The name "ULYSSES" contains seven letters, but there are three S's. For $p = 7$ and $r = 3$:

$$\begin{aligned} \frac{p!}{r!} &= \frac{7!}{3!} \\ &= \frac{7\times6\times5\times4\times \cancel{3\times2\times1}}{\cancel{3\times2\times1}} \\ &= 840 \end{aligned}$$

840 different arrangements are possible.

6. It is given that six members of a school's varsity tennis team will march in a parade. Since order matters, finding the number of different ways the players can be lined up if Angela, the team captain, is always at the front of the line is a permutation problem in which six positions in a line must be filled by six people:

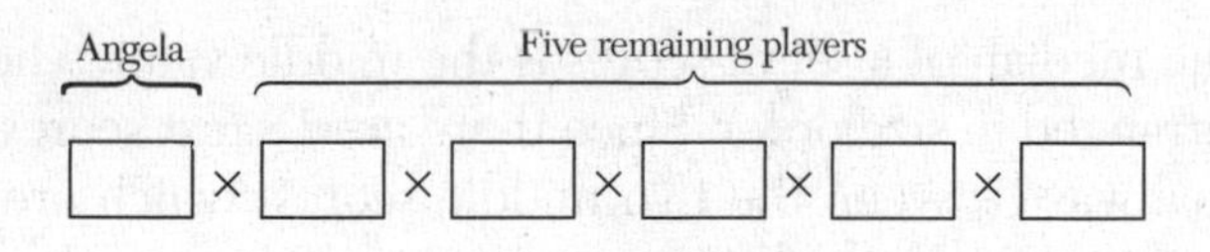

Since Angela is always first in line, the first position can be filled in exactly one way. The second slot can be filled with any of the five remaining players. Once that position is filled, the next slot can be filled with any of the remaining four players, and so forth:

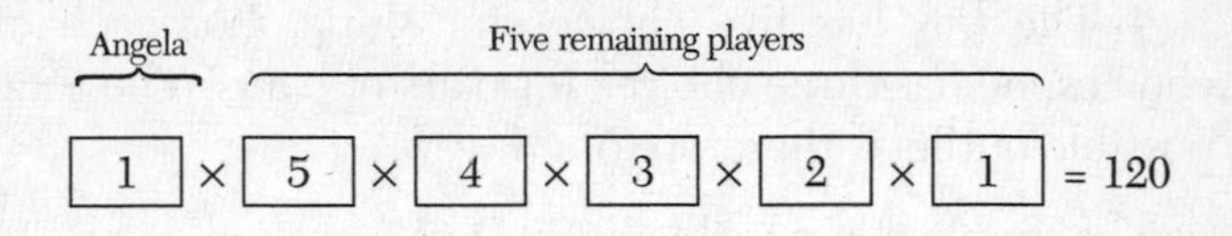

The players can be lined up in **120** ways given that Angela is always at the front of the line.

7. It is given that the accompanying stem-and-leaf plot represents the student scores on Mrs. Frederick's mathematics test.

Stem	Leaves
4	3
6	0 5 5 7 9
7	2 5 6 8 9 9 9
9	0 1 2 5 9

Key: 4 | 3 = 43 points

The key tells us that the numbers to the left of the vertical bar represent the ten's digits of the scores and the numbers to the right of the vertical bar represent the corresponding unit's digits of the scores. Hence, the student scores are:

$$\underbrace{43}_{\text{row 1}};\ \underbrace{60, 65, 65, 67, 69}_{\text{row 2}};\ \underbrace{72, 75, 76, \overset{\text{median}}{\downarrow} 78, 79, 79, 79}_{\text{row 3}};\ \underbrace{90, 91, 92, 95, 99}_{\text{row 4}}$$

The median of a set of scores is the middle value when the scores are arranged in size order. Since there are 18 test scores, the median lies halfway between the two middle scores, which are the 9th and 10th test scores in the list:

$$\text{median} = \frac{76+78}{2} = 77$$

The median of these scores is **77**.

8. The cumulative frequency histogram shows that eight students received scores between 50 and 70 and that 16 students received scores between 50 and 80. Hence, 16 – 8 or 8 students received scores between 71 and 80.

8 students received scores between 71 and 80.

9: It is given that when José surveyed 20 of his friends, he found that 12 have only DVD players, 5 have both DVD players and VCRs, and 2 have neither type of player. The rest of his friends have only VCRs. To find the total number of his friends that have VCRs, organize the facts using a Venn diagram where x represents the number of friends who have only VCRs.

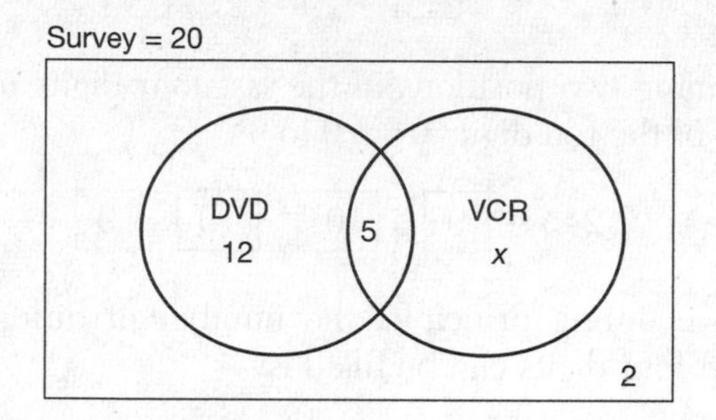

Summarize the number of friends who belong to each of the nonoverlapping sets:

friends who have only DVD players	= 12
friends who have both types of players	= 5
friends who have only VCRs	$= x$
friends who have neither player	= 2

Total $= x + 19 = 20$

$x = 1$

As 1 friend has only a VCR and 5 friends have both VCRs and DVD players, a total of 1 + 5 = **6** friends have VCRs.

10. It is given that all seven-digit telephone numbers in a town begin with 245 and the last four digits do not begin or end in a zero:

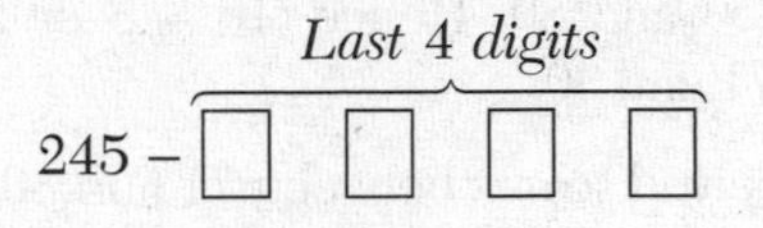

To find the number of telephone numbers that may be assigned in the town, figure out the number of digits that can be used to fill each position of the last four digits. Then use the counting principle:

The last four digits may begin or end with any one of the nine digits from 1 to 9:

$$245 - \boxed{9}\ \boxed{}\ \boxed{}\ \boxed{9}$$

The remaining two positions in the last four digits may be filled with any one of the ten digits from 0 to 9:

$$245 - \boxed{9}\ \boxed{10}\ \boxed{10}\ \boxed{9}$$

Using the counting principle, the number of different ways in which the last four digits can be filled is:

$$\boxed{9} \times \boxed{10} \times \boxed{10} \times \boxed{9} = 8{,}100$$

Hence, **8,100** telephone numbers may be assigned in the town.

11. To find the probability that Selena and Tracy both get a hit the next time at bat, multiply the probabilities of each girl getting a hit the next time at bat.

- Selena has 8 hits out of 20 times at bat, so the probability that she will get a hit her next time at bat is $\frac{8}{20}$.
- Tracey has 6 hits out of 16 times at bat, so the probability that she will get a hit her next time at bat is $\frac{6}{16}$.

- The probability that both girls will get a hit next time at bat is

$$\frac{8}{20} \times \frac{6}{16} = \frac{48}{320}$$

$$= \mathbf{\frac{3}{20}}$$

12. It is given that Tamika's five test scores are integers with 100 the highest-possible score 0 the lowest-possible score. Tamika remembered that the mean (average) score was exactly 80, the median was 81, and the mode was 88.

The median of an odd number of scores is the middle score when the scores are arranged in size order. Hence, when Tamika's five test scores are arranged in size order, the middle score must be 81:

The mode is the score that occurs the most number of times. Since the mode was 88, at least two of the scores must be 88:

Let x represent the lowest score Tamika could have earned.

For a fixed mean (average), the value of x will be as small as possible when each of the remaining scores are as large as possible. Since the second score must be less than 81 and an integer, the largest it can be is 1 less than 81 or 80:

x	80	81	88	88

The mean (average) of a set of scores is the sum of the scores divided by the number of scores. Since the mean (average) of the five test scores was exactly 80:

$$\frac{x+80+81+88+88}{5} = 80$$

$$\frac{x+337}{5} = 80$$

$$x + 337 = 5 \cdot 80$$

$$x = 400 - 337$$

$$x = 63$$

The lowest-possible score Tamika could have received on any one test was **63**.

Glossary of Terms

abscissa See *x-coordinate*.

absolute value The absolute value of a number x, denoted by $|x|$, is its distance from 0 on the number line and, as a result, can never be a negative number. To remove the absolute value sign in $|x|$:

- write x if x is non-negative, as in $|0| = 0$ and $|+3| = 3$.
- write the opposite of x if x is negative, as in $|-3| = -(-3) = 3$.

absolute value function A function whose equation has the form $y = |x|$. It is comprised of the two linear functions, $y = x$ when $x \geq 0$ and $y = -x$ when $x < 0$.

additive inverse The opposite of a real number. The sum of a number and its additive inverse is 0. The additive inverse of +3 is −3 since $(+3) + (-3) = 0$.

algebraic representation The use of an equation or algebraic expression to describe the relationship between variable quantities.

altitude A line segment that is perpendicular to the side to which it is drawn.

angle of depression The angle through which the horizontal line of sight must be lowered in order to view an object below it.

angle of elevation The angle through which the horizontal line of sight must be raised in order to view an object above it.

area The number of square units a region encloses.

associative property The sum or product of three numbers is the same regardless of which two numbers are chosen first to be added or multiplied together:

$$(a + b) + c = a + (b + c)$$
$$(ab)c = a(bc)$$

where a, b, and c are real numbers.

average See *Mean*.

axis of symmetry The line of symmetry of a parabola that contains the turning point (vertex). For the parabola $y = ax^2 + bx + c$, an equation of the axis of symmetry is $-\frac{b}{2a}$.

base of a power The number that is used as the factor when a product is expressed in exponential form. In 3^4, the exponent, 4, tells the number of times the base, 3, is to be used as a factor in the expanded product.

binomial The sum or difference of two unlike terms, as in $2x + 3y$.

bivariate data A data set involving the measurements of two variables.

box-and-whisker plot A diagram that shows how spread out data are about the median using five key values: the lowest score, the lower quartile, the median, the upper quartile, and the highest score.

circumference The distance around a circle. The circumference C of a circle with radius r is given by the formula, $C = 2\pi r$.

closure property A set is closed under an operation if the result of performing that operation on members of the set is also a member of the same set.

coefficient The number that multiplies one or more variable factors in a monomial. The coefficient of $-5xy^2$ is -5.

coefficient of linear correlation A number from -1 to $+1$, denoted by r, that indicates the strength and direction of the relationship between two sets of data. The closer r is to 1, the stronger the relationship between the statistical variables that represents the two sets of data.

commutative property The order in which two numbers are added or multiplied together does not matter:

$$a + b = b + a$$
$$a \times b = b \times a$$

where a, b, and c are real numbers.

complement of a set If A is a subset of S, then the complement of A, denoted by $\overline{A}$, is the set of elements that are in S but *not* in A. If $S = \{1,2,3,4,5,6,7\}$ and $A = \{1,2,3,4\}$ then $\overline{A} = \{5,6,7\}$. The complement of A may also be represented by A' or $\sim A$.

complementary angles Two angles whose degree measures add up to 90.

congruent figures Figures that have the same size and the same shape.

constant A quantity that does not change in value, such as π.

coordinate plane A plane divided into four parts, called quadrants, by a horizontal number line and a vertical number line intersecting at their zero points, called the origin.

correlation coefficient See *Coefficient of Linear Correlation.*

cosine ratio In a right triangle, the ratio of the length of the leg that is adjacent to a given acute angle to the length of the hypotenuse.

counterexample A single, specific instance that contradicts a proposed generalization.

counting principle If event A can occur in m ways and event B can occur in n ways, then the number of ways in which both events can occur is m times .

cumulative frequency The sum of all frequencies from a given data point up to and including another data point.

cumulative frequency histogram A histogram whose bar heights represent the cumulative frequency at stated intervals.

degree of a monomial The sum of the exponents of its variable factors. The degree of $7xy^2$ is 3 since the exponent of x is 1, the exponent of y is 2, and $1 + 2 = 3$.

degree of a polynomial For a polynomial in one variable, the greatest degree of its terms. The degree of $3x^2 + 5x + 11$ is 2.

direct variation When two variables are related so that their ratio remains the same.

distributive property A sum may be multiplied by a number by multiplying each addend separately by the number and then adding the products:

$$a(b + c) = ab + ac$$

where a, b, and c are real numbers.

domain The set of all first members or x-values of the ordered pairs that belong to a relation.

empirical probability An estimate of the probability that an event will happen based on sample data or the results from performing repeated trials of a probability experiment.

empty set A set that has no elements, denoted by { } or $\varnothing$.

equation A sentence in which an equal sign separates two expressions, as in $3x - 7 = x + 1$, thereby indicating the left and right sides have the same value.

equivalent equations Two equations that have the same solution set, as with $2x = 6$ and $x + 2 = 5$.

event A subset of the sample space of a probability experiment.

exponent A number written to the right and a half line above another number, called the base, that tells the number of times the base is used as a factor in a product. Thus, $5^3 = 5 \times 5 \times 5 = 125$.

exponential decay When a quantity decreases by a fixed percent of its current value over successive time periods. If A is the initial amount and r is the rate of decay per time period, then after n successive time periods, the amount that remains, y, is given by $y = A(1 - r)^n$ where $0 < r < 1$.

exponential function A function of the form $y = b^x$ where b is a positive number other than 1. An exponential function, unlike linear and quadratic functions, has a variable in its exponent.

exponential growth When a quantity increases by a fixed percent of its current value over successive time periods. If A is the initial amount and r is the rate of growth per time period, then after n successive time periods, the amount present, y, is given by $y = A(1 = r)^n$ where $r > 0$.

extrapolate The process of estimating a statistical variable *outside* the range of its observed values. The usefulness of an extrapolated value depends on whether the relationship between the data values extends outside the original data set.

factor An exact divisor of a given expression. For example, 3 is a factor of 12. Because $2x(3x + 2) = 6x^2 + 4x$, $2x$ and $3x + 2$ are factors of $6x^2 + 4x$.

factorial *n* Denoted by $n!$ and, for any positive integer n, is the product of the consecutive whole numbers from n to 1. By definition, $0! = 1$. Thus, $4! = 4 \times 3 \times 2 \times 1 = 24$.

factoring Rewriting a polynomial as the product of two or more lower-degree polynomials.

factoring completely Factoring so that each of the factors of a given expression cannot be factored further.

favorable outcomes The set of outcomes for which an event occurs.

five number summary The minimum score, the first quartile, the median, the third quartile, and the maximum score of a data set. These five values are used to construct a box and whisker plot.

FOIL An abbreviation of a rule for multiplying two binomials horizontally by forming the sum of the products of the first terms (F), the outer terms (O), the inner terms (I), and the last terms (L) of the binomial factors, as in:

$$(x+5)(x-2) = \overbrace{x \cdot x}^{\text{F}} + \overbrace{-2x}^{\text{O}} + \overbrace{5x}^{\text{I}} + \overbrace{-2 \cdot 5}^{\text{L}}$$

$$= x^2 + 3x - 10$$

frequency The number of times that a score appears in a set of data values.

function A set of ordered pairs in which no two ordered pairs have the same first member and different second members. The set {(1,3), (2,4), (5,3)} represents a function. The set {(1,4), (2,3), (2,5)} does not represent a function since 2 is paired with two different y-values.

greatest common factor (GCF) For a given polynomial, the greatest monomial that is an exact divisor of each of its terms. The GCF of $8a^2b^2 + 20ab^3$ is $4ab^2$.

half-life The time needed for a substance to decay to one-half of its original amount.

histogram A vertical bar graph whose bars are adjacent to each other. The height of each rectangular bar shows an amount or frequency of a quantity that is indicated at the base of the bar.

hypotenuse In a right triangle, the side opposite the right angle.

independent events When the outcome of one event does not affect the outcome of a second event.

Inequality A sentence that uses an inequality symbol to compare two expressions. The symbols for indicating an inequality relation: $<$ means "is less than"; $\leq$ means "is less than or equal to"; $>$ means "is greater than"; $\geq$ means "is greater than or equal to."

integer An element of the set {. . . –4, –3, –2, –1, 0, 1, 2, 3, 4, . . .}.

interpolate The process of estimating a statistical variable *within* the range of its observed values.

intersection of sets The elements common to the sets. The intersection of sets A and B is denoted by $A \cap B$. If $A = \{1,2,3,4,5\}$ and $B = \{0,2,4,6\}$, then $A \cap B = \{2,4\}$.

irrational number A number that cannot be expressed as the quotient of two integers, such as $\sqrt{3}$ and π. Nonending decimals with no repeating pattern represent irrational numbers.

isosceles trapezoid A trapezoid whose nonparallel sides have the same length.

linear function A function whose graph is a nonvertical line.

line of best fit The line that can be drawn "closest" to a set of data points. Because an equation of this line can be determined using a statistical procedure called regression analysis, it is sometimes referred to as a regression line.

mean For a given set of n numbers, their sum divided by n.

median The middle score when a set of data values are arranged in size order. If the set of data values has an even number of scores, then the median is the average of the two middle scores.

mode The score in a set of data values that has the greatest frequency. A set of data values may have more than one mode or no mode.

monomial A single term that consists of a number, a variable, or the product of a number and one or more variables.

multiplication property of 0 If the product of two expressions is 0, then at least one of these expressions is equal to 0.

multiplicative identity The element of a set that when multiplied by any member of the same set results in that same member. The multiplicative identity for the set of real numbers is 1.

multiplicative inverse For any given member of a set, the element that it must be multiplied by to produce the multiplicative identity for that set. For the set of real numbers, the multiplicative inverse of each nonzero number is its reciprocal because their product is 1.

mutually exclusive events Events from the same sample space that have no outcomes in common.

null set See *Empty set*.

ordered pair Two numbers that are written in a definite order.

ordinate See *y-coordinate*.

origin The zero point on a number line.

parabola The graph of a quadratic equation in which either x or y, but not both, are squared. The graph of $y = ax^2 + bx + c$ $(a \neq 0)$ is a parabola that has a vertical line of symmetry, an equation of which is $x = -\frac{b}{2a}$.

parallel lines Lines that lie in the same plane that do not intersect.

parallelogram A quadrilateral in which both pairs of opposite sides are parallel. The diagonals of a parallelogram bisect each other.

perfect square A rational expression whose square root is also rational, such as $4(\sqrt{4} = 2)$ and $\frac{9}{25}\left(\sqrt{\frac{9}{25}} = \frac{3}{5}\right)$.

perimeter The distance around a figure.

permutation An arrangement of objects in which order matters.

permutation notation The notation ${}_nP_n$ represents the arrangement of n objects in n positions so ${}_nP_n = n!$ The arrangement of n objects when fewer than n positions are available is denoted by ${}_nP_r$ where $r \leq n$. To evaluate ${}_nP_r$, multiply together the r greatest factors of n, as in ${}_5P_3 = 5 \times 4 \times 3 = 60$.

point-slope form An equation of a line of the form $y - b = m(x - a)$, where m is the slope of the line and (a,b) is a point on the line.

polynomial A monomial or the sum of two or more monomials.

power A number written with an exponent, as in 2^4 which is read "2 raised to the fourth power."

prime number A positive integer greater than 1 whose only positive factors are itself and 1. The number 2 is the only even integer that is a prime number.

principal square root The positive square root of a positive number. Although 49 has two square roots, +7 and –7, the principal square root is +7. The radical sign, $\sqrt{}$, indicates the principal square root of the number underneath it. Thus, $\sqrt{49} = 7$.

probability of an event A measure of the likelihood that an event will occur expressed as a number from 0 to 1. If all the outcomes of an event are equally likely to occur, then the probability that the event will occur is the number of ways in which it can occur divided by the total number of equally likely outcomes.

proportion An equation that states that two ratios are equal.

***p*th percentile** The score at or below which $p\%$ of the scores in a set of data values lie.

Pythagorean theorem The relationship that states that in a right triangle, the square of the length of the hypotenuse is equal to the sum of the squares of the lengths of the two legs.

quadrant One of four rectangular-like regions into which the coordinate plane is divided.
quadratic equation An equation in which a variable has an exponent of 2 and no variable has an exponent greater than 2.
quadratic polynomial A polynomial whose degree is 2.
quadrilateral A polygon with four sides.

radical sign The symbol $\sqrt{}$ that denotes the principal root of the number underneath it, as in $\sqrt{49} = 7$.
radicand The number that appears underneath a radical sign. In $\sqrt{49}$, the radicand is 49.
range The set of all second members or y-values of the ordered pairs that comprise a relation.
range of a set of scores The difference between the highest and the lowest scores.
rate A comparison of two quantities measured in different units by division.
ratio A comparison of two quantities measured in the same units by division. The ratio of a to b (a:b) is the fraction $\frac{a}{b}$, provided $b \neq 0$.
rational number A number that can be written as a fraction having an integer in the numerator and a nonzero integer in the denominator. The set of rational numbers includes decimals in which a set of digits endlessly repeat, as in 0.25000… $\left(=\frac{1}{4}\right)$ and 0.333. . . $\left(=\frac{1}{3}\right)$.
real number A number that is either rational or irrational. The set of all points on the number line corresponds in one-to-one fashion to the set of real numbers.
reciprocal For a nonzero number, the number by which it must be multiplied in order to produce 1. The reciprocal of $\frac{a}{b}$ is $\frac{b}{a}$ provided $a,b \neq 0$.
rectangle A parallelogram with four right angles. The diagonals of a rectangle have the same length.

regression line See *Line of best fit*.
relation A set of ordered pairs.
replacement set The set whose elements may be substituted for a variable. Also referred to as the domain of the variable.
rhombus A parallelogram whose four sides have the same length. The diagonals of a rhombus intersect at right angles.
right angle An angle that measures 90°.
right triangle A triangle that contains a right angle.
root A number from the replacement set that when substituted for a variable, makes an equation a true statement. An equation may have more than one root or may have no roots.

sample space The set of all possible outcomes of a probability experiment.
scientific notation A number expressed as the product of a number greater than or equal to 1 but less than 10 and a power of 10. In scientific notation, 81,000 is written as 8.1×10^4 and 0.0072 is written as 7.2×10^{-3}.
set A collection of objects usually described within a set of braces, { }. The set of integers greater than 2 and less than 6 is {3,4,5}.
similar triangles Two triangles are similar when the angles of one triangle are congruent to the corresponding angles of the other triangle. The lengths of corresponding sides of similar triangles are in proportion.
sine ratio In a right triangle, the ratio of the length of the leg that is opposite a given acute angle to the length of the hypotenuse.
slope A measure of steepness. A line that rises as x increases has a positive slope. A line that falls as x increases has a negative slope. The slope of a horizontal line is 0, and the slope of a vertical line is undefined.
slope formula The slope, m, of a nonvertical line that contains the points $A(x_A,y_A)$ and $B(x_B,y_B)$ can be determined using the formula,

$$m = \frac{y_B - y_A}{x_B - x_A}.$$

slope-intercept form An equation of a line of the form $y = mx + b$, where m is the slope of the line and b is the y-intercept.
solution set The set consisting of those members of the replacement set that when substituted for the variable in an equation or inequality, results in a true statement.

square A parallelogram with four right angles and whose sides have the same length. The diagonals of a square have the same length and intersect at right angles.

square root One of two identical factors of a nonnegative number. Every positive number has two square roots. The square root of 9 is +3 or –3.

subset A set each of whose members are elements of another set. If $A = \{1,2,3,4,5\}$ and $C = \{1,3,5\}$, then C is a subset of A, denoted by $C \subset A$.

successes See *Favorable outcomes*.

supplementary angles Two angles whose degree measures add up to 180.

system of equations A set of equations with the same variables whose solution is the set of values that make each of the equations true at the same time.

tangent ratio In a right triangle, the ratio of the length of the leg that is opposite a given acute angle to the length of the leg that is adjacent to the same angle.

theoretical probability A probability ratio that can be calculated based on a logical analysis of a probability experiment without the need actually to perform it.

trapezoid A quadrilateral that has one pair of parallel sides and one pair of sides that are not parallel.

tree diagram A diagram whose branches describe the different possible outcomes in a probability experiment.

trinomial A polynomial with three unlike terms.

turning point of a parabola See *Vertex of a parabola*.

union of two sets The set each of whose members belong to either or both of two other sets. The union of sets P and Q is denoted by $P \cup Q$. If $P = \{1,3\}$ and $Q = \{0,3,5\}$, then $P \cup Q = \{0,1,3,5\}$.

variable A symbol, usually a single lowercase letter, that represents an unspecified member of a given set called the replacement set or domain of the variable.

Venn diagram A diagram in which circles are used to represent the logical relationships among two or more sets that may have members in common.

vertex of a parabola The point at which the axis of symmetry intersects a parabola. At this point, the parabola achieves either its maximum or minimum y-value. Also referred to as the turning point.

vertical line test A graph represents a function if it is not possible to draw a vertical line that intersects the graph in more than one point.

volume A measure of capacity that gives the number of unit cubes a solid figure can hold.

***x*-axis** The horizontal axis in the coordinate plane.

***x*-coordinate** The first member (abscissa) in an ordered pair of numbers that indicates the location of a point in the coordinate plane. The x-coordinate of (3,2) is 3.

***x*-intercept** The x-coordinate of the point at which a graph intersects the x-axis.

***y*-axis** The vertical axis in the coordinate plane.

***y*-coordinate** The second member (ordinate) in an ordered pair of numbers that indicates the location of a point in the coordinate plane. The y-coordinate of (3,2) is 2.

***y*-intercept** The y-coordinate of the point at which a graph intersects the y-axis.

zero product rule If the product of two expressions is 0, then at least one of the expressions must be equal to 0.

Regents Examinations, Answers, and Self-Analysis Charts

Important Note

The first sample test that follows is *not* an actual Regents Examination. It is a practice test that has the same format and similar topic coverage of an actual Regents Examination in Integrated Algebra.

An official Test Sampler from the New York State Education Department begins on page 212.

The last four exams in the book (starting on page 261) were actually administered in June and August 2008 and June and August 2009. These are the first four actual tests available for your review, so they should be very helpful for you as you prepare for the Regents.

Examination Sample Test
Integrated Algebra

FORMULAS

Trigonometric ratio $\sin A = \dfrac{\text{opposite}}{\text{hypotenuse}}$

$\cos A = \dfrac{\text{adjacent}}{\text{hypotenuse}}$

$\tan A = \dfrac{\text{opposite}}{\text{adjacent}}$

Area Trapezoid $A = \frac{1}{2}h(b_1 + b_2)$

Volume Cylinder $V = \pi r^2 h$

Surface area Rectangular prism $SA = 2lw + 2hw + 2lh$

Cylinder $SA = 2\pi r^2 + 2\pi rh$

Coordinate geometry $m = \dfrac{\Delta y}{\Delta x} = \dfrac{y_2 - y_1}{x_2 - x_1}$

PART I

Answer all questions in this part. Each correct answer will receive 2 credits. No partial credit will be allowed. For each question, write in the space provided the numeral preceding the word or expression that best completes the statement or answers the question. [60]

1 When given the true statements: "t is a multiple of 3" and "t is even," what could be a value of t?

(1) 8 (3) 15
(2) 9 (4) 24 1 _____

2 The accompanying circle graph shows how Joan invested her money.

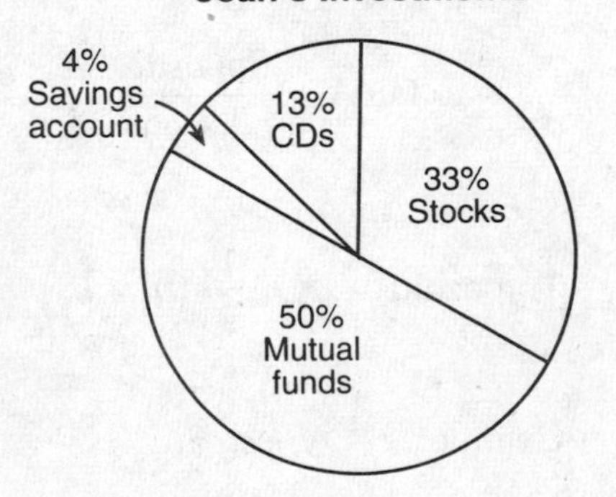

If she invested a total of $12,000, how much money did she invest in CDs?

(1) $1,560 (3) $15,600
(2) $9,230 (4) $92,308 2 _____

3 Super Painters charges $1.00 per square foot plus an additional fee of $25.00 to paint a living room. If x represents the area of the walls of Francesca's living room, in square feet, and y represents the cost, in dollars, which graph best represents the cost of painting her living room?

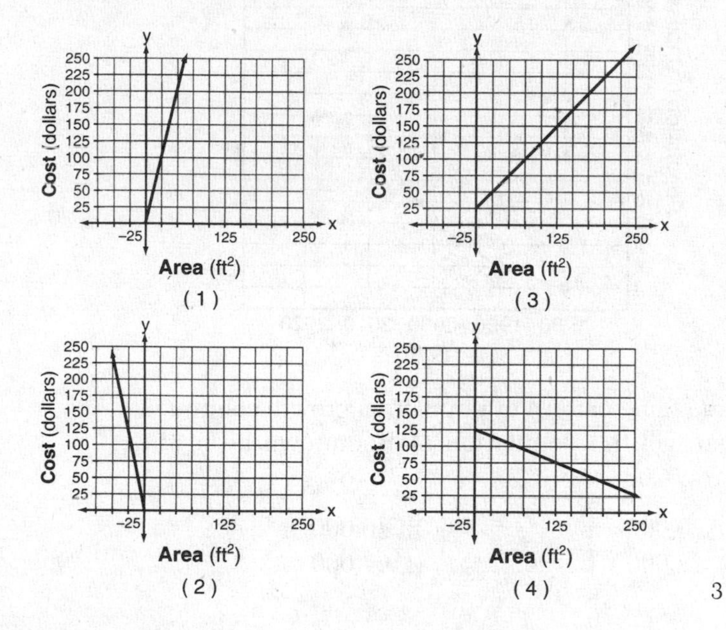

3 _____

4 Jen and Barry's ice cream stand has three types of cones, six flavors of ice cream, and four kinds of sprinkles. If a serving consists of a cone, one flavor of ice cream, and one kind of sprinkles, how many different servings are possible?

(1) 90
(2) 72
(3) 13
(4) ${}_{13}P_3$

4 _____

5 The population growth of Boomtown is shown in the accompanying graph.

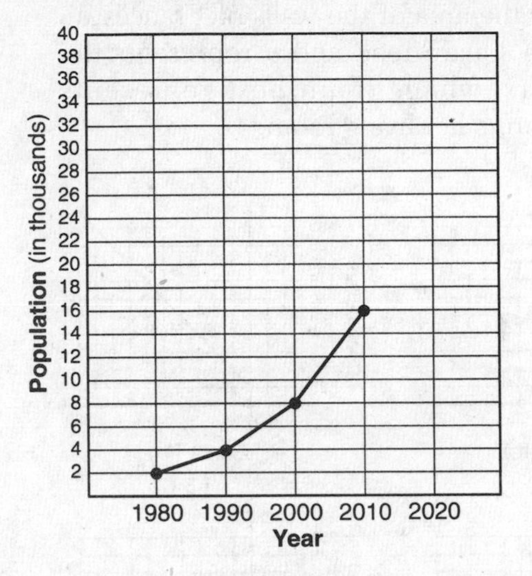

If the same pattern of population growth continues, what will the population of Boomtown be in the year 2020?

(1) 20,000 (3) 40,000
(2) 32,000 (4) 64,000 5 _____

6 If $a + 3b = 13$ and $a + b = 5$, the value of b is

(1) 1 (3) 4.5
(2) 7 (4) 4 6 _____

7 A cable 20 feet long connects the top of a flagpole to a point on the ground that is 16 feet from the base of the pole. How tall is the flagpole?

(1) 8 ft. (3) 12 ft.
(2) 10 ft. (4) 26 ft. 7 _____

8 In the equation $\frac{1}{4}n + 5 = 5\frac{1}{2}$, n is equal to

(1) 8
(2) 2
(3) $\frac{1}{2}$
(4) $\frac{1}{8}$

8 _____

9 Which equation represents the function shown in the accompanying graph?

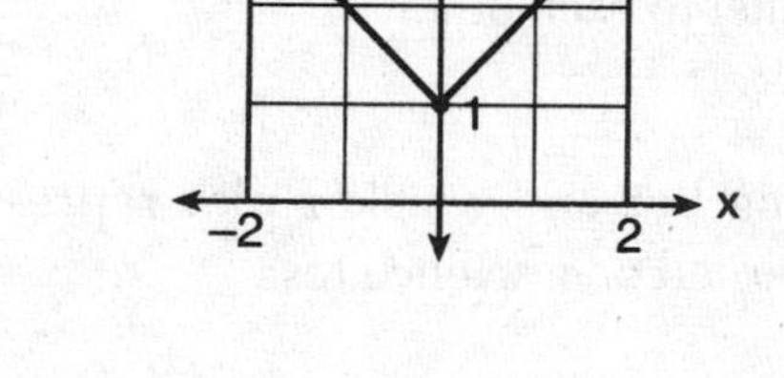

(1) $y = |x| + 1$
(2) $y = |x| - 1$
(3) $y = |x + 1|$
(4) $y = |x - 1|$

9 _____

10 The sum of $8x^2 - x + 4$ and $x - 5$ is

(1) $8x^2 + 9$
(2) $8x^2 - 1$
(3) $8x^2 - 2x + 9$
(4) $8x^2 - 2x - 1$

10 _____

11 One factor of the expression $x^2y^2 - 16$ is

(1) $xy - 4$
(2) $xy - 8$
(3) $x^2 - 4$
(4) $x^2 + 8$

11 _____

12 What is the sum of $\sqrt{50}$ and $\sqrt{8}$?

(1) $\sqrt{58}$ (3) $9\sqrt{2}$

(2) $7\sqrt{2}$ (4) $29\sqrt{2}$ 12 ____

13 Which statement describes the lines whose equations are $y = \frac{1}{3}x + 12$ and $6y = 2x + 6$?

(1) They are segments.
(2) They are perpendicular to each other.
(3) They intersect each other.
(4) They are parallel to each other. 13 ____

14 The accompanying box-and-whisker plot represents the scores earned on a science test.

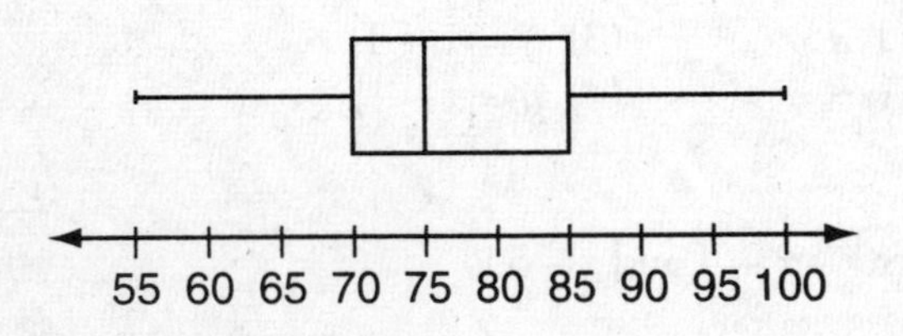

What is the median score?

(1) 73 (3) 75
(2) 79 (4) 81 14 ____

15 The video of the movie *Star Wars* earned \$193,500,000 in rental fees during its first year. When expressed in scientific notation, the number of dollars earned is

(1) 1935×10^8 (3) 1.935×10^6
(2) 193.5×10^6 (4) 1.935×10^8 15 _____

16 In the Ambrose family, the ages of the three children are three consecutive even integers. If the age of the youngest child is represented by $x + 3$, which expression represents the age of the oldest child?

(1) $x + 5$ (3) $x + 7$
(2) $x + 6$ (4) $x + 8$ 16 _____

17 If $t < \sqrt{t}$, t could be

(1) 0 (3) $\frac{1}{2}$
(2) 2 (4) 4 17 _____

18 Which number is irrational?

(1) $\frac{5}{4}$ (3) $\sqrt{121}$
(2) $0.\overline{3}$ (4) π 18 _____

19 Which type of function could be used to model the data shown in the accompanying graph?

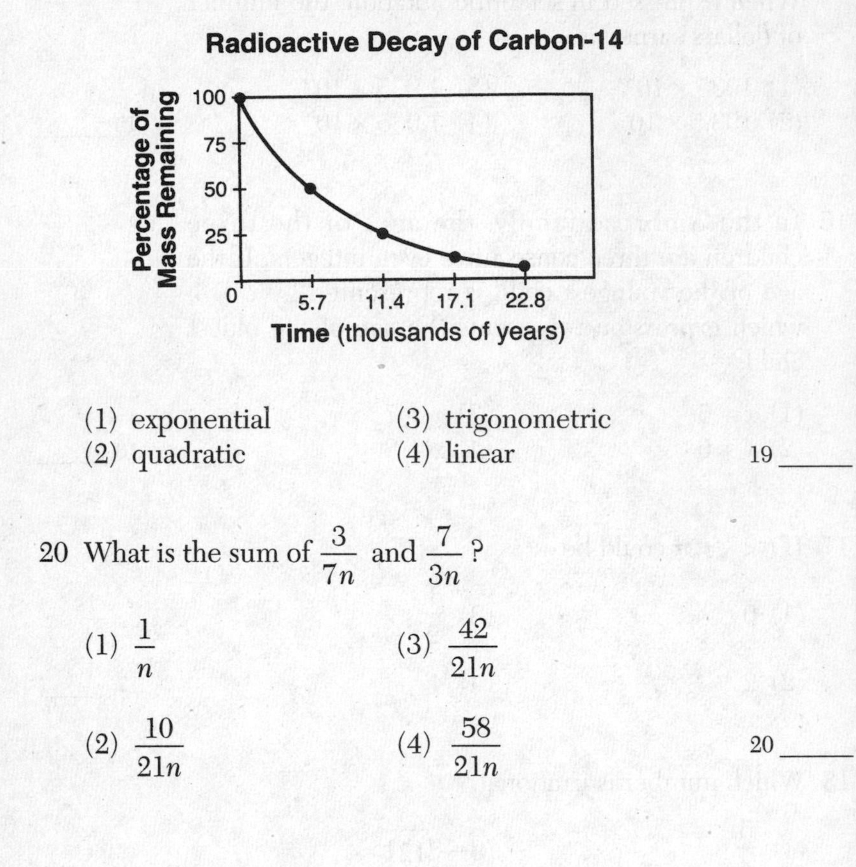

(1) exponential
(2) quadratic
(3) trigonometric
(4) linear

19 ______

20 What is the sum of $\frac{3}{7n}$ and $\frac{7}{3n}$?

(1) $\frac{1}{n}$
(2) $\frac{10}{21n}$
(3) $\frac{42}{21n}$
(4) $\frac{58}{21n}$

20 ______

21 What could be the approximate value of the correlation coefficient for the accompanying scatter plot?

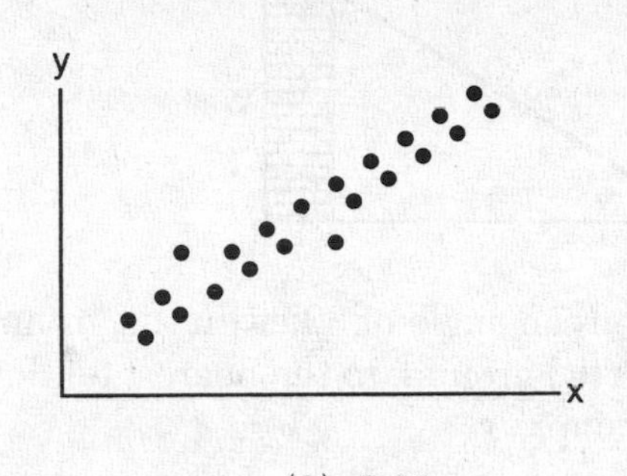

(1) –0.85
(2) –0.16
(3) 0.21
(4) 0.90

21 ______

22 Which equation is equivalent to $3x + 4y = 15$?

(1) $y = \frac{15 - 3x}{4}$
(2) $y = \frac{3x - 15}{4}$
(3) $y = 15 - 3x$
(4) $y = 3x - 15$

22 ______

23 Which set of ordered pairs does *not* represent a function?

(1) {(3,–2), (–2,3), (4,–1), (–1,4)}
(2) {(3,–2), (3,–4), (4,–1), (4,–3)}
(3) {(3,–2), (4,–3), (5,–4), (6,–5)}
(4) {(3,–2), (5,–2), (4,–2), (–1,–2)}

23 ______

24 The accompanying diagram shows a ramp 30 feet long leaning against a wall at a construction site.

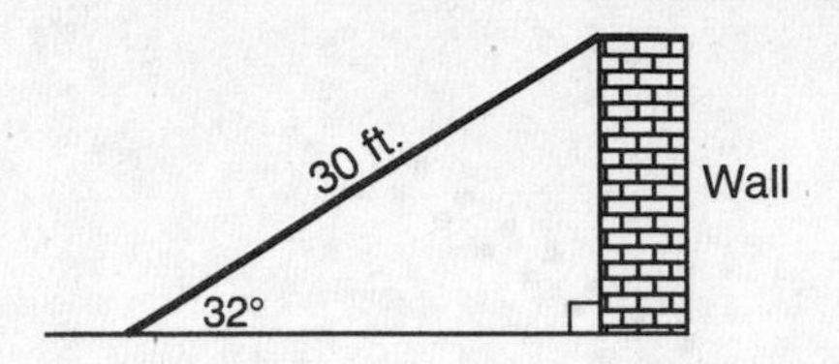

If the ramp forms an angle of 32° with the ground, how high above the ground, to the nearest tenth, is the top of the ramp?

(1) 15.9 ft. (3) 25.4 ft.
(2) 18.7 ft. (4) 56.5 ft. 24 ______

25 Which equation illustrates the associative property?

(1) $a(1) = a$
(2) $a + b = b + a$
(3) $a(b + c) = (ab) + (ac)$
(4) $(a + b) + c = a + (b + c)$ 25 ______

26 The graph of the equation $2x + 6y = 4$ passes through point $(x,-2)$. What is the value of x?

(1) -4 (3) 16
(2) 8 (4) 4 26 ______

27 Which expression represents the number of different 8-letter arrangements that can be made from the letters of the word "SAVANNAH" if each letter is used only once?

(1) $\frac{8!}{5!}$ (3) $_8P_5$

(2) $\frac{8!}{3!2!}$ (4) $8!$ 27 ______

28 Line segment AB has a slope of $\frac{3}{4}$. If the coordinates of point A are (2,5), the coordinates of point B could be

(1) (6,8) (3) (–1,1)
(2) (5,9) (4) (6,2) 28 ______

29 Which is *not* a property of all similar triangles?

(1) The corresponding angles are congruent.
(2) The corresponding sides are congruent.
(3) The perimeters are in the same ratio as the corresponding sides.
(4) The altitudes are in the same ratio as the corresponding sides. 29 ______

30 The expression $\left(\frac{3}{4}\right)^2 \bullet \left(\frac{1}{4}\right)^{-2}$ is equivalent to

(1) $\frac{9}{16}$ (3) 3

(2) $\frac{9}{256}$ (4) 9 30 ______

PART II

Answer all questions in this part. Each correct answer will receive 2 credits. Clearly indicate the necessary steps, including appropriate formula substitutions, diagrams, graphs, charts, etc. For all questions in this part, a correct numerical answer with no work shown will receive only 1 credit. [6]

31 A population of wolves in a county is represented by the equation $w = 80(0.98)^t$, where t is the number of years since 1998. Using this equation, what is the best approximation for the number of wolves, w, that will be in the wolf population in the year 2008?

32 Thelma and Laura start a lawn-mowing business and buy a lawn mower for \$225. They plan to charge \$15 to mow one lawn. What is the minimum number of lawns they need to mow if they wish to earn a profit of at least \$750?

33 The accompanying table shows the enrollment of a preschool from 1980 through 2000. If the data were plotted as points of the form (x,y), what would be an equation of the line of best fit where x represents the number of years since 1980?

Year (x)	Enrollment (y)
1980	14
1985	20
1990	22
1995	28
2000	37

PART III

Answer all questions in this part. Each correct answer will receive 3 credits. Clearly indicate the necessary steps, including appropriate formula substitutions, diagrams, graphs, charts, etc. For all questions in this part, a correct numerical answer with no work shown will receive only 1 credit. [9]

34 The trip from Manhattan to Montauk Point is 120 miles by train or by car. A train makes the trip in 2 hours, while a car makes the trip in $2\frac{1}{2}$ hours. How much faster, in miles per hour, is the average speed of the train than the average speed of the car?

35 Tracey has two empty, cube-shaped containers with sides of 5 inches and 7 inches, as shown in the accompanying diagram. She fills the smaller container completely with water and then pours all the water from the smaller container into the larger container. How deep, to the *nearest tenth of an inch*, will the water be in the larger container?

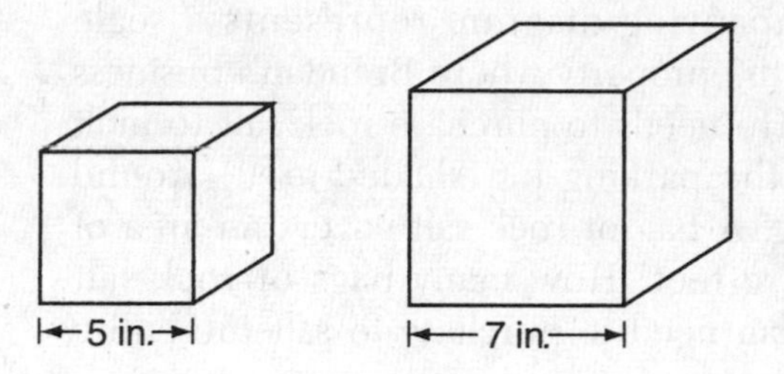

36 The tickets for a dance recital cost $5.00 for adults and $2.00 for children. If the total number of tickets sold was 295 and the total amount collected was $1,220, how many adult tickets were sold? [Only an algebraic solution can receive full credit.]

PART IV

Answer all questions in this part. Each correct answer will receive 4 credits. Clearly indicate the necessary steps, including appropriate formula substitutions, diagrams, graphs, charts, etc. For all questions in this part, a correct numerical answer with no work shown will receive only 1 credit. [12]

37 The accompanying diagram represents a scale drawing of the property where Brendan's business is located. He needs to purchase rock salt to melt the ice on the parking lot (shaded area) around his building. A bag of rock salt covers an area of 1,500 square feet. How many bags of rock salt does Brendan need to purchase to salt the entire parking lot?

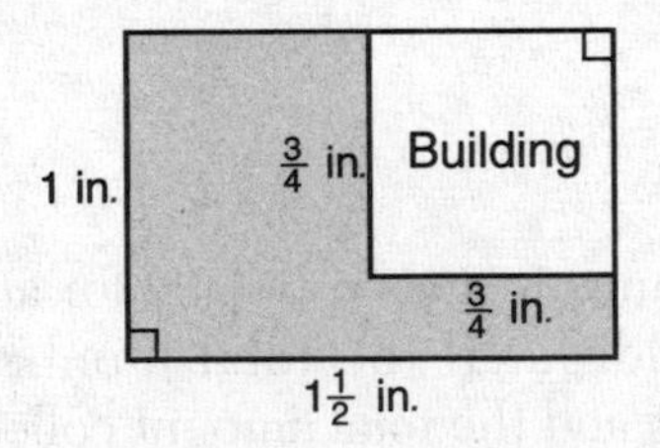

38 The path of a rocket fired during a fireworks display is given by the equation $s = 64t - 16t^2$, where t is the time, in seconds, and s is the height, in feet.

a What is the maximum height, in feet, the rocket will reach?

b In how many seconds will the rocket hit the ground?

39 In the time trials for the 400-meter run at the state sectionals, the 15 runners recorded times shown in the table below.

400-Meter Run	
Time (sec.)	**Frequency**
50.0–50.9	
51.0–51.9	11
52.0–52.9	~~1111~~ 1
53.0–53.9	111
54.0–54.9	1111

a What percent of the runners completed the time trial between 52.0 and 53.9 seconds?

b Complete the accompanying cumulative frequency table.

400-Meter Run	
Time (sec.)	**Cumulative Frequency**
50.0–50.9	
50.0–51.9	
50.0–52.9	
50.0–53.9	
50.0–54.9	

c Using the data from the cumulative frequency table, draw a cumulative frequency histogram.

Answers Sample Test
Integrated Algebra

Answer Key

PART I

1. (4)
2. (1)
3. (3)
4. (2)
5. (2)
6. (4)
7. (3)
8. (2)
9. (1)
10. (2)
11. (1)
12. (2)
13. (4)
14. (3)
15. (4)
16. (3)
17. (3)
18. (4)
19. (1)
20. (4)
21. (4)
22. (1)
23. (2)
24. (1)
25. (4)
26. (2)
27. (2)
28. (1)
29. (2)
30. (4)

PART II

31. 66
32. 65
33. $y = 1.08x + 13.4$

PART III

34. 12 miles per hour
35. 2.6 inches
36. 210

PART IV

37. 4
38. **a.** 64 feet **b.** 4 seconds
39. **a.** 60% **b. and c.** See *Answers Explained* section.

In **PARTS II–IV** you are required to show how you arrived at your answers. For sample methods of solutions, see the *Answers Explained* section.

Answers Explained

PART I

1. It is given that "t is a multiple of 3" and "t is even" are true statements. To determine which answer choice gives a correct value of t, substitute each of the proposed values of t into both statements:

- Choice (1): If $t = 8$, the first statement is false and the second statement is true since 8 is divisible by 2. ✘
- Choice (2): If $t = 9$, the first statement is true since $3 \times 3 = 9$ and the second statement is false since 9 is not divisible by 2. ✘
- Choice (3): If $t = 15$, the first statement is true since $3 \times 5 = 15$ and the second statement is false since 15 is not divisible by 2. ✘
- Choice (4): If $t = 24$, the first statement is true since $3 \times 8 = 24$ and the second statement is also true since 24 is divisible by 2. ✔

The correct choice is **(4)**.

2. According to the circle graph, Joan invested 13% of her money in CDs. Since she invested a total of \$12,000, the amount Joan invested in CDs is:

$$13\% \times \$12{,}000 = 0.13 \times \$12{,}000$$
$$= \$1{,}560$$

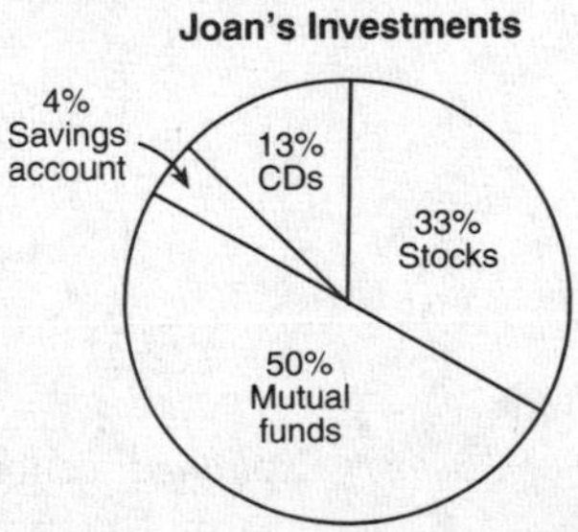

Joan's Investments

4% Savings account
13% CDs
33% Stocks
50% Mutual funds

The correct choice is **(1)**.

3. It is given that Super Painters charges \$1.00 per square foot plus an additional fee of \$25.00 to paint a living room. If x represents the number of square feet in the area of the walls of Francesa's living room and y represents the cost, in dollars, of painting her living room, then $y = 1.00x + 25$.

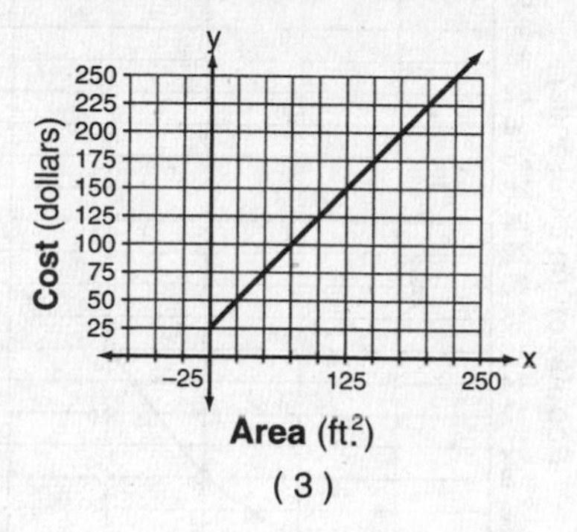

(3)

The graph of $y = 1.00x + 25$ has a slope of 1, the coefficient of x, and a y-intercept of +25. The only graph that has 25 as its y-intercept is the graph in choice (3).

The correct choice is **(3)**.

4. It is given that an ice cream stand has three types of cones, six flavors of ice cream, and four kinds of sprinkles. A serving consists of a cone, one flavor of ice cream, and one kind of sprinkles. According to the fundamental principle of counting, the number of different servings that are possible is the product of the number of types of cones, the number of flavor choices, and the number of kinds of sprinkles, which is $3 \times 6 \times 4 = 72$.

The correct choice is **(2)**.

5. It is given that the accompanying graph shows the population growth of Boomtown.

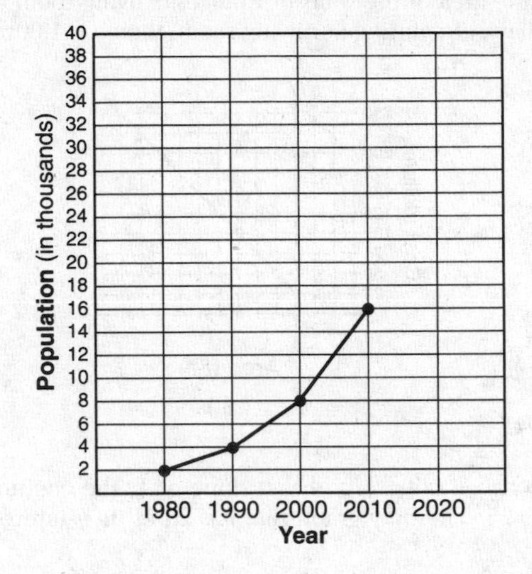

Use the graph to find the populations for the years 1980, 1990, 2000, and 2010. Since you will need to discover a pattern, organize the data in a table:

Year	Population
1980	2,000
1990	4,000
2000	8,000
2010	16,000
2020	?

- In 1980, the population was 2,000. In 1990, the population was double that number, or 4,000. From 1990 to 2000, the population doubled again, and from 2000 to 2010 the population is projected to double as well.
- Assuming the same pattern of population growth continues, the population of Boomtoom in the year 2020 will be double its population in 2010 or $2 \times 16{,}000 = 32{,}000$.

The correct choice is **(2)**.

6. It is given that $a + 3b = 13$ and $a + b = 5$. Write the second equation underneath the first equation:

$$a + 3b = 13$$
$$a + b = 5$$

To solve the system of equations for variable b, eliminate variable a by subtracting corresponding sides of the two equations:

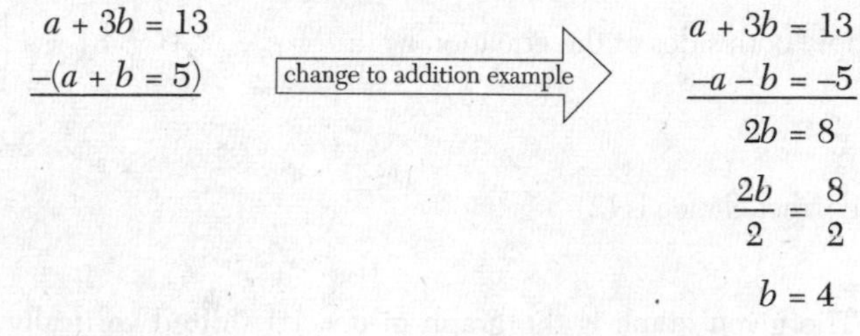

The correct choice is **(4)**.

7. It is given that a cable 20 feet long connects the top of a flagpole to a point on the ground that is 16 feet from the base of the pole. You are required to find the height of the flagpole.

- Draw a right triangle, as shown in the accompanying figure.
- Represent the height of the flagpole by x.

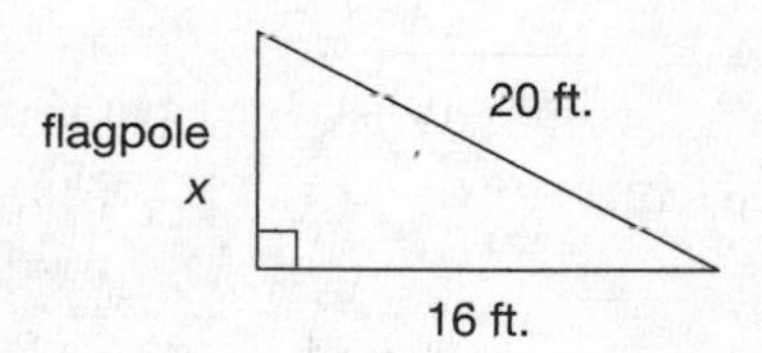

- Use the Pythagorean theorem to find x:

$$x^2 + 16^2 = 20^2$$
$$x^2 + 256 = 400$$
$$x^2 = 400 - 256$$
$$x = \sqrt{144}$$
$$= 12 \text{ feet}$$

The correct choice is **(3)**.

8. The given equation is: $\frac{1}{4}n + 5 = 5\frac{1}{2}$

Subtract 5 from each side: $-5 = -5$

$$\frac{1}{4}n = \frac{1}{2}$$

Multiply both sides of the equation by 4: $\overset{1}{\cancel{4}}\left(\frac{1}{\cancel{4}}n\right) = \overset{2}{\cancel{4}}\left(\frac{1}{\cancel{2}}\right)$

$$n = 2$$

The correct choice is **(2)**.

9. The given graph is the graph of $y = |x|$ shifted vertically up 1 unit. Hence, its equation is $y = |x| + 1$.

The correct choice is **(1)**.

10. To find the sum of $8x^2 - x + 4$ and $x - 5$, write the second polynomial underneath the first polynomial, aligning like terms in the same column. Then combine like terms:

$$\begin{array}{r} 8x^2 - x + 4 \\ x - 5 \\ \hline 8x^2 + 0x - 1 \end{array} \quad \text{or, equivalently, } 8x^2 - 1$$

The correct choice is **(2)**.

11. The given expression, $x^2y^2 - 16$, represents the difference of two squares:

$$x^2y^2 - 16 = (xy)^2 - (4)^2$$

The difference of two squares can be factored as the product of the sum and difference of the two terms that are being squared:

$$= (xy - 4)(xy + 4)$$

The factor $xy - 4$ is choice (1).

The correct choice is **(1)**.

12. To find the sum of $\sqrt{50}$ and $\sqrt{8}$, change each radical to an equivalent radical with the same radicand.

Factor each radicand such that one of the two factors is its greatest perfect square factor: $\sqrt{50} + \sqrt{8} = \sqrt{25 \cdot 2} + \sqrt{4 \cdot 2}$

Write the radical over each factor of the radicand: $= \sqrt{25} \cdot \sqrt{2} + \sqrt{4} \cdot \sqrt{2}$

Evaluate the square root of each perfect square factor: $= 5\sqrt{2} + 2\sqrt{2}$

Combine like radicals: $= 7\sqrt{2}$

The correct choice is **(2)**.

13. The given equations are $y = \frac{1}{3}x + 12$ and $6y = 2x + 6$.

Write the second equation in $y = mx + b$ form where m is the slope of the line and b is its y-intercept.

$$6y = 2x + 6$$

$$\frac{6y}{6} = \frac{2x}{6} + \frac{6}{6}$$

$$y = \frac{1}{3}x + 1$$

Compare the slope-intercept forms of the equations of the two lines. Since the two equations have the same slope, $\frac{1}{3}$, the lines are parallel to each other.

The correct choice is **(4)**.

14. As shown in the accompanying figure, a box-and-whisker plot uses five key data values: lowest score (L), highest score (H), first quartile ($Q1$), second quartile or median ($Q2$), and third quartile ($Q3$).

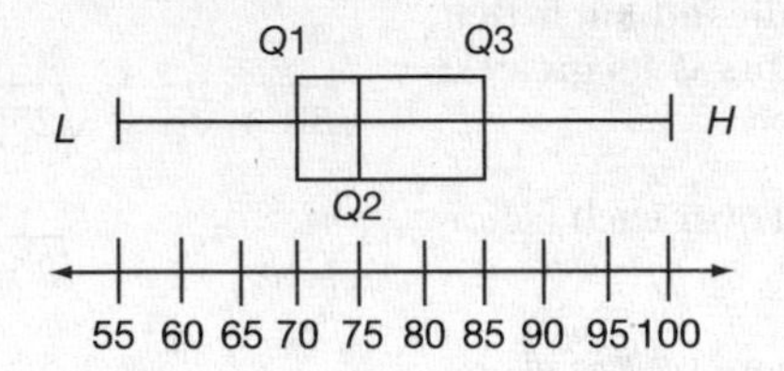

Since $Q2$ is aligned with 75 on the number line, the median score is 75.

The correct choice is **(3)**.

15. A number is expressed in scientific notation when it has the form $N \times 10^b$ where N is a number greater than or equal to 1 but less than 10 and where b is a nonzero integer. To express \$193,500,000 in scientific notation, write the given number with an implied decimal point:

$$\$193{,}500{,}000.$$

Determine the number of decimal places, b, the decimal point must be moved so that the resulting number is between 1 and 10:

$$\$1\ 9\ 3{,}\ 5\ 0\ 0{,}\ 0\ 0\ 0\ .$$

$$b = 8$$

Since the decimal point must be moved to the *left*, b is positive.

Write the number in scientific notation using $N = 1.935$ and $b = 8$:

$$\$193{,}500{,}000 = \$1.935 \times 10^8$$

The correct choice is **(4)**.

16. It is given that the ages of three children can be represented by three consecutive even integers and that the age of the youngest child is represented by $x + 3$. Since consecutive even integers differ by 2, the age of the next oldest child is $x + 3 + 2 = x + 5$. The age of the oldest child is, therefore, $x + 5 + 2 = x + 7$.

The correct choice is **(3)**.

17. It is given that $t < \sqrt{t}$. To find which answer choice has a value of t that satisfies the inequality, test each answer choice in turn:

- Choice (1): If $t = 0$, then $\sqrt{0} = 0$ so $t + \sqrt{t}$. ✘
- Choice (2): If $t = 2$, then $\sqrt{2} \approx 1.4$. Since $2 > 1.4$, $t > \sqrt{t}$. ✘
- Choice (3): If $t = \frac{1}{2}$, then $\sqrt{\frac{1}{2}} \approx 0.7$. Since $\frac{1}{2} < 0.7$, $t < \sqrt{t}$. ✔
- Choice (4): If $t = 4$, then $\sqrt{4} = 2$ so $t > \sqrt{t}$. ✘

The correct choice is **(3)**.

18. An irrational number is a real number that cannot be expressed as the quotient of two integers. To find which answer choice gives an irrational number, consider each answer choice in turn:

- Choice (1): $\frac{5}{4}$ is a rational number since it is the quotient of two integers. ✘
- Choice (2): Since $0.\overline{3} = 0.3333\ldots = \frac{1}{3}$, $0.\overline{3}$ is a rational number. In general, any decimal number that ends in a repeating set of digits is rational. ✘
- Choice (3): Since $\sqrt{121} = 11$, $\sqrt{121}$ is a rational number. ✘
- Choice (4): Although the number π is *approximated* by $\frac{22}{7}$ or 3.14, it does not have an exact fractional or decimal equivalent. When written as a decimal, the value of π takes the form of a nonending, nonrepeating decimal number. Hence, π is an irrational number. ✔

The correct choice is **(4)**.

19. You are required to identify the type of function that could be used to model the data shown in the accompanying graph.

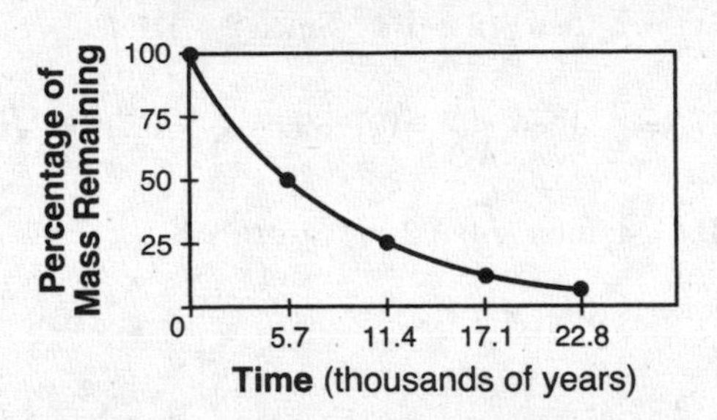

List the key points indicated on the curve:

Time (thousands of years)	Percentage of Mass Remaining
0	0%
5.7	50%
11.4	25%
17.1	12.5%
22.8	6.25%

Notice that after every 5,700 years,the percentage of mass that remains is one-half of the previous percentage. This process is an example of exponential decay. Hence, the data shown in the accompanying graph could be modeled by an exponential function.

The correct choice is **(1)**.

20. To find the sum of $\frac{3}{7n}$ and $\frac{7}{3n}$, change each fraction to an equivalent fraction that has the LCD of the two fractions as its denominator.

The LCD is $21n$ since $21n$ is the smallest expression into which $7n$ and $3n$ divide evenly.

To change the first fraction into an equivalent fraction that has $21n$ as its denominator, multiply the fraction by 1 in the form of $\frac{3}{3}$. Similarly, change the second fraction so that its denominator is also $21n$ by multiplying it by 1 in the form of $\frac{7}{7}$:

$$\begin{aligned}\frac{3}{7n}+\frac{7}{3n}&=\frac{3}{7n}\cdot\left(\frac{3}{3}\right)+\frac{7}{3n}\cdot\left(\frac{7}{7}\right)\\&=\frac{9}{21n}+\frac{49}{21n}\\&=\frac{9+49}{21n}\\&=\frac{58}{21n}\end{aligned}$$

The correct choice is **(4)**.

21. The scatter plot shows that as x increases, y increases. Thus, the correlation coefficient is positive. This eliminates choices (1) and (2). Since the data points appear to be closely clustered about a line, the correlation coefficient must be closer to 1 than to 0, which makes 0.90 a better approximation than 0.21.

The correct choice is **(4)**.

22. To determine which answer choice contains an expression equivalent to $3x + 4y = 15$, solve for y in terms of x by isolating y in the usual way:

The given equation is: $3x + 4y = 15$

Subtract $3x$ from both sides: $-3x \quad = \quad -3x$

$$4y = 15 - 3x$$

Divide both sides by 4: $\frac{4y}{4} = \frac{15-3x}{4}$

$$y = \frac{15-3x}{4}$$

The correct choice is **(1)**.

23. A set of ordered pairs represents a function when no two ordered pairs have the same x-value but different y-values. Since choice (2) includes the ordered pairs (3,–2) and (3,–4), it does not represent a function.

The correct choice is **(2)**.

24. It is given that in the accompanying diagram a ramp 30 feet long forms an angle of 32° with the ground. Let x represent how high above the ground, in feet, is the top of the ramp. Since x is the length of the side opposite the 32° angle in a right triangle in which the 30-foot ramp is the hypotenuse, use the sine ratio to find x:

$$\sin 32° = \frac{\text{length of side opposite given angle}}{\text{hypotenuse}}$$

$$= \frac{x}{30}$$

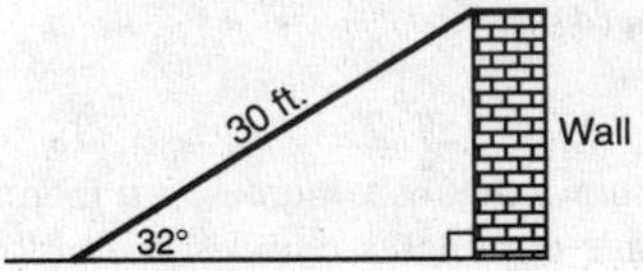

Solve for x before evaluating sin 32°:

$$x = 30 \cdot \sin 32°$$

Using your calculator, multiply 30 by the stored value of sin 32°:

$$x = 15.89757793$$

The height of the top of the ramp, correct to the *nearest tenth* of a foot, is 15.9 feet.

The correct choice is **(1)**.

25. The associative property states that the sum or product of three real numbers is the same regardless of which two numbers are chosen to be added or multiplied together first. The equation $(a + b) + c = a + (b + c)$ illustrates the associative property of addition. On the left side of the equation, a and b are added together first, and on the right side of the equation, b and c are added together first.

The correct choice is **(4)**.

26. The coordinates of any point a line passes through must satisfy the equation of the line. It is given that the graph of the equation $2x + 6y = 4$ passes through $(x,-2)$. To find the value of x, replace y with -2 and solve for x:

$$2x + 6(-2) = 4$$

$$2x - 12 = 4$$

$$2x = 4 + 12$$

$$\frac{2x}{2} = \frac{16}{2}$$

$$x = 8$$

The correct choice is **(2)**.

27. The number of different ways in which n objects can be arranged when p of the n objects are identical and q of the n objects are identical is $\frac{n!}{p!\,q!}$. The word "SAVANNAH" consists of 8 letters of which there are 3 identical As and 2 identical Ns. The number of different 8-letter arrangements that can be made from the letters of this word is $\frac{n!}{p!\,q!}$ where $n = 8$, $p = 3$, and $q = 2$ or, equivalently, $\frac{8!}{3!\,2!}$.

The correct choice is **(2)**.

28. It is given that the slope of $\overline{AB}$ is $\frac{3}{4}$ and that the coordinates of point A are (2,5). To find the possible coordinates of point B, evaluate the slope formula using the coordinates of point A and the coordinates of point B given in each answer choice. Then compare each slope value to $\frac{3}{4}$.

The formula for the slope, m, of a line determined by two points is $m = \frac{\Delta y}{\Delta x}$ where Δx represents the difference in the x-coordinates of the two points and Δy represents the corresponding difference in their y-coordinates:

- Choice (1): Calculate the slope for $A(2,5)$ and $B(6,8)$:

$$\begin{aligned} m &= \frac{\Delta y}{\Delta x} \\ &= \frac{8-5}{6-2} \\ &= \frac{3}{4} \quad ✔ \end{aligned}$$

Thus, (6,8) could represent the coordinates of point B.

- Choice (2): Calculate the slope for $A(2,5)$ and $B(5,9)$:

$$\begin{aligned} m &= \frac{\Delta y}{\Delta x} \\ &= \frac{9-5}{5-2} \\ &= \frac{4}{3} \quad ✘ \end{aligned}$$

- Choice (3): Calculate the slope for $A(2,5)$ and $B(-1,1)$:

$$\begin{aligned} m &= \frac{\Delta y}{\Delta x} \\ &= \frac{1-5}{-1-2} \\ &= \frac{-4}{-3} \\ &= \frac{4}{3} \quad ✘ \end{aligned}$$

- Choice (4): Calculate the slope for $A(2,5)$ and $B(6,2)$:

$$m = \frac{\Delta y}{\Delta x}$$

$$= \frac{2-5}{6-2}$$

$$= -\frac{3}{4} \quad ✘$$

The correct choice is **(1)**.

29. If two triangles are similar, their corresponding angles are congruent and the lengths of their corresponding sides are in proportion. Since corresponding sides of similar triangles are not necessarily congruent, choice (2) is not a property of all similar triangles.

The correct choice is **(2)**.

30. The given expression is: $\left(\frac{3}{4}\right)^2 \cdot \left(\frac{1}{4}\right)^{-2}$

Evaluate the first fraction by raising both the numerator and the denominator to the second power: $\frac{9}{16} \cdot \left(\frac{1}{4}\right)^{-2}$

Change the exponent of the second fraction from negative to positive by inverting it: $\frac{9}{16} \cdot \left(\frac{4}{1}\right)^2$

Raise both the numerator and the denominator of the second fraction to the second power: $\frac{9}{16} \cdot \frac{16}{1}$

Multiply: $\frac{9}{\cancel{16}} \cdot \frac{\overset{1}{\cancel{16}}}{1}$

$$= 9$$

The correct choice is **(4)**.

PART II

31. To find the number of wolves in 2008, use your calculator to evaluate $w = 80(0.98)^t$ for $t = 10$:

$$= 80(0.98)^{10}$$
$$\approx 65.36582455$$

Since a fractional wolf is not possible, the best approximation is **66** wolves.

32. It is given that Thelma and Laura buy a lawn mower for $225 and plan to charge $15 to mow one lawn. If x represents the number of lawns they mow, then:

- $\$15x$ represents their income from mowing lawns.
- $225 represents their cost.
- $15x - 225$, the difference between income and cost, represents their profit in dollars.

You are required to find the number of lawns that they must mow to earn a profit of at least (≥) $750. Thus:

$$15x - 225 \geq 750$$
$$+225 = +225$$
$$15x \geq 975$$
$$\frac{15x}{15} \geq \frac{975}{15}$$
$$x \geq 65$$

They must mow a minimum of **65** lawns if they wish to earn a profit of at least $750.

33. Enter the paired data values in the given table into the statistics editor of your graphing calculator. The data in the first column of the table are entered under "L1," and the corresponding data values in the second column are entered under "L2." Use the regression feature of your graphing calculator to calculate the line of best fit or the regression line. The line of best fit is $\boldsymbol{y = 1.08x + 13.4}$.

PART III

34. It is given that the trip from Manhattan to Montauk Point is 120 miles by train or by car and that a train makes the trip in 2 hours while a car makes the trip in $2\frac{1}{2}$ hours. You are required to find how much faster, in miles per hour, is the average speed of the train compared with the average speed of the car.

Use the formula rate × time = distance to find the average speeds of the train and car:

- If x represents the average speed of the train in miles per hour, then $2x = 120$ so $x = \frac{120}{2} = 60$ miles per hour.
- If y represents the average speed of the car in miles per hour, then $\left(2\frac{1}{2}\right)y = 120$ so $y = \frac{120}{2.5} = 48$ miles per hour.
- The difference in average speeds is $60 - 48 = 12$ miles per hour

The average speed of the train is **12 miles per hour** faster than the average speed of the car.

35. It is given that Tracey fills a cube-shaped container that has a 5-inch side completely with water and then pours all the water from it into a cube-shaped container with a 7-inch side.

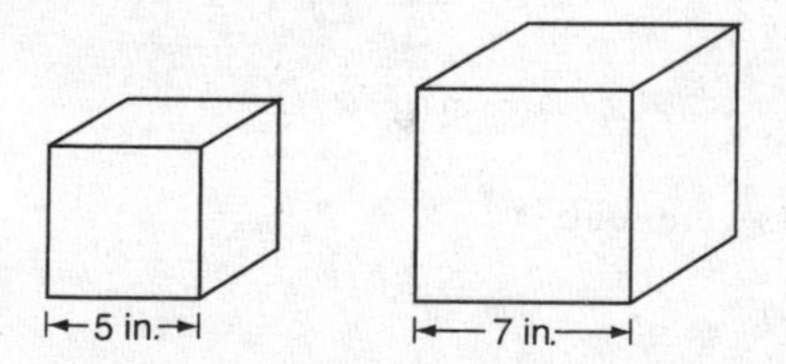

You are required to find how deep, to the *nearest tenth of an inch*, the water will be in the larger container.

- The volume of the water in the smaller cube-shaped container is $5^3 = 5 \times 5 \times 5 = 125$ cubic inches.
- If h represents the depth of the water in the larger cube-shaped container, then the volume of the water in this container is $7 \times 7 \times h = 49h$ cubic inches.
- Since the volumes of water in the two containers are equal:

$$49h = 125$$

$$\frac{49h}{49} = \frac{125}{49}$$

$$h \approx 2.55\ldots$$

The water will be **2.6 inches** deep, to the *nearest tenth of an inch.*

36. It is given that tickets for a dance recital cost $5.00 for adults and $2.00 for children. The total number of tickets sold was 295, and the total amount collected was $1,220. You are required to find the number of adult tickets that were sold. If x represents the number of adult tickets that were sold, then $295 - x$ represents the number of children's tickets that were sold. Hence:

$$5x + 2(295 - x) = 1{,}220$$

$$5x + 590 - 2x = 1{,}220$$

$$3x = 1{,}220 - 590$$

$$\frac{3x}{3} = \frac{630}{3}$$

$$x = 210$$

210 adult tickets were sold.

PART IV

37. It is given that the accompanying diagram is a scale drawing of the property where Brendan's business is located. The shaded region represents a parking lot covered with ice. It is also given that one bag of rock salt covers an area of 1,500 square feet. You are required to find the number of bags of rock salt Brendan needs to purchase in order to salt the entire parking lot.

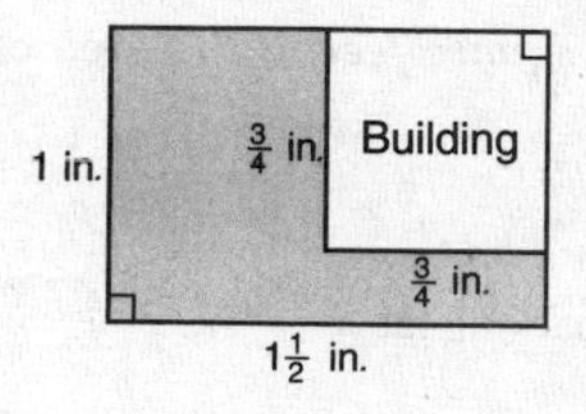

Scale: $\frac{1}{4}$ in. = 18 ft.

The area of the parking lot is the difference between the areas of the rectangle and the building:

$$\text{scaled area of parking lot} = \left(1 \times 1\frac{1}{2}\right) - \left(\frac{3}{4} \times \frac{3}{4}\right)$$

$$= 1\frac{1}{2} - \frac{9}{16}$$

$$= \frac{24}{16} - \frac{9}{16}$$

$$= \frac{15}{16} \text{ in.}^2$$

Express the scale conversion factor in terms of square feet. Because it is given that $\frac{1}{4}$ in. = 18 ft.:

$$\left(\frac{1}{4}\text{ in.}\right)^2 = (18\text{ ft.})^2$$

$$\frac{1}{16}\text{ in.}^2 = 324\text{ ft.}^2$$

Find the number of *square feet* in the area of the parking lot. Since $\frac{1}{16}$ in.^{2} = 324 ft.^{2},

$$\begin{aligned}\text{area of parking lot} &= \frac{15}{16}\text{ in.}^2\\ &= 15\left(\frac{1}{16}\text{ in.}^2\right)\\ &= 15 \times 324\text{ ft.}^2\\ &= 4{,}860\text{ ft.}^2\end{aligned}$$

Find the number of bags of rock salt that are needed. As one bag of rock salt covers an area of 1,500 square feet, $\frac{4{,}860}{1{,}500} = 3.24$ bags of rock salt are needed.

Since a fractional part of a bag of rock salt cannot be purchased, Brendan needs to purchase **4** bags of rock salt.

38. It is given that the equation $x = 64t - 16t^2$ represents the path of a rocket fired during a fireworks display in which t is the time, in seconds, and s is the height, in feet. The equation $s = 64t - 16t^2$ describes a parabola with a maximum turning point since the coefficient of the quadratic term is negative.

a. The maximum height of the rocket is the y-coordinate of the vertex or turning point of the parabola that describes its path.

The x-coordinate of the turning point of a parabola whose equation has the form $y = ax^2 + bx + c$ is $x = \frac{-b}{2a}$. For the parabola equation $s = 64t - 16t^2$, $a = -16$ and $b = 64$. Hence:

$$x = -\frac{64}{2(-16)} = 2$$

Find the y-coordinate of the turning point of the parabola by replacing t with 2 in its equation:

$$x = 64(2) - 16(2)^2$$
$$= 64$$

The maximum height of the rocket is **64 feet**.

b. When the rocket hits the ground, $s = 0$. Hence:

$$s = 64t - 16t^2 = 0$$
$$16t(4 - t) = 0$$
$$16t = 0 \quad \text{or} \quad 4 - t = 0$$
$$t = 0 \quad \Big| \quad 4 = t$$

At $t = 0$, the rocket is at ground level when it is fired. When $t = 4$, the rocket hits the ground after it has been fired.

The rocket will hit the ground in **4 seconds**.

39. a. There are 6 runners in the interval from 52.0 to 52.9 seconds and 3 runners in the interval from 53.0 to 53.9 seconds.

Hence, 6 + 3 = 9 of the 15 runners, or $\frac{9}{15}$ = **60%** of the runners, completed the time trial between 52.0 and 53.9 seconds.

b. See the accompanying table.

400-Meter Run	
Time (sec.)	**Cumulative Frequency**
50.0–50.9	0
50.0–51.9	0 + 2 = **2**
50.0–52.9	2 + 6 = **8**
50.0–53.9	8 + 3 = **11**
50.0–54.9	11 + 4 = **15**

c. See the accompanying graph.

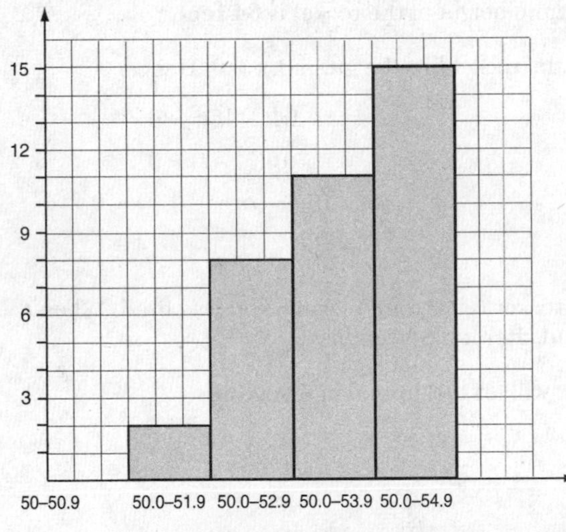

Topic	Question Numbers	Number of Points	Your Points	Your Percentage
1. Sets and Numbers; Interval Notation; Properties of Real Numbers; Percent	1, 17, 18, 25	2 + 2 + 2 + 2 = 8		
2. Operations on Rat'l. Numbers & Monomials	20	2		
3. Laws of Exponents for Integer Exponents; Scientific Notation	15, 30	2 + 2 = 4		
4. Operations on Polynomials	10	2		
5. Square Root; Operations with Radicals	12	2		
6. Evaluating Formulas & Algebraic Expressions	—	—		
7. Solving Linear Eqs. & Inequalities	8, 32	2 + 2 = 4		
8. Solving Literal Eqs. & Formulas for a Particular Letter	22	2		
9. Alg. Operations (including factoring)	11	2		
10. Quadratic Equations (incl. alg. solution; parabolas)	38	4		
11. Coordinate Geometry (eq. of a line; graphs of linear eqs.; slope)	3, 13, 26, 28	2 + 2 + 2 +2 = 8		
12. Systems of Linear Eqs. & Inequalities (algebraic & graphical solutions)	6	2		
13. Mathematical Modeling (using: eqs.; tables; graphs).	5, 16, 36	2 + 2 + 3 = 7		
14. Linear-Quadratic systems	—	—		
15. Properties of Triangles & Parallelograms	—	—		
16. Perimeter; Circumference; Area of Common Figures	37	4		
17. Volume and Surface Area of Common Figures; Relative Error in Measurement	35	3		
18. Angle & Line Relationships (suppl., compl., vertical angles; parallel lines; congruence)	—	—		
19. Ratio & Proportion (incl. similar polygons, scale drawings, & rates)	29, 34	2 + 3 = 5		
20. Pythagorean Theorem	7	2		

Topic	Question Numbers	Number of Points	Your Points	Your Percentage
21. Right Triangle Trigonometry	24	2		
22. Functions (def.; domain and range; vertical line test; absolute value)	9, 23	2 + 2 = 4		
23. Exponential Functions (properties; growth and decay)	19, 31	2 + 2 = 4		
24. Probability (incl. tree diagrams & sample spaces)	—	—		
25. Permutations and Counting Methods (incl. Venn diagrams)	4, 27	2 + 2 = 4		
26. Statistics (mean, percentiles, quartiles; freq. dist., histograms; box-and-whisker plots; causality; bivariate data; circle graphs)	2, 14, 39	2 + 2 + 4 = 8		
27. Line of Best Fit (including linear regression, scatter plots, and linear correlation)	21, 33	2 + 2 = 4		
28. Nonroutine Word Problems Requiring Arith. or Alg. Reasoning	—	—		

HOW TO CONVERT YOUR RAW SCORE TO YOUR INTEGRATED ALGEBRA REGENTS EXAMINATION SCORE

Below is a conversion chart that can be used to determine your final score on the Sample Test in Integrated Algebra Regents Examination. To find your final exam score, locate in the column labeled "Raw Score" the total number of points you scored out of a possible 87 points. Since partial credit is allowed in Parts II, III, and IV of the test, you may need to approximate the credit you would receive for a solution that is not completely correct. Then locate in the adjacent column to the right the scaled score that corresponds to your raw score. The scaled score is your final sample Regents Examination score.

Regents Examination in Integrated Algebra—Sample Test Chart for Converting Total Test Raw Scores to Final Examination Scores (Scaled Scores)

Raw Score	Scaled Score	Raw Score	Scaled Score	Raw Score	Scaled Score
87	**100**	57	**79**	27	**54**
86	**99**	56	**79**	26	**53**
85	**99**	55	**78**	25	**52**
84	**98**	54	**77**	24	**52**
83	**97**	53	**77**	23	**51**
82	**97**	52	**76**	22	**50**
81	**96**	51	**75**	21	**49**
80	**95**	50	**75**	20	**48**
79	**94**	49	**74**	19	**47**
78	**93**	48	**73**	18	**46**
77	**93**	47	**73**	17	**45**
76	**92**	46	**72**	16	**43**
75	**91**	45	**71**	15	**40**
74	**91**	44	**71**	14	**38**
73	**90**	43	**70**	13	**36**
72	**89**	42	**70**	12	**33**
71	**89**	41	**69**	11	**31**
70	**88**	40	**68**	10	**29**
69	**87**	39	**68**	9	**27**
68	**87**	38	**67**	8	**25**
67	**86**	37	**66**	7	**23**
66	**85**	36	**65**	6	**21**
65	**85**	35	**64**	5	**18**
64	**84**	34	**63**	4	**16**
63	**83**	33	**61**	3	**13**
62	**83**	32	**60**	2	**8**
61	**82**	31	**59**	1	**4**
60	**81**	30	**57**	0	**0**
59	**81**	29	**56**		
58	**80**	28	**55**		

Official Test Sampler Fall 2007

Integrated Algebra

FORMULAS

Trigonometric ratio $\sin A = \frac{\text{opposite}}{\text{hypotenuse}}$

$\cos A = \frac{\text{adjacent}}{\text{hypotenuse}}$

$\tan A = \frac{\text{opposite}}{\text{adjacent}}$

Area Trapezoid $A = \frac{1}{2}h(b_1 + b_2)$

Volume Cylinder $V = \pi r^2 h$

Surface area Rectangular prism $SA = 2lw + 2hw + 2lh$

Cylinder $SA = 2\pi r^2 + 2\pi rh$

Coordinate geometry $m = \frac{\Delta y}{\Delta x} = \frac{y_2 - y_1}{x_2 - x_1}$

PART I

Answer all questions in this part. Each correct answer will receive 2 credits. No partial credit will be allowed. For each question, write in the space provided the numeral preceding the word or expression that best completes the statement or answers the question. [60]

1 For 10 days, Romero kept a record of the number of hours he spent listening to music. The information is shown in the table below.

Day	1	2	3	4	5	6	7	8	9	10
Hours	9	3	2	6	8	6	10	4	5	2

Which scatter plot shows Romero's data graphically?

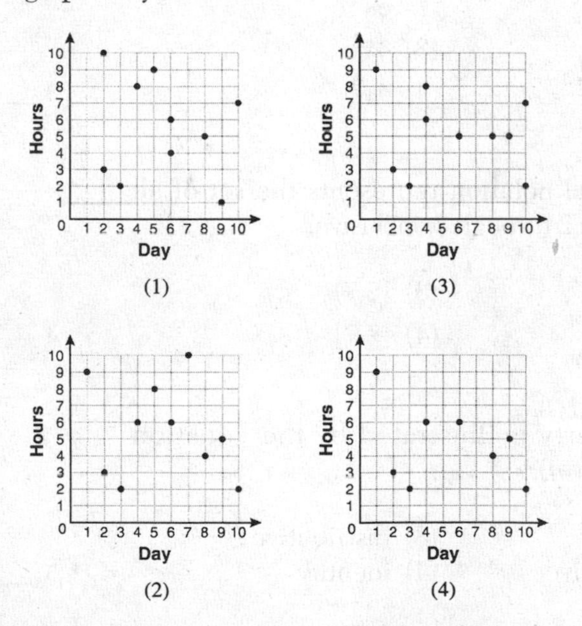

1 ______

2 Throughout history, many people have contributed to the development of mathematics. These mathematicians include Pythagoras, Euclid, Hypatia, Euler, Einstein, Agnesi, Fibonacci, and Pascal. What is the probability that a mathematician's name selected at random from those listed will start with either the letter E or the letter A?

(1) $\frac{2}{8}$ (3) $\frac{4}{8}$

(2) $\frac{3}{8}$ (4) $\frac{6}{8}$ 2 ____

3 Which expression represents $\frac{(2x^3)(8x^5)}{4x^6}$ in simplest form?

(1) x^2 (3) $4x^2$

(2) x^9 (4) $4x^9$ 3 ____

4 Which interval notation represents the set of all numbers from 2 through 7, inclusive?

(1) (2,7] (3) [2,7)

(2) (2,7) (4) [2,7] 4 ____

5 Which property is illustrated by the equation $ax + ay = a(x + y)$?

(1) associative (3) distributive

(2) commutative (4) identity 5 ____

6 The expression $x^2 - 16$ is equivalent to

(1) $(x + 2)(x - 8)$
(2) $(x - 2)(x + 8)$
(3) $(x + 4)(x - 4)$
(4) $(x + 8)(x - 8)$

6 ____

7 Which situation describes a correlation that is *not* a causal relationship?

(1) The rooster crows, and the Sun rises.
(2) The more miles driven, the more gasoline needed.
(3) The more powerful the microwave, the faster the food cooks.
(4) The faster the pace of a runner, the quicker the runner finishes.

7 ____

8 The equations $5x + 2y = 48$ and $3x + 2y = 32$ represent the money collected from school concert ticket sales during two class periods. If x represents the cost for each adult ticket and y represents the cost for each student ticket, what is the cost for each adult ticket?

(1) \$20
(2) \$10
(3) \$8
(4) \$4

8 ____

9 The data set 5, 6, 7, 8, 9, 9, 9, 10, 12, 14, 17, 17, 18, 19, 19 represents the number of hours spent on the Internet in a week by students in a mathematics class. Which box-and-whisker plot represents the data?

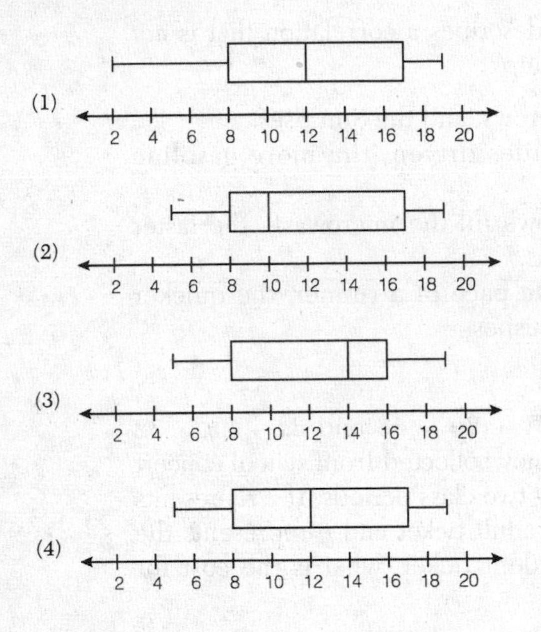

9 ______

10 Given:

Set $A = \{(-2,-1), (-1,0), (1,8)\}$
Set $B = \{(-3,-4), (-2,-1), (-1,2), (1,8)\}$.

What is the intersection of sets A and B?

(1) $\{(1,8)\}$
(2) $\{(-2,-1)\}$
(3) $\{(-2,-1), (1,8)\}$
(4) $\{(-3,-4), (-2,-1), (-1,2), (-1,0), (1,8)\}$

10 ______

11 Tanya runs diagonally across a rectangular field that has a length of 40 yards and a width of 30 yards, as shown in the diagram below.

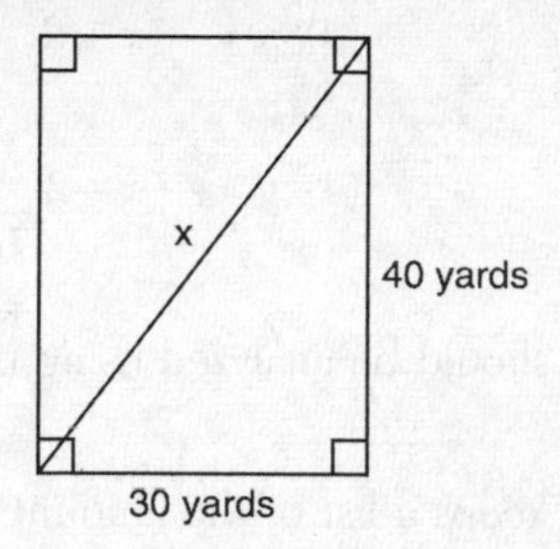

What is the length of the diagonal, in yards, that Tanya runs?

(1) 50
(2) 60
(3) 70
(4) 80

11 ______

12 A cylindrical container has a diameter of 12 inches and a height of 15 inches, as illustrated in the diagram below.

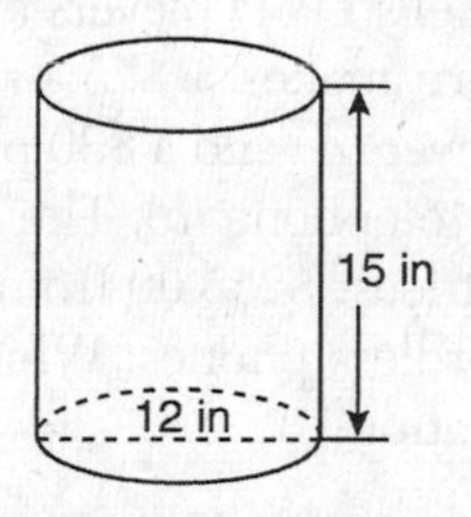

(Not drawn to scale)

What is the volume of this container to the *nearest tenth* of a cubic inch?

(1) 6,785.8
(2) 4,241.2
(3) 2,160.0
(4) 1,696.5

12 ______

13 What is an equation for the line that passes through the coordinates (2,0) and (0,3)?

(1) $y = -\frac{3}{2}x + 3$ (3) $y = -\frac{2}{3}x + 2$

(2) $y = -\frac{3}{2}x - 3$ (4) $y = -\frac{2}{3}x - 2$

13 ______

14 Which situation should be analyzed using bivariate data?

(1) Ms. Saleem keeps a list of the amount of time her daughter spends on her social studies homework.
(2) Mr. Benjamin tries to see if his students' shoe sizes are directly related to their heights.
(3) Mr. DeStefan records his customers' best video game scores during the summer.
(4) Mr. Chan keeps track of his daughter's algebra grades for the quarter.

14 ______

15 An electronics store sells DVD players and cordless telephones. The store makes a $75 profit on the sale of each DVD player (d) and a $30 profit on the sale of each cordless telephone (c). The store wants to make a profit of at least $255.00 from its sales of DVD players and cordless phones. Which inequality describes this situation?

(1) $75d + 30c < 255$ (3) $75d + 30c > 255$
(2) $75d + 30c \leq 255$ (4) $75d + 30c \geq 255$

15 ______

16 What is the slope of the line containing the points (3,4) and (–6,10)?

(1) $\frac{1}{2}$ (3) $-\frac{2}{3}$

(2) 2 (4) $-\frac{3}{2}$

16 ______

17 Which type of graph is shown in the diagram below?

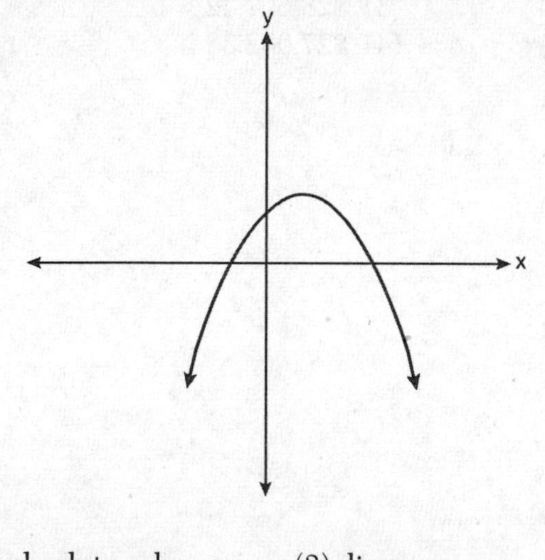

(1) absolute value (3) linear
(2) exponential (4) quadratic

17 ______

18 The expression $\dfrac{9x^4 - 27x^6}{3x^3}$ is equivalent to

(1) $3x(1 - 3x)$
(2) $3x(1 - 3x^2)$
(3) $3x(1 - 9x^5)$
(4) $9x^3(1 - x)$

18 _____

19 Daniel's Print Shop purchased a new printer for $35,000. Each year it depreciates (loses value) at a rate of 5%. What will its approximate value be at the end of the fourth year?

(1) $33,250.00
(2) $30,008.13
(3) $28,507.72
(4) $27,082.33

19 _____

20 Which inequality is represented by the graph below?

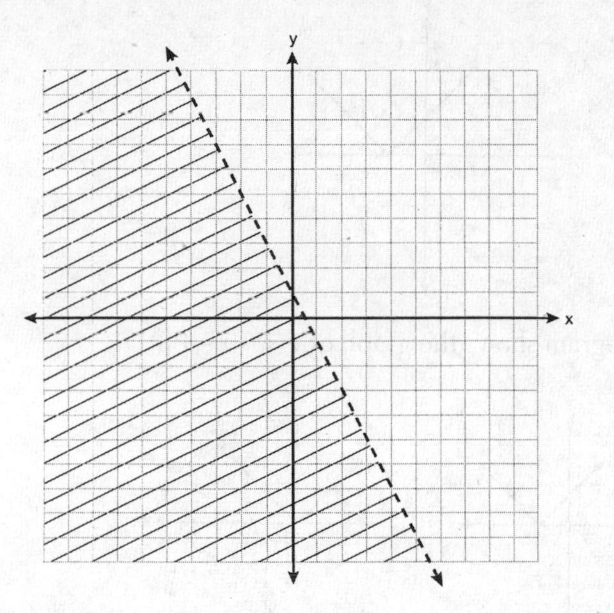

(1) $y < 2x + 1$

(2) $y < -2x + 1$

(3) $y < \frac{1}{2}x + 1$

(4) $y < -\frac{1}{2}x + 1$

20 ______

21 In triangle MCT, the measure of $\angle T = 90°$, $MC = 85$ cm, $CT = 84$ cm, and $TM = 13$ cm. Which ratio represents the sine of $\angle C$?

(1) $\frac{13}{85}$

(2) $\frac{84}{85}$

(3) $\frac{13}{84}$

(4) $\frac{84}{13}$

21 ______

22 The diagram below shows the graph of $y = |x - 3|$.

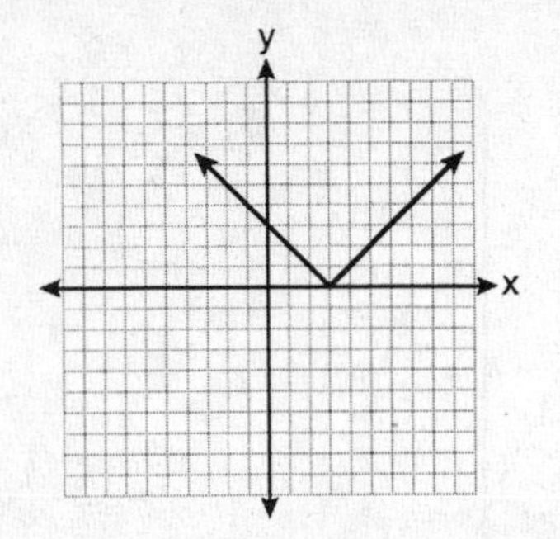

Which diagram shows the graph of $y = -|x - 3|$?

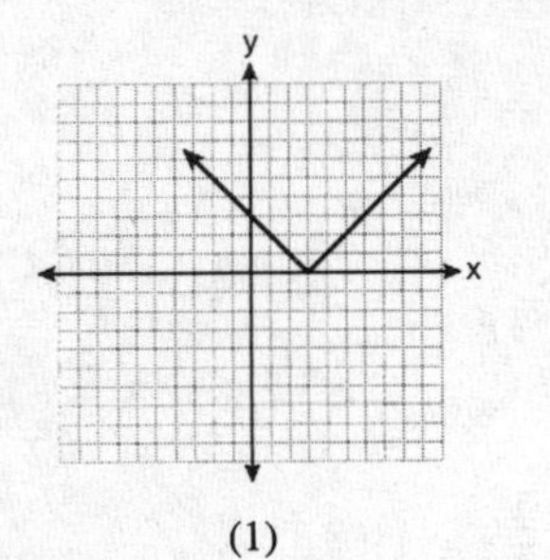

(1)

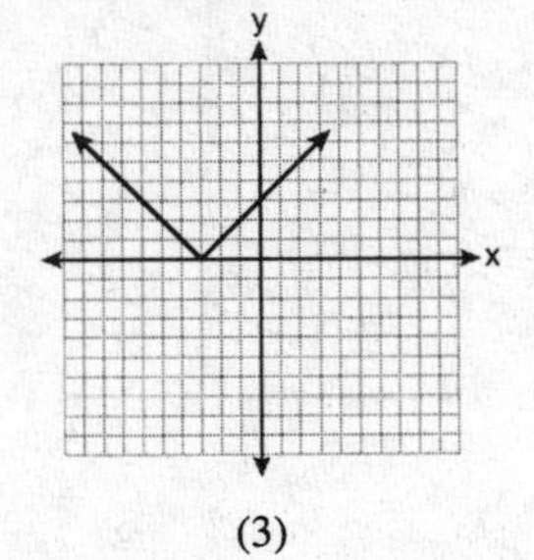

(3)

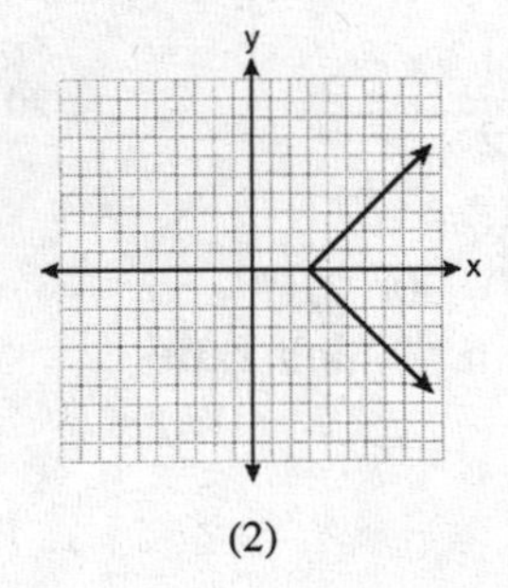

(2)

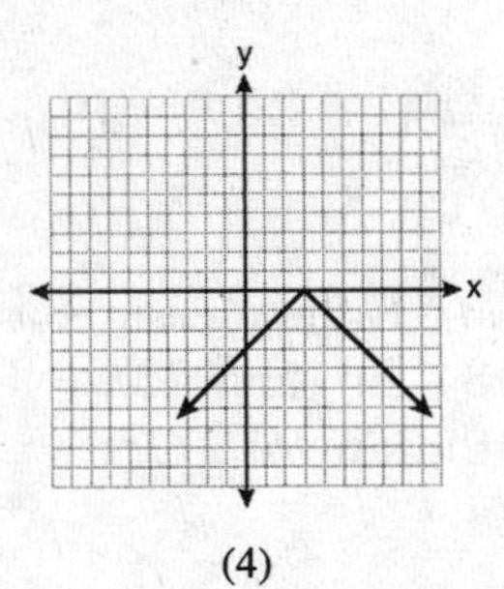

(4)

22 ______

23 The groundskeeper is replacing the turf on a football field. His measurements of the field are 130 yards by 60 yards. The actual measurements are 120 yards by 54 yards. Which expression represents the relative error in the measurement?

(1) $\frac{(130)(60)-(120)(54)}{(120)(54)}$

(2) $\frac{(120)(54)}{(130)(60)-(120)(54)}$

(3) $\frac{(130)(60)-(120)(54)}{(130)(60)}$

(4) $\frac{(130)(60)}{(130)(60)-(120)(54)}$

23 _____

24 Which value of x is in the solution set of the inequality $-2x + 5 > 17$?

(1) –8
(2) –6
(3) –4
(4) 12

24 _____

25 What is the quotient of 8.05×10^6 and 3.5×10^2?

(1) 2.3×10^3
(2) 2.3×10^4
(3) 2.3×10^8
(4) 2.3×10^{12}

25 _____

26 The length of a rectangular window is 5 feet more than its width, w. The area of the window is 36 square feet. Which equation could be used to find the dimensions of the window?

(1) $w^2 + 5w + 36 = 0$
(2) $w^2 - 5w - 36 = 0$
(3) $w^2 - 5w + 36 = 0$
(4) $w^2 + 5w - 36 = 0$

26 ______

27 What is the sum of $\frac{d}{2}$ and $\frac{2d}{3}$ expressed in simplest form?

(1) $\frac{3d}{5}$
(2) $\frac{3d}{6}$
(3) $\frac{7d}{5}$
(4) $\frac{7d}{6}$

27 ______

28 For which value of x is $\frac{x-3}{x^2-4}$ undefined?

(1) −2
(2) 0
(3) 3
(4) 4

28 ______

29 Which verbal expression represents $2(n - 6)$?

(1) two times n minus six
(2) two times six minus n
(3) two times the quantity n less than six
(4) two times the quantity six less than n

29 ______

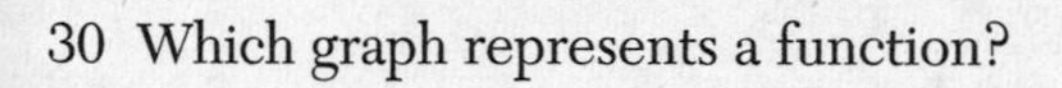

30 Which graph represents a function?

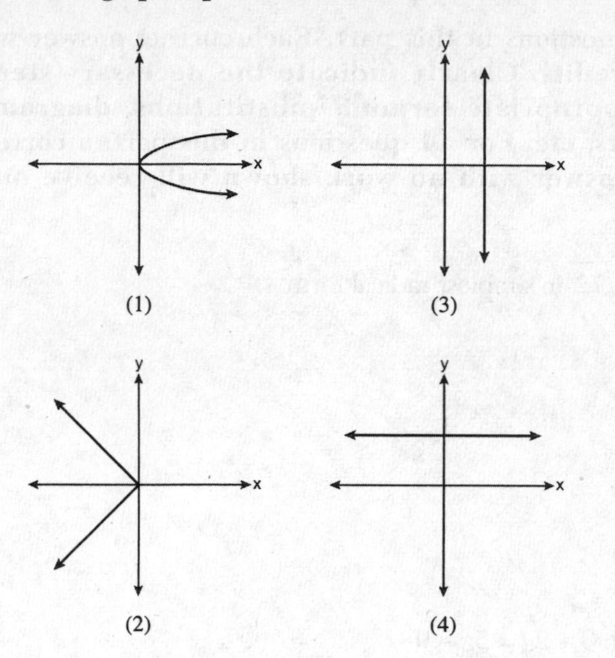

30 ______

PART II

Answer all questions in this part. Each correct answer will receive 2 credits. Clearly indicate the necessary steps, including appropriate formula substitutions, diagrams, graphs, charts, etc. For all questions in this part, a correct numerical answer with no work shown will receive only 1 credit. [6]

31 Express $5\sqrt{72}$ in simplest radical form.

32 Solve for g: $3 + 2g = 5g - 9$

33 Serena's garden is a rectangle joined with a semicircle, as shown in the diagram below. Line segment AB is the diameter of semicircle P. Serena wants to put a fence around her garden.

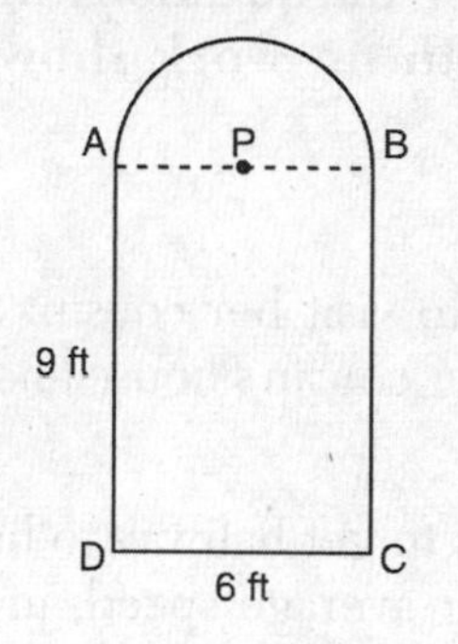

Calculate the length of fence Serena needs to the *nearest tenth of a foot*.

PART III

Answer all questions in this part. Each correct answer will receive 3 credits. Clearly indicate the necessary steps, including appropriate formula substitutions, diagrams, graphs, charts, etc. For all questions in this part, a correct numerical answer with no work shown will receive only 1 credit. [9]

34 Hannah took a trip to visit her cousin. She drove 120 miles to reach her cousin's house and the same distance back home.

It took her 1.2 hours to get halfway to her cousin's house. What was her average speed, in miles per hour, for the first 1.2 hours of the trip?

Hannah's average speed for the remainder of the trip to her cousin's house was 40 miles per hour. How long, in hours, did it take her to drive the remaining distance?

Traveling home along the same route, Hannah drove at an average rate of 55 miles per hour. After 2 hours her car broke down. How many miles was she from home?

35 A prom ticket at Smith High School is $120. Tom is going to save money for the ticket by walking his neighbor's dog for $15 per week. If Tom already has saved $22, what is the minimum number of weeks Tom must walk the dog to earn enough to pay for the prom ticket?

36 Mr. Laub has three children: two girls (Sue and Karen) and one boy (David). After each meal, one child is chosen at random to wash dishes.

If the same child can be chosen for both lunch and dinner, construct a tree diagram or list a sample space of all the possible outcomes of who will wash dishes after lunch and dinner on Saturday.

Determine the probability that one boy and one girl will wash dishes after lunch and dinner on Saturday.

PART IV

Answer all questions in this part. Each correct answer will receive 4 credits. Clearly indicate the necessary steps, including appropriate formula substitutions, diagrams, graphs, charts, etc. For all questions in this part, a correct numerical answer with no work shown will receive only 1 credit. [12]

37 The values of 11 houses on Washington St. are shown in the table below.

Value per House	Number of Houses
$100,000	1
$175,000	5
$200,000	4
$700,000	1

Find the mean value of these houses in dollars.

Find the median value of these houses in dollars.

State which measure of central tendency, the mean or the median, *best* represents the values of these 11 houses. Justify your answer.

38 Solve the following systems of equations graphically, on the set of axes below, and state the coordinates of the point(s) in the solution set.

$$y = x^2 - 6x + 5$$
$$2x + y = 5$$

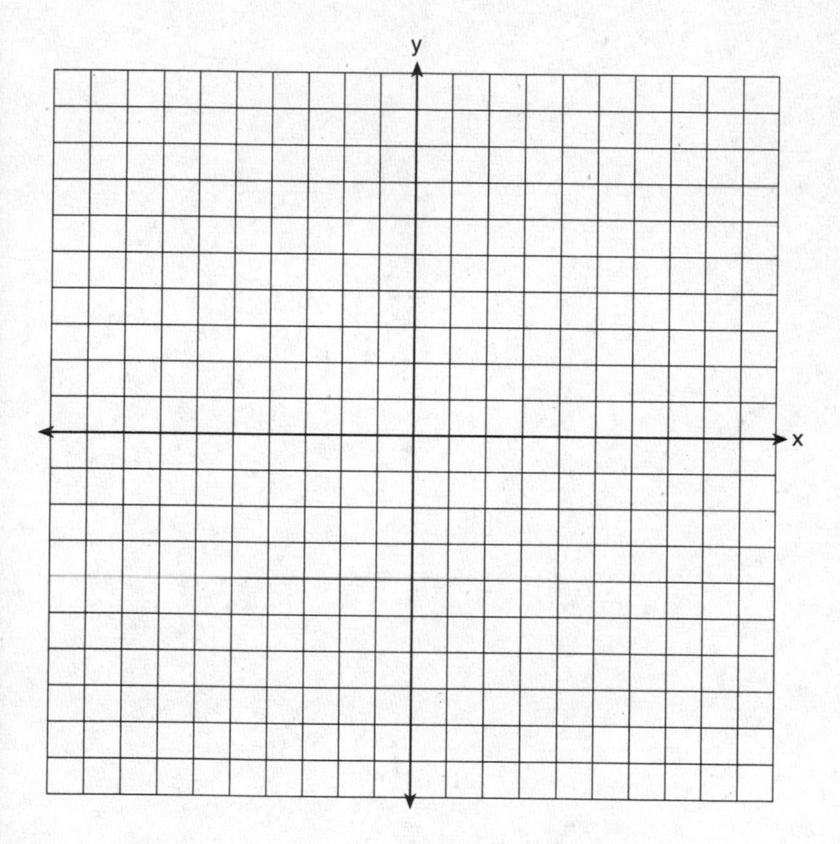

39 Solve for x: $\dfrac{x+1}{x} = \dfrac{-7}{x-12}$

Answers Official Test Sampler

Integrated Algebra

Answer Key

PART I

1. (2)	**6.** (3)	**11.** (1)	**16.** (3)	**21.** (1)	**26.** (4)
2. (3)	**7.** (1)	**12.** (4)	**17.** (4)	**22.** (4)	**27.** (4)
3. (3)	**8.** (3)	**13.** (1)	**18.** (2)	**23.** (1)	**28.** (1)
4. (4)	**9.** (2)	**14.** (2)	**19.** (3)	**24.** (1)	**29.** (4)
5. (3)	**10.** (3)	**15.** (4)	**20.** (2)	**25.** (2)	**30.** (4)

PART II

31. $30\sqrt{2}$
32. 4
33. 33.4

PART III

34. 50; 1.5; 10
35. 7
36. See *Answers Explained*; $\frac{4}{9}$

PART IV

37. mean = 25,000; median = 175,000; See *Answers Explained*
38. (0,5) and (4,–3)
39. 6 and –2

In **PARTS II–IV** you are required to show how you arrived at your answers. For sample methods of solutions, see the *Answers Explained* section.

Answers Explained

PART I

1. It is given that for 10 days Romero kept a record of the number of hours he spent listening to music. The information is shown in the accompanying table.

Day	1	2	3	4	5	6	7	8	9	10
Hours	9	3	2	6	8	6	10	4	5	2

A scatter plot of the data in the table displays the data graphically as a set of 10 points whose coordinates have the form (day,hours): (1,9), (2,3), (3,2), (4,6), (5,8), (6,6), (7,10), (8,4), (9,5), and (10,2). Examine each of the answer choices in turn for the scatter plot that contains each of the 10 data points.

- The scatter plot for choice (1) does not include the data point (1,9), so this choice can be eliminated.
- The scatter plot for choice (3) does not include the data point (5,8), so this choice can be eliminated.
- The scatter plot for choice (4) does not include the data point (5,8) so this choice can be eliminated.
- The scatter plot for choice (2) includes all 10 of the data points in the table.

The correct choice is **(2)**.

2. The names of 8 mathematicians are given: Pythagoras, Euclid, Hypatia, Euler, Einstein, Agnesi, Fibonacci, and Pascal.

- Three names start with the letter *E*: Euclid, Euler, and Einstein.
- One name starts with the letter *A*: Agnesi.
- Since 4 of the 8 mathematician's names start with either the letter *E* or the letter *A*, the probability that a name selected at random will start with either of these letters is $\frac{4}{8}$.

The correct choice is **(3)**.

3. The given expression is: $\frac{(2x^3)(8x^5)}{4x^6}$

Find the product of the terms in the numerator by multiplying their numerical coefficients and by multiplying powers of the same variable by adding their exponents:

$$\frac{(2 \cdot 8)(x^{3+5})}{4x^6}$$

$$\frac{16x^8}{4x^6}$$

Find the quotient by dividing the numerical coefficients and by dividing powers of the same base by subtracting their exponents:

$$(16 \div 4)(x^{8-6})$$

$$4x^2$$

The correct choice is **(3)**.

4. An interval of numbers from a to b may include one, both, or neither of its endpoints. The accompanying table lists the possibilities:

Description	Interval Notation
Interval includes a and b	$[a,b]$
Interval includes a but not b	$[a,b)$
Interval includes b but not a	$(a,b]$
Interval includes neither endpoint	(a,b)

When using interval notation, a bracket next to an endpoint indicates that the endpoint is included in the interval. A left or right parenthesis indicates that the endpoint is *not* included in the interval.

The given interval is set of all numbers from 2 through 7, *inclusive*. Because both 2 and 7 are included, this interval is represented in interval notation as [2,7].

The correct choice is **(4)**.

5. It is given that $ax + ay = a(x + y)$. According to the distributive property, a sum may be multiplied by a number, as in $a(x + y)$, by multiplying each addend separately by that number and then adding the two products, as in $ax + ay$.

The correct choice is **(3)**.

6. The given expression, $x^2 - 16$, represents the difference of two squares and can be factored by writing the product of the sum and difference of the terms that are being squared:

$$\begin{aligned} x^2 - 16 &= (x)^2 - (4)^2 \\ &= (x + 4)(x - 4) \end{aligned}$$

The correct choice is **(3)**.

7. A causal or cause-and-effect relationship exists between two events when the occurrence of the first event causes the second event. For the statement in choice (1),

$$\underbrace{\text{The rooster crows}}_{\text{cause}}\text{, and }\underbrace{\text{the Sun rises}}_{\text{effect}}$$

the two events are "the rooster crows" and "the sun rises." Since a crowing rooster does not cause the Sun to rise, there is no causal relationship between the two events.

The correct choice is **(1)**.

8. It is given that the equations $5x + 2y = 48$ and $3x + 2y = 32$ represent the money collected from school concert ticket sales during two class periods where x represents the cost for each adult ticket and y represents the cost for each student ticket. To find the cost for each adult ticket, solve the system of equations for x:

$$\left.\begin{aligned} 5x + 2y &= 48 \\ -\ 3x + 2y &= 32 \end{aligned}\right\}\ \text{linear system}$$

Subtract corresponding sides: $\quad 2x = 16$

Solve for x: $\quad \dfrac{2x}{2} = \dfrac{16}{2}$

$$x = 8$$

The correct choice is **(3)**.

9. You are required to find the answer choice that shows the box-and-whisker plot for the data set 5, 6, 7, 8, 9, 9, 9, 10, 12, 14, 17, 17, 18, 19, 19. A box-and-whisker plot is constructed using these five key values:

- The smallest data value in the set.
- The median of the data set. When listed in ascending order, the median is the middle data value in the list.
- The lower quartile of the data set, which is the median of the bottom half of the data.
- The upper quartile of the data set, which is the median of the upper half of the data.
- The greatest data value in the set.

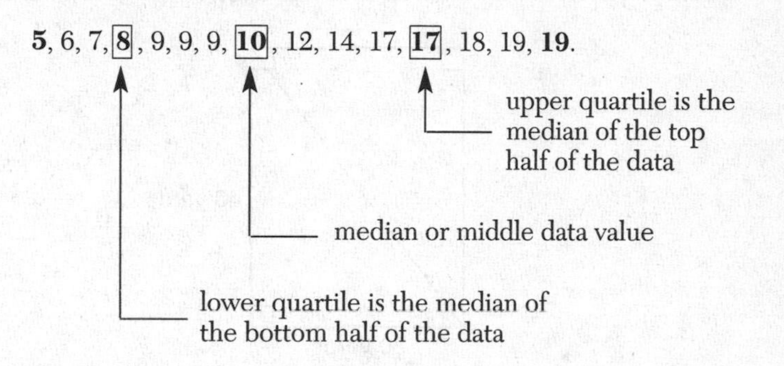

Hence, the five key values for this data set are **5** (the smallest value), **8** (the lower quartile), **10** (the median), **17** (the upper quartile), and **19** (the greatest value).

- Examine each of the answer choices for a box-and-whisker plot whose whiskers extend from 5 to 19. You can eliminate choice (1) since its left whisker starts at 2 rather than at 5.
- Examine each of the remaining answer choices for a median mark (vertical segment) of 10 in the rectangular box. The box-and-whisker plots in choices (3) and (4) have their median marks at different values, so they can be eliminated.

The correct choice is **(2)**.

10. The intersection of two sets is the set that contains those elements, if any, that are contained in both sets. It is given that:

$$\text{Set } A = \{(\mathbf{-2,-1}), (-1,0), (\mathbf{1,8})\}$$

$$\text{Set } B = \{(-3,-4), (\mathbf{-2,-1}), (-1,2), (\mathbf{1,8})\}$$

The intersection of sets A and B is the set of ordered pairs common to both A and B: $\{(-2,-1), (1,8)\}$.

The correct choice is **(3)**.

11. It is given that Tanya runs diagonally across a rectangular field that has a length of 40 yards and a width of 30 yards, as shown in the diagram below. You are required to find the length of the diagonal, in yards, that Tanya runs.

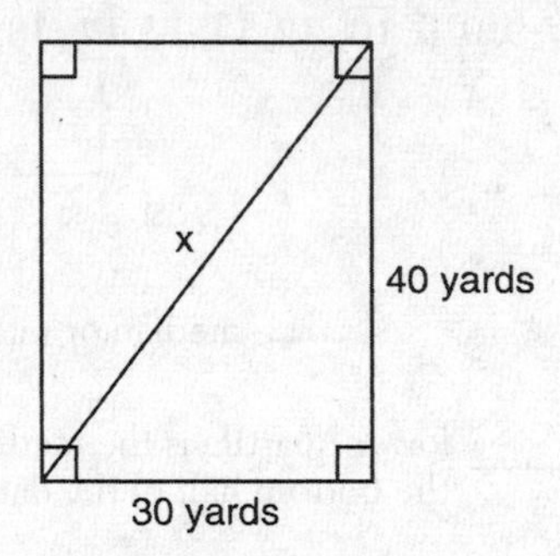

The diagonal whose length is labeled as x is the hypotenuse of a right triangle whose side lengths form a multiple of a 3-4-5 Pythagorean triple:

$$\underbrace{3\times 10}_{30},\ \underbrace{4\times 10}_{40},\ \underbrace{5\times 10}_{x=50}$$

The correct choice is **(1)**.

12. It is given that a cylindrical container has a diameter of 12 inches and a height of 15 inches, as illustrated in the accompanying diagram.

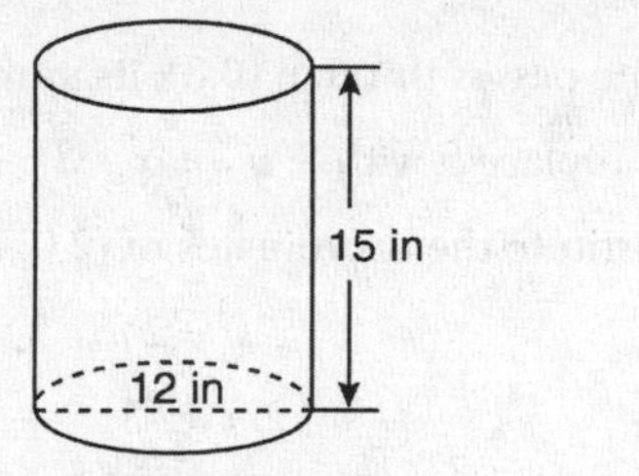

(Not drawn to scale)

The volume, V, of the cylinder is given by the formula $V = \pi r^2 h$ where $h = 15$ and $r = \frac{1}{2} \times 12 = 6$:

$$\begin{aligned} V &= \pi r^2 h \\ &= \pi(6^2)(15) \\ &= 540\pi \end{aligned}$$

Use your calculator to multiply 540 by the stored value of π: $= 1{,}696.460033$

Round off to the *nearest tenth*: $= 1{,}696.5$

The correct choice is **(4)**.

13. The equation $y = mx + b$ is the general equation of a line whose slope is m and y-intercept is b. It is given that the required line passes through the points (2,0) and (0,3).

- Because the line passes through (0,3), its y-intercept is 3. Hence, $b = 3$.
- In $y = mx + b$, replace b with 3: $y = mx + 3$.
- To find m, substitute the coordinates of (2,0) in $y = mx + 3$:

$$0 = 2m + 3$$
$$2m = -3$$
$$m = -\frac{3}{2}$$

Because $b = 3$ and $m = -\frac{3}{2}$, an equation of the line is $y = -\frac{3}{2}x + 3$.

The correct choice is **(1)**.

14. A set of bivariate data includes paired data values for two different statistical variables. Examine each of the answer choices in turn until you find a situation in which two different sets of data would need to be collected, as in choice (2):

> Mr. Benjamin tries to see if his students' shoe sizes are directly related to their heights.

There are two distinct variables: students' shoes sizes and students' heights. Data would need to be collected for students' shoe sizes and for the corresponding heights of the students. Both data values for each student would then need to be paired.

The correct choice is **(2)**.

15. It is given that an electronics store makes a \$75 profit on the sale of each DVD player (d) and a \$30 profit on the sale of each cordless telephone (c).

- $75d$ represents the profit from the sales of DVD players, and $30c$ represents the profit from the sales of cordless phones.
- $75d + 30c$ represents the total profit from the sales of both items.
- In order for the total profit to be at least (equal to or greater than) \$255,

$$75d + 30c \geq 255.$$

The correct choice is **(4)**.

16. To find the slope of the line containing the points (3,4) and (–6,10), use the slope formula provided on the reference sheet in the test booklet:

$$m = \frac{\Delta y}{\Delta x} = \frac{y_2 - y_1}{x_2 - x_1}$$

For the given points, $y_2 - y_1 = 10 - 4 = 6$ and $x_2 - x_1 = -6 - 3 = -9$. Thus:

$$m = \frac{\Delta y}{\Delta x} = \frac{6}{-9} = -\frac{2}{3}.$$

The correct choice is **(3)**.

17. The graph in the accompanying figure is given:

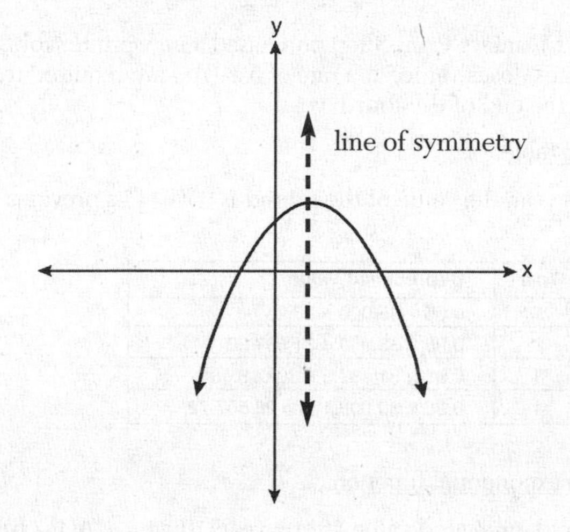

The graph is not linear since it is curved. It is not an absolute value graph since it is not V shaped. An exponential graph always increases or always decreases but not both, so it is not an exponential graph. The given graph has the shape of a parabola whose equation is quadratic since it rises to a maximum and then falls such that it has a vertical line of symmetry.

The correct choice is **(4)**.

18. The given expression is: $\frac{9x^4 - 27x^6}{3x^3}$

Divide each term of the numerator by the monomial denominator $\frac{9x^4}{3x^3} - \frac{27x^6}{3x^3}$

$3x^{4-3} - 9x^{6-3}$

$3x - 9x^3$

Factor out the greatest common monomial factor, which is $3x$: $3x(1 - 3x^2)$

The correct choice is **(2)**.

19. It is given that Daniel's Print Shop purchased a new printer for $35,000. Each year it depreciates (loses value) at a rate of 5%. You are required to find its approximate value at the end of the fourth year.

Method 1: Make a table.

At the end of each year, the value of the printer is 95% of its previous value.

Year	Depreciated Value
1	0.95 × 35,000 = 33,250
2	0.95 × 33,250 = 31,587.50
3	0.95 × 31,587.5 = 30,008.13
4	0.95 × 30,008.125 = 28,507.72

Method 2: Use an exponential function.

If y represents the depreciated value after x years, then $y = A(1 - b)^x$ where A is the initial value and b is the rate of decay or depreciation. Using your calculator, evaluate $y = A(1 - b)^x$ for A = 35,000, b = 5% = 0.05, and $x = 4$:

$$y = 35,000(1 - 0.05)^4$$
$$= 35,000(0.95)^4$$
$$= 28,507.72$$

The correct choice is **(3)**.

20. You are required to identify the inequality whose graph is given:

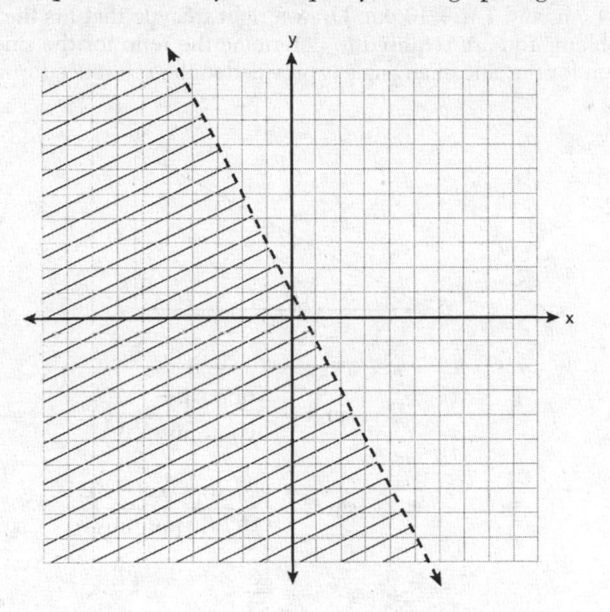

- The broken line has a negative slope since it falls as x increases. You can, therefore, eliminate answer choices (1) and (3). These graphs have a positive slope since the coefficient of x in each of the inequalities is positive.
- Pick any point that lies in the solution set, such as (–1,2). Check which of the remaining inequalities results in a true statement when $x = -1$ and $y = 2$:

Choice (2): $y < -2x + 1$

$2 \boxed{?} -2(-1) + 1$

$2 \boxed{?} 2 + 1$

$2 < 3$ ✔

Choice (4): $y < -\frac{1}{2}x + 1.$

$2 \boxed{?} -\frac{1}{2}(-1) + 1$

$2 \boxed{?} \frac{1}{2} + 1$

$2 \not< 1\frac{1}{2}$

The correct choice is **(2)**.

21. It is given that in triangle MCT, the measure of $\angle T = 90°$, $MC = 85$ cm, $CT = 84$ cm, and $TM = 13$ cm. Draw a right triangle that fits the conditions of the problem. You are required to determine the ratio for the sine of $\angle C$. The definition for the sine of an angle is provided in the reference formula sheet:

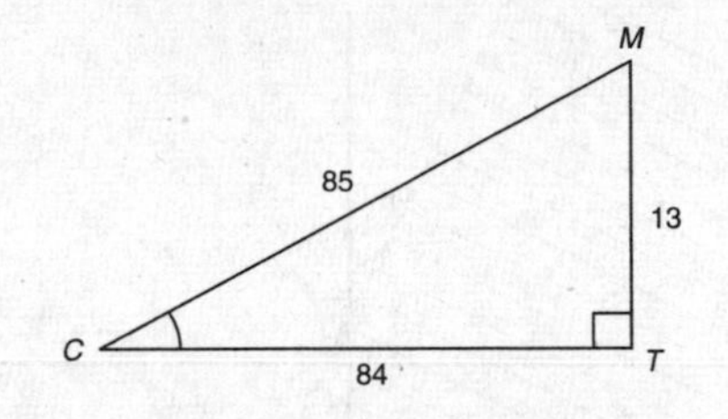

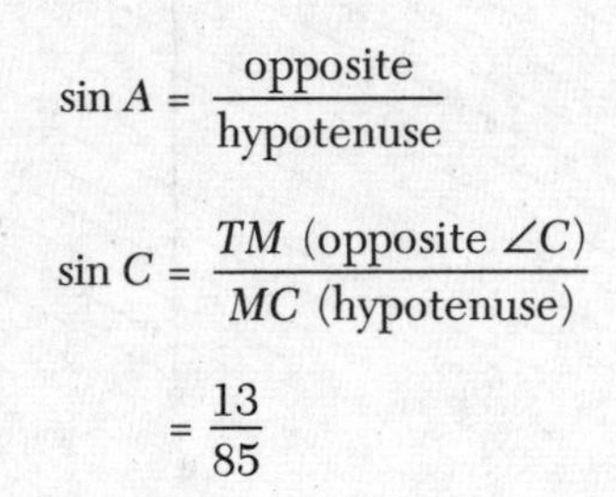

$$\sin A = \frac{\text{opposite}}{\text{hypotenuse}}$$

$$\sin C = \frac{TM \text{ (opposite } \angle C)}{MC \text{ (hypotenuse)}}$$

$$= \frac{13}{85}$$

The correct choice is **(1)**.

22. The graph of $y = |x - 3|$ is given:

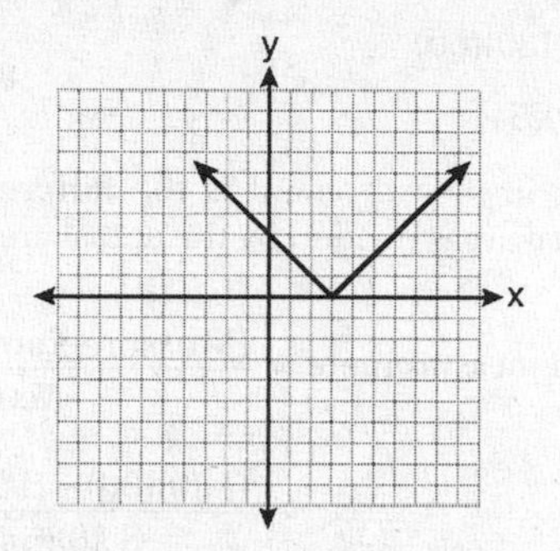

You are required to identify the graph of $y = -|x - 3|$. The effect of placing a negative sign in front of the absolute value sign changes the sign of the y-coordinate of each point on the graph to its opposite, which flips the graph over the x-axis:

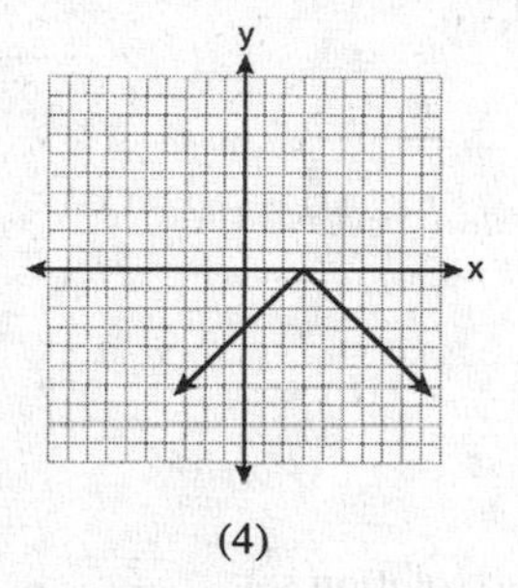

(4)

The correct choice is **(4)**.

23. It is given that the measurements taken of a football field were 130 yards by 60 yards but the actual measurements are 120 yards by 54 yards. Thus:

- Measured area = (130)(60)
- Actual area = (120)(54)
- The relative error in measurement is the difference between the measured and the actual areas divided by the actual area:

$$\text{Relative error in measurement} = \frac{(\text{Measured area}) - (\text{Actual area})}{\text{Actual area}}$$

$$= \frac{(130)(60) - (120)(54)}{(120)(54)}$$

The correct choice is **(1)**.

24. The given inequality is: $-2x + 5 > 17$

Subtract 5 from both sides: $\quad -5 = -5$

$$-2x > 12$$

Divide each side by −2 *and* reverse the direction of the inequality:

$$\frac{-2x}{-2} \boxed{>} \frac{12}{-2}$$

$$x < -6$$

Since −8 < −6, −8 is in the solution set.

The correct choice is **(1)**.

25. To find the quotient of 8.05×10^6 and 3.5×10^2, divide:

$$\frac{8.05 \times 10^6}{3.5 \times 10^2} = \frac{8.05}{3.5} \times 10^{6-2}$$

$$= 2.3 \times 10^4$$

The correct choice is **(2)**.

26. It is given that the length of a rectangular window is 5 feet more than its width, w, and that the area of the window is 36 square feet.

- The length can be represented by $w + 5$. Since the area of a rectangle is length times width:

$$w(w + 5) = 36$$

- Remove the parentheses by multiplying each term inside the parentheses by w:

$$w^2 + 5w = 36$$

- Put the equation in standard form by subtracting 36 from each side:

$$w^2 + 5w - 36 = 0$$

The correct choice is **(4)**.

27. The given sum is $\frac{d}{2} + \frac{2d}{3}$. The lowest common denominator is 6. Change each fraction into an equivalent fraction having 6 as its denominator by multiplying the first fraction by 1 in the form of $\frac{3}{3}$ and the second fraction by 1 in the form of $\frac{2}{2}$:

$$\frac{d}{2} + \frac{2d}{3} = \left(\frac{3}{3}\right)\left(\frac{d}{2}\right) + \left(\frac{2}{2}\right)\left(\frac{2d}{3}\right)$$

$$= \frac{3d}{6} + \frac{4d}{6}$$

$$= \frac{3d + 4d}{6}$$

$$= \frac{7d}{6}$$

The correct choice is **(4)**.

28. A fraction with a variable denominator is undefined for each value of the variable that makes the denominator equal to 0. The given fraction is

$$\frac{x-3}{x^2-4} = \frac{x-3}{(x+2)(x-2)}$$

When $x = -2$ or $x = +2$, the denominator equals 0 and the fraction is undefined. Look for either or both of these values of x among the answer choices.

The correct choice is **(1)**.

29. The given algebraic expression, $2(n - 6)$, represents two times the quantity inside the parentheses, which is six less than n.

The correct choice is **(4)**.

30. A graph represents a function if no vertical line intersects it at more than one point. The graph in choice (4) is the only graph among the answer choices that passes the vertical line test:

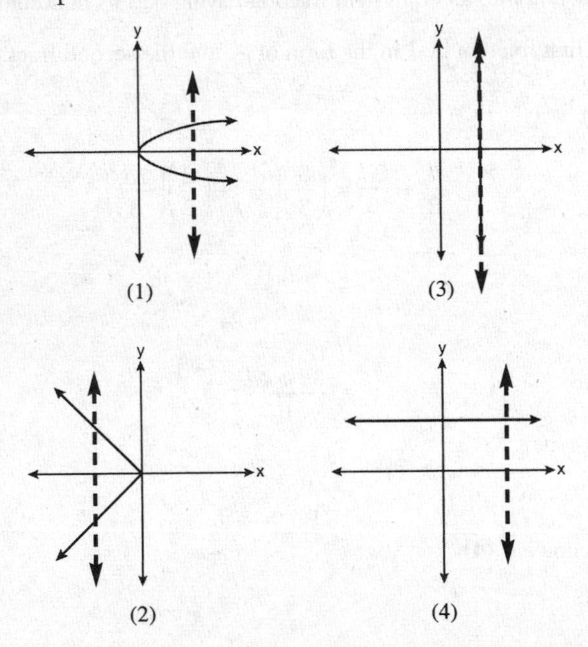

The correct choice is **(4)**.

PART II

31. The given radical expression is: $5\sqrt{72}$

Factor the radicand so that one of the factors is the greatest possible perfect square factor of 72: $5\sqrt{36 \cdot 2}$

Write the radical over each factor: $5 \cdot \sqrt{36} \cdot \sqrt{2}$

Evaluate the square root of the perfect square factor: $5 \cdot 6 \cdot \sqrt{2}$

$30\sqrt{2}$

In simplest radical form, $5\sqrt{72}$ is $\mathbf{30\sqrt{2}}$.

32. The given equation is: $3 + 2g = 5g - 9$

Subtract 3 from each side: $2g = 5g - 9 - 3$

Subtract 5g from each side: $2g - 5g = -12$

Combine like terms: $-3g = -12$

Divide each side by –3: $\frac{-3g}{-3} = \frac{-12}{-3}$

$g = 4$

The value of g is **4**.

33. It is given that Serena's garden is a rectangle joined with a semicircle, as shown in the diagram below where line segment AB is the diameter of semicircle P. You are required to find the perimeter of the composite figure to the *nearest tenth of a foot*.

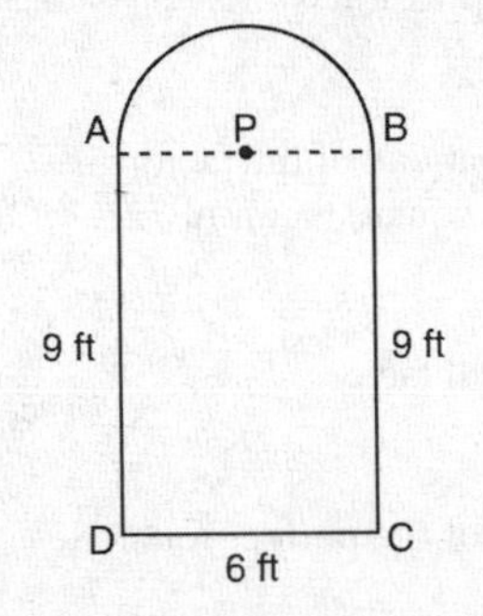

The distance around the figure is comprised of the circumference of semicircle P plus the sum of the lengths of CD, AD, and BC.

- The circumference of semicircle P is $\frac{1}{2} \times (2\pi r)$. The radius, r, of the semicircle is one-half of its diameter. The diameter of the semicircle is $AB = CD = 6$ feet, so r is $\frac{1}{2} \times 6 = 3$ feet. Thus:

$$\text{Circumference of semicircle} = \frac{1}{2} \times (2\pi r)$$

$$= \frac{1}{2} \times (2\pi) \times 3$$

$$= 3\pi$$

- Perimeter of figure = $CD + AD + BC$ + circumference of semicircle

$$= 6 + 9 + 9 + 3\pi$$

$$= 24 + 3\pi$$

$$= 33.42477796$$

The length of fence Serena needs to the *nearest tenth of a foot* is **33.4**.

PART III

34. It is given that Hannah drove 120 miles to reach her cousin's house and drove the same distance back home. It took her 1.2 hours to get halfway to her cousin's house.

- Since it took her 1.2 hours to travel $\frac{1}{2} \times 120 = 60$ miles, her average speed for this part of the trip was:

$$\frac{60 \text{ miles}}{1.2 \text{ hours}} = 50 \text{ miles per hour}$$

Her average speed was **50** miles per hour.

- It is also given that Hannah's average speed for the remaining 60 miles of the trip to her cousin's house was 40 miles per hour. Since rate × time = distance:

$$40 \times t = 60$$

$$\frac{40t}{40} = \frac{60}{40}$$

$$t = 1.5 \text{ hours}$$

It took her **1.5** hours to drive the remaining distance.

- It is also given that traveling while home along the same route, Hannah drove at an average rate of 55 miles per hour and that after 2 hours her car broke down. You are required to find the number of miles she was from her home. By traveling 2 hours at an average rate of 55 miles per hour, Hannah drove a distance of $55 \times 2 = 110$ miles. Since the total distance of the return trip is 120 miles, she was 10 miles from her home.

Hannah was **10** miles from her home when her car broke down.

35. It is given that Tom is going to save \$120 to pay for a prom ticket by walking his neighbor's dog for \$15 per week. It is also given that Tom already has saved \$22. If x represents the number of weeks Tom walks his neighbor's dog, then $15x + 22$ is the amount of money he has after x weeks. To save at least \$120:

$$15x + 22 \geq 120$$

$$15x \geq 120 - 22$$

$$15x \geq 98$$

$$\frac{15x}{15} \geq \frac{98}{15}$$

$$x \geq 6.5333...$$

Since x must be an integer, $x = 7$.

Tom must walk the dog a minimum of **7** weeks to earn enough to pay for the prom ticket.

36. Mr. Laub has three children: two girls (**S**ue and **K**aren) and one boy (**D**avid). After each meal, one child is chosen at random to wash dishes. The same child can be chosen for both lunch and dinner. The sample space can be represented either as a tree diagram or as a set of ordered pairs:

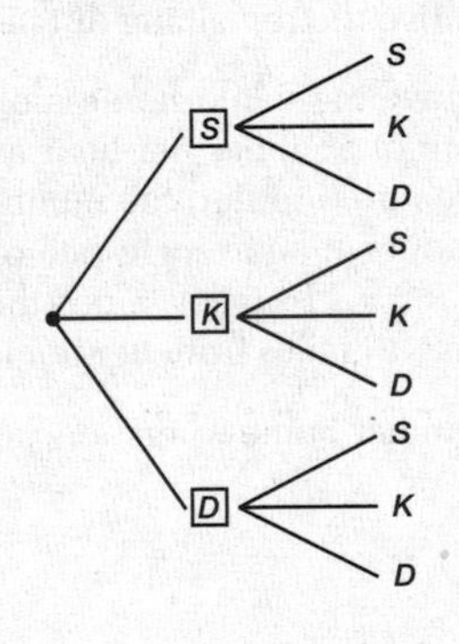

or

(S,S)	(K,S)	(D,S) ✔
(S,K)	(K,K)	(D,K) ✔
(S,D) ✔	(K,D) ✔	(D,D)

Four out of the nine ordered pairs are comprised of one boy and one girl: (S,D), (K,D), (D,S), and (D,K). Thus, the probability that one boy and one girl will wash dishes after lunch and dinner on Saturday is $\frac{4}{9}$.

The probability is $\mathbf{\frac{4}{9}}$.

PART IV

37. The values of 11 houses on Washington St. are shown in the table below.

Value per House	Number of Houses
$100,000	1
$175,000	5
$200,000	4
$700,000	1

- The mean (average) value of these houses in dollars is the sum of the values of the houses divided by the number of houses:

$$\frac{(1\times 100{,}000)+(5\times 175{,}000)+(4\times 200{,}000)+(1\times 700{,}000)}{1+5+4+1} = \frac{2{,}475{,}000}{11}$$

$$= 225{,}000$$

The mean value of the houses in dollars is **225,000**.

- The median value is the middle value in an ordered list of numbers. Since 11 houses are in the table, the price of the 6th house in the table represents the median value of the houses in dollars:

Value per House	Number of Houses	
$100,000	1	
$175,000	5	← 1 + 5 = 6
$200,000	4	
$700,000	1	

The median value of these houses in dollars is **175,000**.

- The data contains two outliers ($100,000 and $700,000), which vary greatly from the other house values. As a result, these outliers distort the mean. Since most of the house prices are clustered in the range of $175,000 to $200,000, the median is more representative of the values of the houses than the mean.

The **median** *best* represents the values of the 11 houses.

38. To solve the given system of equations graphically, graph each equation on the same set of axes and then read the coordinates of their points of intersection. Graph the parabola $y = x^2 - 6x + 5$ by plotting its vertex (turning point) and three points on either side of it.

- To find the x-coordinate of the vertex of $y = x^2 - 6x + 5$:

$$x = -\frac{b}{2a}$$

$$= -\frac{(-6)}{2(1)}$$

$$= 3$$

- A table of values can be obtained by entering the equation $Y_1 = x^2 - 6x + 5$ into your calculator and then using its TABLE feature, as shown in the accompanying table. If necessary, scroll up or down the table until you locate $x = 3$, the x-coordinate of the vertex.

$$Y_1 = x^2 - 6x + 5$$

X	Y1	
0	5	
1	0	
2	-3	
3	-4	
4	-3	
5	0	
6	5	

X=0

- On the set of axes provided, plot these points: (0,5), (1,0), (2,–3), (3,–4), (4,–3), (5,0), and (6,5). Connect these points with a smooth U-shaped curve, as shown in the accompanying figure.

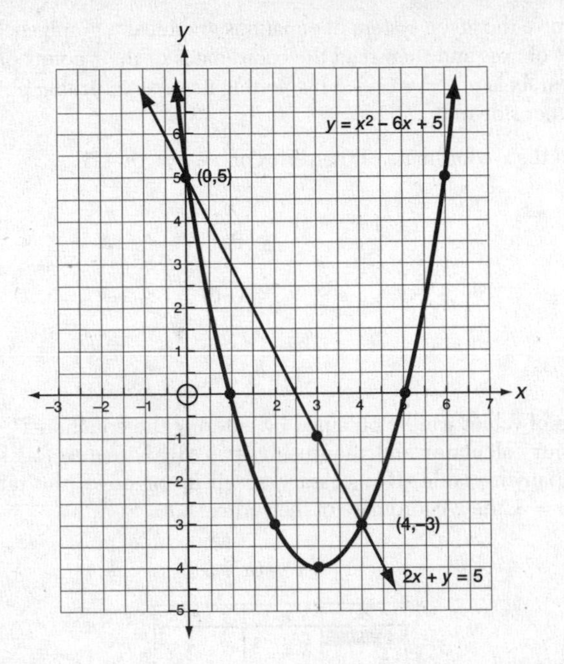

Graph $2x + y = 5$ by finding and then plotting at least two points that satisfy the equation.

- Use your calculator to create a table of values for $2x + y = 5$ by entering the equation in the form $y = -2x + 5$.
- Plot any two convenient points from the table, such as (1,3) and (3,–1). Then draw a line through these points, as shown in the accompanying figure. The solution consists of the set of points where the two graphs intersect.

$$Y_2 = -2x + 5$$

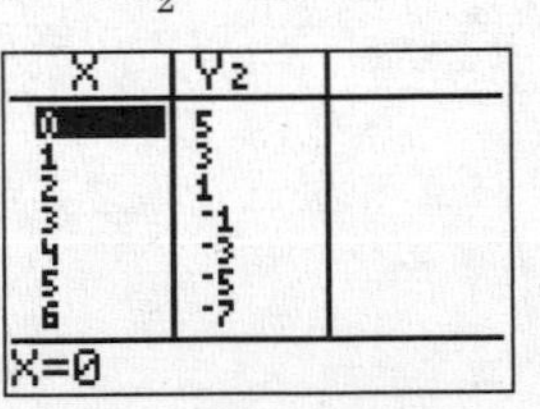

X	Y2	
0	5	
1	3	
2	1	
3	-1	
4	-3	
5	-5	
6	-7	

X=0

The points in the solution set are **(0,5)** and **(4,–3)**.

39. The given equation is: $\frac{x+1}{x} = \frac{-7}{x-12}$

Eliminate the fractions by setting the cross-products equal: $(x + 1)(x - 12) = -7x$

Multiply the binomials together using FOIL: $x^2 - 12x + 1x - 12 = -7x$

Simplify: $x^2 - 11x - 12 = -7x$

Collect all the nonzero terms on the left side of the equation so that the right side is 0: $x^2 - 11x + 7x - 12 = 0$

Combine like terms: $x^2 - 4x - 12 = 0$

Factor the quadratic trinomial as the product of two binomials: $(x + ?)(x + ?) = 0$

The missing terms of the binomial factors are the two integers whose product is –12 and whose sum is –4. Since (–6)(+2) = –12 and (–6) + (+2) = –4, the missing terms are –6 and +2: $(x - 6)(x + 2) = 0$

Set each factor equal to 0: $x - 6 = 0$ or $x + 2 = 0$

Solve for x: $x = 6$ or $x = -2$

The solutions for x are **6** and **–2**.

Topic	Question Numbers	Number of Points	Your Points	Your Percentage
1. Sets and Numbers; Interval Notation; Properties of Real Numbers; Percent	4, 5, 10, 28	2 + 2 + 2 + 2 = 8		
2. Operations on Rat'l. Numbers & Monomials	3, 27	2 + 2 = 4		
3. Laws of Exponents for Integer Exponents; Scientific Notation	25	2		
4. Operations on Polynomials	18	2		
5. Square Root; Operations with Radicals	31	2		
6. Evaluating Formulas & Algebraic Expressions	—	—		
7. Solving Linear Eqs. & Inequalities	20, 24, 32, 35	2 + 2 +2 + 3 = 9		
8. Solving Literal Eqs. & Formulas for a Particular Letter	—	—		
9. Alg. Operations (including factoring)	6	2		
10. Quadratic Equations (incl. alg. solution; parabolas)	17, 39	2 + 4 = 6		
11. Coordinate Geometry (eq. of a line; graphs of linear eqs.; slope)	13, 16	2 + 2 = 4		
12. Systems of Linear Eqs. & Inequalities (algebraic & graphical solutions)	8	2		
13. Mathematical Modeling (using: eqs.; tables; graphs).	15, 26, 29	2 + 2 + 2 = 6		
14. Linear-Quadratic systems	38	4		
15. Properties of Triangles & Parallelograms	—	—		
16. Perimeter; Circumference; Area of Common Figures	33	2		
17. Volume and Surface Area of Common Figures; Relative Error in Measurement	12, 23	2 + 2 = 4		
18. Angle & Line Relationships (suppl., compl., vertical angles; parallel lines; congruence)	—	—		
19. Ratio & Proportion (incl. similar polygons, scale drawings, & rates)	—	—		
20. Pythagorean Theorem	11	2		

Topic	Question Numbers	Number of Points	Your Points	Your Percentage
21. Right Triangle Trigonometry	21	2		
22. Functions (def.; domain and range; vertical line test; absolute value)	22, 30	2 + 2 = 4		
23. Exponential Functions (properties; growth and decay)	19	2		
24. Probability (incl. tree diagrams & sample spaces)	2, 36	2 + 3 = 5		
25. Permutations and Counting Methods (incl. Venn diagrams)	—	—		
26. Statistics (mean, percentiles, quartiles; freq. dist., histograms; box-and-whisker plots; causality, bivariate data; circle graphs)	7, 9, 14, 37	2 + 2 + 2 +4 = 10		
27. Line of Best Fit (including linear regression, scatter plots, and linear correlation)	1	2		
28. Nonroutine Word Problems Requiring Arith. or Alg. Reasoning	34	3		

Raw Versus Scaled Score

No conversion chart was provided by the New York State Education Department, so it is not possible to convert your final raw score on the Test Sampler into a scaled score.

MAP TO LEARNING STANDARDS

The table below shows which content strand each item is aligned to. The numbers in the table represent the question numbers on the test.

Content Strand	Multiple-Choice Item Number	2-Credit Item Number	3-Credit Item Number	4-Credit Item Number
Number Sense and Operations	5, 25	31		
Algebra	3, 4, 6, 8, 10, 11, 13, 15, 16, 18, 19, 21, 24, 26, 27, 28, 29	32	35	39
Geometry	12, 17, 20, 22, 30	33		38
Measurement	23		34	
Probability and Statistics	1, 2, 7, 9, 14		36	37

Examination June 2008
Integrated Algebra

FORMULAS

Trigonometric ratio

$\sin A = \dfrac{\text{opposite}}{\text{hypotenuse}}$

$\cos A = \dfrac{\text{adjacent}}{\text{hypotenuse}}$

$\tan A = \dfrac{\text{opposite}}{\text{adjacent}}$

Area Trapezoid $A = \frac{1}{2}h(b_1 + b_2)$

Volume Cylinder $V = \pi r^2 h$

Surface area Rectangular prism $SA = 2lw + 2hw + 2lh$

Cylinder $SA = 2\pi r^2 + 2\pi rh$

Coordinate geometry $m = \dfrac{\Delta y}{\Delta x} = \dfrac{y_2 - y_1}{x_2 - x_1}$

PART I

Answer all questions in this part. Each correct answer will receive 2 credits. No partial credit will be allowed. For each question, write in the space provided the numeral preceding the word or expression that best completes the statement or answers the question. [60]

1 Which graph represents a linear function?

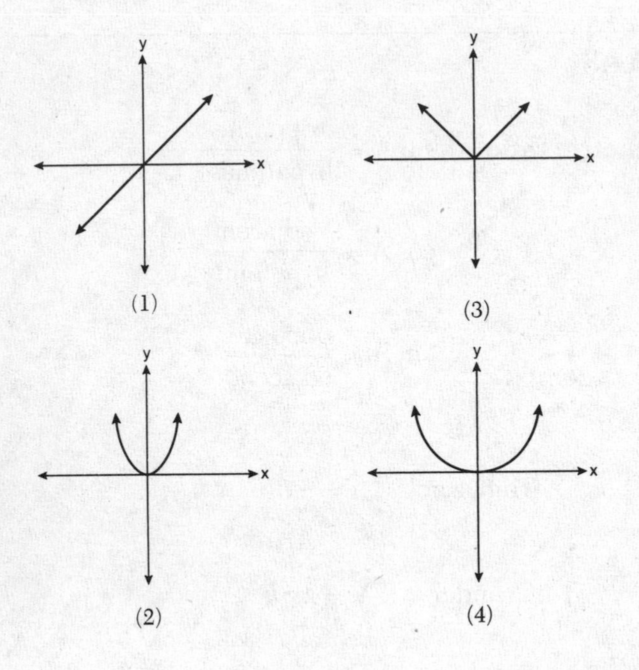

1 ______

2 A spinner is divided into eight equal regions as shown in the diagram below.

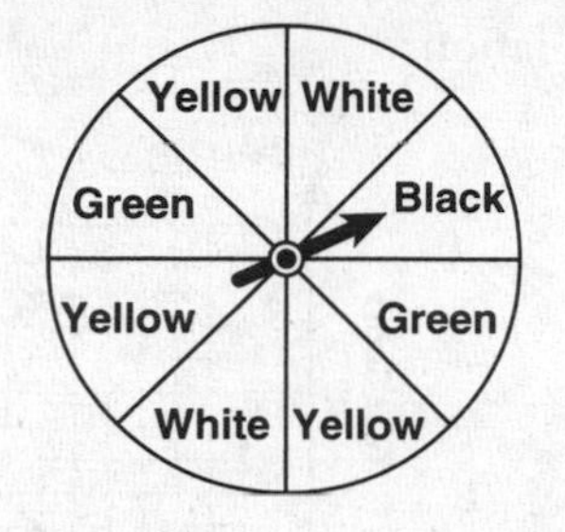

Which event is most likely to occur in one spin?

(1) The arrow will land in a green or white area.
(2) The arrow will land in a green or black area.
(3) The arrow will land in a yellow or black area.
(4) The arrow will land in a yellow or green area. 2 _____

3 A school wants to add a coed soccer program. To determine student interest in the program, a survey will be taken. In order to get an unbiased sample, which group should the school survey?

(1) every third student entering the building
(2) every member of the varsity football team
(3) every member in Ms. Zimmer's drama classes
(4) every student having a second-period French class 3 _____

4 Factored, the expression $16x^2 - 25y^2$ is equivalent to

(1) $(4x - 5y)(4x + 5y)$ (3) $(8x - 5y)(8x + 5y)$
(2) $(4x - 5y)(4x - 5y)$ (4) $(8x - 5y)(8x - 5y)$ 4 _____

5 There is a negative correlation between the number of hours a student watches television and his or her social studies tests score. Which scatter plot displays this correlation?

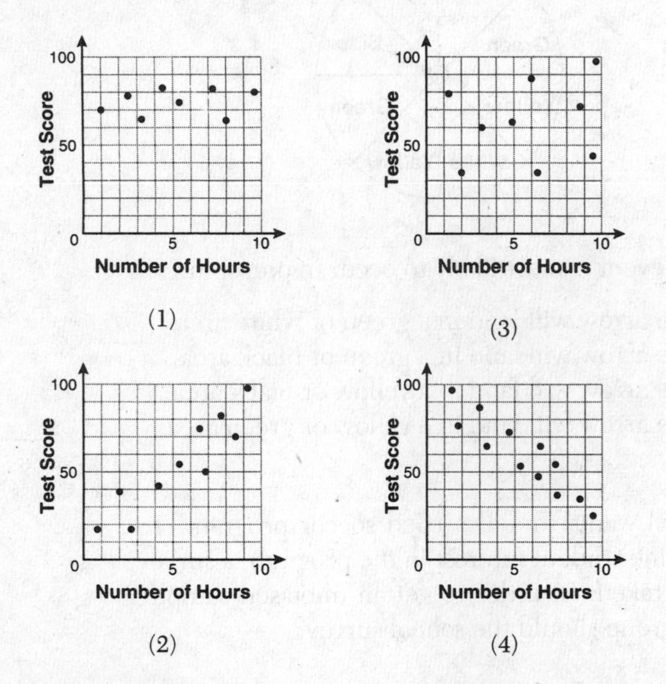

5 _____

6 Jack bought 3 slices of cheese pizza and 4 slices of mushroom pizza for a total cost of \$12.50. Grace bought 3 slices of cheese pizza and 2 slices of mushroom pizza for a total cost of \$8.50. What is the cost of one slice of mushroom pizza?

(1) \$1.50
(2) \$2.00
(3) \$3.00
(4) \$3.50

6 _____

7 What is the product of $-3x^2y$ and $(5xy^2 + xy)$?

(1) $-15x^3y^3 - 3x^3y^2$
(2) $-15x^3y^3 - 3x^3y$
(3) $-15x^2y^2 - 3x^2y$
(4) $-15x^3y^3 - xy$

7 _____

8 The bowling team at Lincoln High School must choose a president, vice president, and secretary. If the team has 10 members, which expression could be used to determine the number of ways the officers could be chosen?

(1) ${}_3P_{10}$
(2) ${}_7P_3$
(3) ${}_{10}P_3$
(4) ${}_{10}P_7$

8 _____

9 Lenny made a cube in technology class. Each edge measured 1.5 cm. What is the volume of the cube in cubic centimeters?

(1) 2.25
(2) 3.375
(3) 9.0
(4) 13.5

9 _____

10 Which ordered pair is a solution to the system of equations $y = x$ and $y = x^2 - 2$?

(1) $(-2,-2)$
(2) $(-1,1)$
(3) $(0,0)$
(4) $(2,2)$

10 _____

11 What are the vertex and the axis of symmetry of the parabola shown in the diagram below?

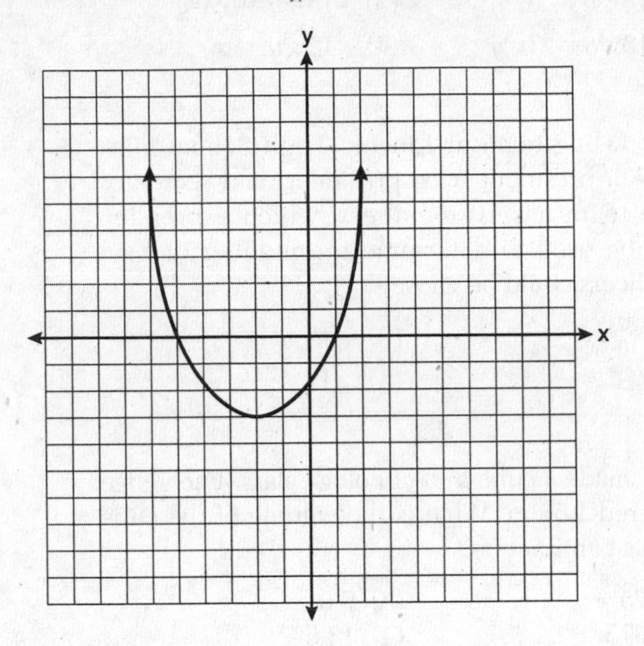

(1) The vertex is (–2,–3), and the axis of symmetry is $x = -2$.
(2) The vertex is (–2,–3), and the axis of symmetry is $y = -2$.
(3) The vertex is (–3,–2), and the axis of symmetry is $y = -2$.
(4) The vertex is (–3,–2), and the axis of symmetry is $x = -2$.

11 ______

12 Pam is playing with red and black marbles. The number of red marbles she has is three more than twice the number of black marbles she has. She has 42 marbles in all. How many red marbles does Pam have?

(1) 13
(2) 15
(3) 29
(4) 33

12 ____

13 What is half of 2^6?

(1) 1^3
(2) 1^6
(3) 2^3
(4) 2^5

13 ____

14 Which equation represents a line that is parallel to the line $y = -4x + 5$?

(1) $y = -4x + 3$
(2) $y = -\frac{1}{4}x + 5$
(3) $y = \frac{1}{4}x + 3$
(4) $y = 4x + 5$

14 ____

15 What is the product of $\frac{x^2 - 1}{x + 1}$ and $\frac{x + 3}{3x - 3}$ expressed in simplest form?

(1) x
(2) $\frac{x}{3}$
(3) $x + 3$
(4) $\frac{x + 3}{3}$

15 ____

16 The center pole of a tent is 8 feet long, and a side of the tent is 12 feet long as shown in the diagram below.

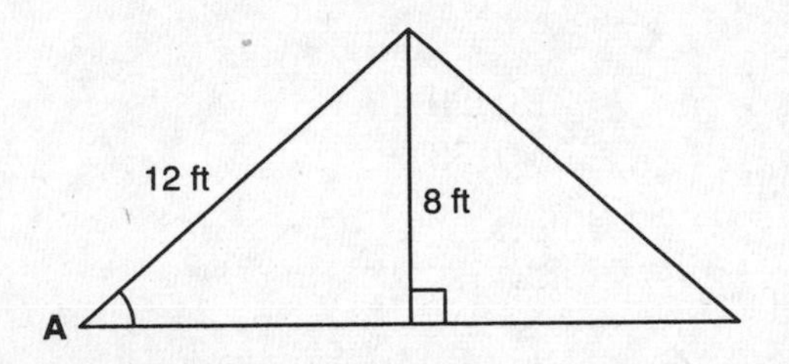

If a right angle is formed where the center pole meets the ground, what is the measure of angle A to the *nearest degree*?

(1) 34 (3) 48
(2) 42 (4) 56

16 ______

17 Which value of x makes the expression $\frac{x+4}{x-3}$ undefined?

(1) −4 (3) 3
(2) −3 (4) 0

17 ______

18 Consider the set of integers greater than −2 and less than 6. A subset of this set is the positive factors of 5. What is the complement of this subset?

(1) {0, 2, 3, 4}
(2) {−1, 0, 2, 3, 4}
(3) {−2, −1, 0, 2, 3, 4, 6}
(4) {−2, −1, 0, 1, 2, 3, 4, 5, 6}

18 ______

19 Which data set describes a situation that could be classified as qualitative?

(1) the elevations of the five highest mountains in the world
(2) the ages of presidents at the time of their inauguration
(3) the opinions of students regarding school lunches
(4) the shoe sizes of players on the basketball team

19 _____

20 What is the slope of the line that passes through the points (–6,1) and (4,–4)?

(1) –2
(2) 2
(3) $-\frac{1}{2}$
(4) $\frac{1}{2}$

20 _____

21 Students in a ninth grade class measured their heights, h, in centimeters. The height of the shortest student was 155 cm, and the height of the tallest student was 190 cm. Which inequality represents the range of heights?

(1) $155 < h < 190$
(2) $155 \leq h \leq 190$
(3) $h \geq 155$ or $h \leq 190$
(4) $h > 155$ or $h < 190$

21 _____

22 The table below shows a cumulative frequency distribution of runners' ages.

Cumulative Frequency Distribution of Runners' Ages

Age Group	Total
20–29	8
20–39	18
20–49	25
20–59	31
20–69	35

According to the table, how many runners are in their forties?

(1) 25 (3) 7
(2) 10 (4) 6 22 ______

23 Mr. Turner bought x boxes of pencils. Each box holds 25 pencils. He left 3 boxes of pencils at home and took the rest to school. Which expression represents the total number of pencils he took to school?

(1) $22x$ (3) $25 - 3x$
(2) $25x - 3$ (4) $25x - 75$ 23 ______

24 Which expression represents $\frac{2x^2 - 12x}{x - 6}$ in simplest form?

(1) 0 (3) $4x$
(2) $2x$ (4) $2x + 2$ 24 ______

25 Don placed a ladder against the side of his house as shown in the diagram below.

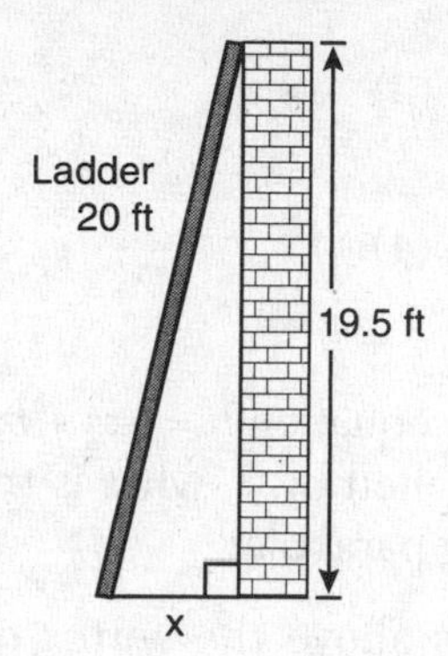

Which equation could be used to find the distance, x, from the foot of the ladder to the base of the house?

(1) $x = 20 - 19.5$
(2) $x = 20^2 - 19.5^2$
(3) $x = \sqrt{20^2 - 19.5^2}$
(4) $x = \sqrt{20^2 + 19.5^2}$

25 _____

26 Which value of x is a solution of $\frac{5}{x} = \frac{x+13}{6}$?

(1) −2
(2) −3
(3) −10
(4) −15

26 _____

27 Mrs. Ayer is painting the outside of her son's toy box, including the top and bottom. The toy box measures 3 feet long, 1.5 feet wide, and 2 feet high. What is the total surface area she will paint?

(1) 9.0 ft^2
(2) 13.5 ft^2
(3) 22.5 ft^2
(4) 27.0 ft^2

27 _____

28 What is $\frac{\sqrt{32}}{4}$ expressed in simplest radical form?

(1) $\sqrt{2}$ (3) $\sqrt{8}$

(2) $4\sqrt{2}$ (4) $\frac{\sqrt{8}}{2}$ 28 ______

29 Consider the graph of the equation $y = ax^2 + bx + c$, when $a \neq 0$. If a is multiplied by 3, what is true of the graph of the resulting parabola?

(1) The vertex is 3 units above the vertex of the original parabola.
(2) The new parabola is 3 units to the right of the original parabola.
(3) The new parabola is wider than the original parabola.
(4) The new parabola is narrower than the original parabola. 29 ______

30 Kathy plans to purchase a car that depreciates (loses value) at a rate of 14% per year. The initial cost of the car is $21,000. Which equation represents the value, v, of the car after 3 years?

(1) $v = 21{,}000(0.14)^3$ (3) $v = 21{,}000(1.14)^3$
(2) $v = 21{,}000(0.86)^3$ (4) $v = 21{,}000(0.86)(3)$ 30 ______

PART II

Answer all questions in this part. Each correct answer will receive 2 credits. Clearly indicate the necessary steps, including appropriate formula substitutions, diagrams, graphs, charts, etc. For all questions in this part, a correct numerical answer with no work shown will receive only 1 credit. [6]

31 Tom drove 290 miles from his college to home and used 23.2 gallons of gasoline. His sister, Ann, drove 225 miles from her college to home and used 15 gallons of gasoline. Whose vehicle had better gas mileage? Justify your answer.

32 A designer created the logo shown below. The logo consists of a square and four quarter-circles of equal size.

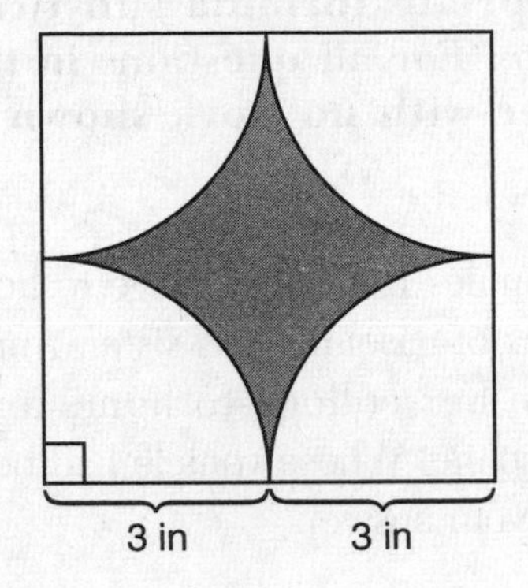

Express, in terms of π, the exact area, in square inches, of the shaded region.

33 Maureen tracks the range of outdoor temperatures over three days. She records the following information?

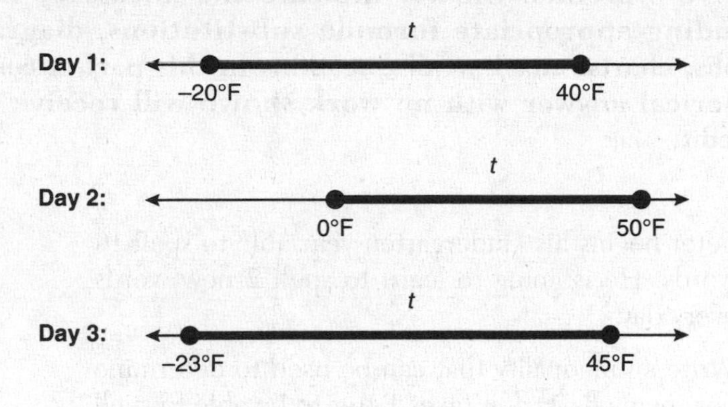

Express the intersection of the three sets as an inequality in terms of temperature, t.

PART III

Answer all questions in this part. Each correct answer will receive 3 credits. Clearly indicate the necessary steps, including appropriate formula substitutions, diagrams, graphs, charts, etc. For all questions in this part, a correct numerical answer with no work shown will receive only 1 credit. [9]

34 Peter begins his kindergarten year able to spell 10 words. He is going to learn to spell 2 new words every day.

Write an inequality that can be used to determine how many days, d, it takes Peter to be able to spell *at least* 75 words.

Use this inequality to determine the minimum number of whole days it will take for him to be able to spell *at least* 75 words.

35 The Hudson Record Store is having a going-out-of-business sale. CDs normally sell for $18.00. During the first weeks of the sale, all CDs will sell for $15.00.

Written as a fraction, what is the rate of discount?

What is this rate expressed as a percent? Round your answer to the *nearest hundredth of a percent.*

During the second week of the sale, the same CDs will be on sale for 25% off the *original* price. What is the price of a CD during the second week of the sale?

36 Graph the equation $y = x^2 - 2x - 3$ on the accompanying set of axes. Using the graph, determine the roots of the equation $x^2 - 2x - 3 = 0$.

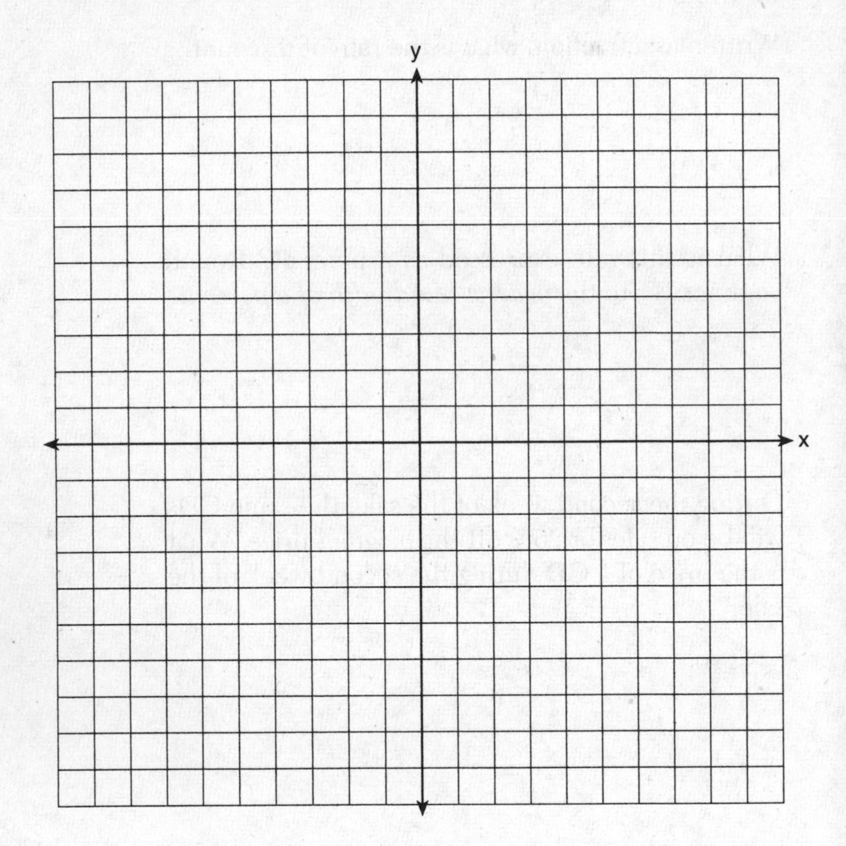

PART IV

Answer all questions in this part. Each correct answer will receive 4 credits. Clearly indicate the necessary steps, including appropriate formula substitutions, diagrams, graphs, charts, etc. For all questions in this part, a correct numerical answer with no work shown will receive only 1 credit. [12]

37 A contractor needs 54 square feet of brick to construct a rectangular walkway. The length of the walkway is 15 feet more than the width.

Write an equation that could be used to determine the dimensions of the walkway. Solve this equation to find the length and width, in feet, of the walkway.

38 Sophie measured a piece of paper to be 21.7 cm by 28.5 cm. The piece of paper is actually 21.6 cm by 28.4 cm.

Determine the number of square centimeters in the area of the piece of paper using Sophie's measurements.

Determine the number of square centimeters in the actual area of the piece of paper.

Determine the relative error in calculating the area. Express your answer as a decimal to the *nearest thousandth*.

Sophie does not think there is a significant amount of error. Do you agree or disagree? Justify your answer.

39 The prices of seven race cars sold last week are listed in the table below.

Price per Race Car	Number of Race Cars
$126,000	1
$140,000	2
$180,000	1
$400,000	2
$819,000	1

What is the mean value of these race cars, in dollars?

What is the median value of these race cars, in dollars?

State which of these measures of central tendency best represents the value of the seven race cars. Justify your answer.

Answers June 2008

Integrated Algebra

Answer Key

PART I

1. (1)	**6.** (2)	**11.** (1)	**16.** (2)	**21.** (2)	**26.** (4)
2. (4)	**7.** (1)	**12.** (3)	**17.** (3)	**22.** (3)	**27.** (4)
3. (1)	**8.** (3)	**13.** (4)	**18.** (2)	**23.** (4)	**28.** (1)
4. (1)	**9.** (2)	**14.** (1)	**19.** (3)	**24.** (2)	**29.** (4)
5. (4)	**10.** (4)	**15.** (4)	**20.** (3)	**25.** (3)	**30.** (2)

PART II

31. Ann's vehicle

32. $36 - 9\pi$ or $36 - 3^2\pi$

33. $0 \leq t \leq 40$

PART III

34. $10 + 2d \geq 75$; 33

35. $\frac{3}{18}$; 16.67%; $13.50

36. –1; 3; see *Answers Explained* section.

PART IV

37. $w^2 + 15w - 54 = 0$ where w = width; width = 3; length = 18.

38. 618.45; 613.44; 0.008. The amount of error is *not* significant since the error is less than 1%.

39. Mean = $315,000; median = $180,000. The median is the best measure of central tendency because the outlier car price of $819,000 distorts the mean but does not affect the median.

In **PARTS II–IV** you are required to show how you arrived at your answers. For sample methods of solutions, see the *Answers Explained* section.

Answers Explained

PART I

1. The graph of a linear function is a straight line.

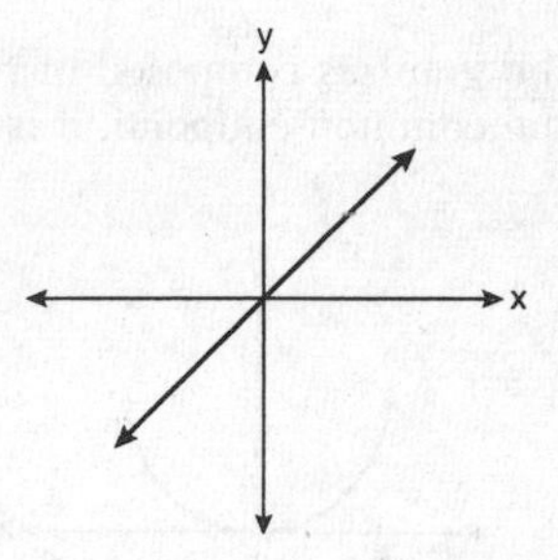

- Choice (1): Since the graph is a continuous line, it represents a linear function. ✔

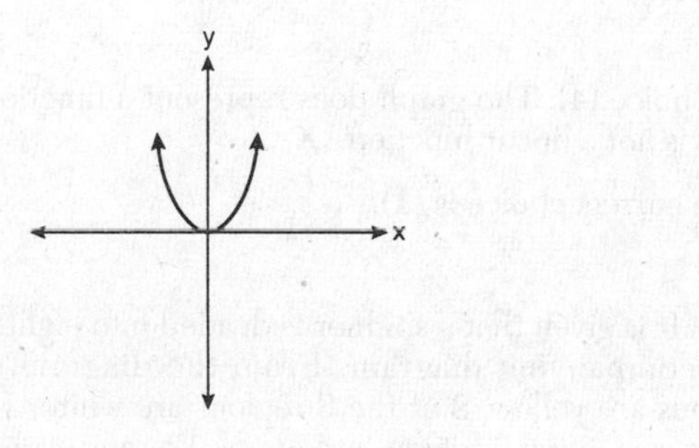

- Choice (2): The graph does represent a function. It is curved, though, so it is *not* a linear function. ✗

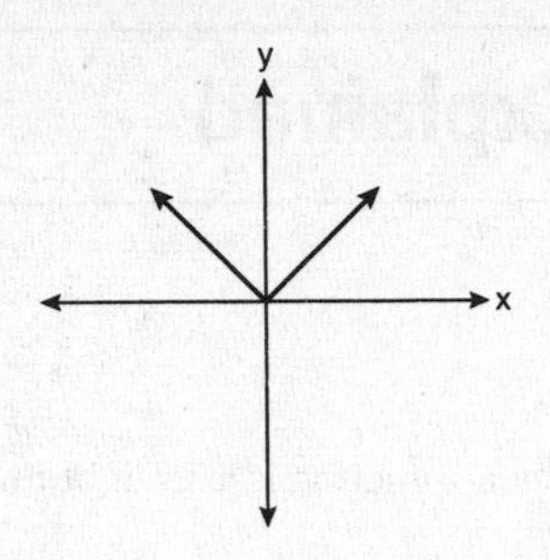

- Choice (3): Since the graph is composed of rays in Quadrants I and II with the origin as the common endpoint, it is the graph of the absolute value function. ✗

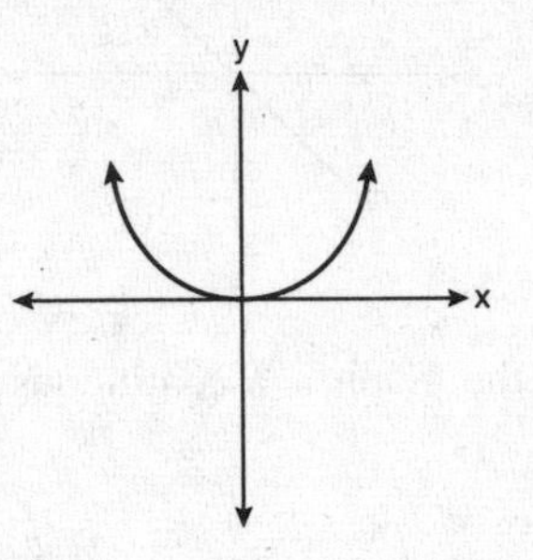

- Choice (4): The graph does represent a function. It is curved, though, so it is *not* a linear function. ✗

The correct choice is **(1)**.

2. It is given that a spinner is divided into eight equal regions as shown in the accompanying diagram. From the diagram, you know that 3 of the 8 regions are yellow, 2 of the 8 regions are white, 1 of the 8 regions is black, and the remaining 2 regions are green. For any event E,

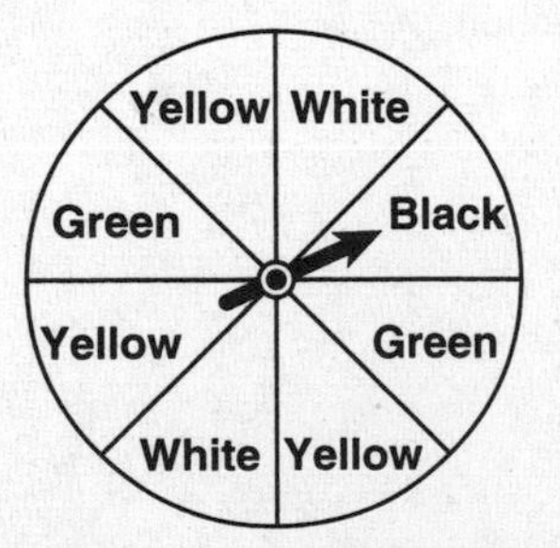

$$P(E) = \frac{\text{Number of favorable outcomes}}{\text{Total number of possible outcomes}}$$

$$P(\text{yellow area}) = \frac{3}{8}$$

$$P(\text{white area}) = \frac{2}{8}$$

$$P(\text{black area}) = \frac{1}{8}$$

$$P(\text{green area}) = \frac{2}{8}$$

The probability that the spinner will land in either of two different-colored regions is the sum of the probabilities of the spinner landing in each colored region.

- Choice (1):

$$P(\text{green or white areas}) = P(\text{green area}) + P(\text{white area})$$
$$= \frac{2}{8} + \frac{2}{8}$$
$$= \frac{4}{8}$$

- Choice (2):

$$P(\text{green or black areas}) = P(\text{green area}) + P(\text{black area})$$
$$= \frac{2}{8} + \frac{1}{8}$$
$$= \frac{3}{8}$$

- Choice (3):

$$P(\text{yellow or black areas}) = P(\text{yellow area}) + P(\text{black area})$$
$$= \frac{3}{8} + \frac{1}{8}$$
$$= \frac{4}{8}$$

- Choice (4):

$$P(\text{yellow or green areas}) = P(\text{yellow area}) + P(\text{green area})$$
$$= \frac{3}{8} + \frac{2}{8}$$
$$= \frac{5}{8}$$

Since the event that is most likely to occur is the event with the greatest probability of occurring, the arrow landing in a yellow or green area is most likely to occur.

The correct choice is **(4)**.

3. To obtain an unbiased sample, each student in the school must have an equal chance of being selected to participate in the survey. Choices (2), (3), and (4) survey special subsets of students in the school and, as a result, represent biased samples. Surveying every third student entering the building, as proposed in choice (1), approximates a random selection process that comes closest to meeting the goal of ensuring that each student has an equal chance of being included in the survey.

The correct choice is **(1)**.

4. The given expression, $16x^2 - 25y^2$, represents the difference of two perfect squares:

$$16x^2 - 25y^2 = (4x)^2 - (5y)^2$$

The difference of two perfect squares can be factored as the product of the sum and the difference of the two terms being squared. Because the two terms that are being squared are $4x$ and $5y$:

$$16x^2 - 25y^2 = (4x - 5y)(4x + 5y).$$

The correct choice is **(1)**.

5. When two quantities have a negative correlation, such as the number of hours a student watches television and his or her social studies test score, as one of the quantities increases, the other decreases. The direction of linear correlation corresponds to the slope of the line that best fits the points in the scatter plot. A best-fit line with a positive slope indicates a positive correlation between the two quantities in the scatter plot. A best-fit line with a negative slope means that a negative correlation exists between the two quantities.

For each scatter plot, draw the approximate line of best fit, if possible:

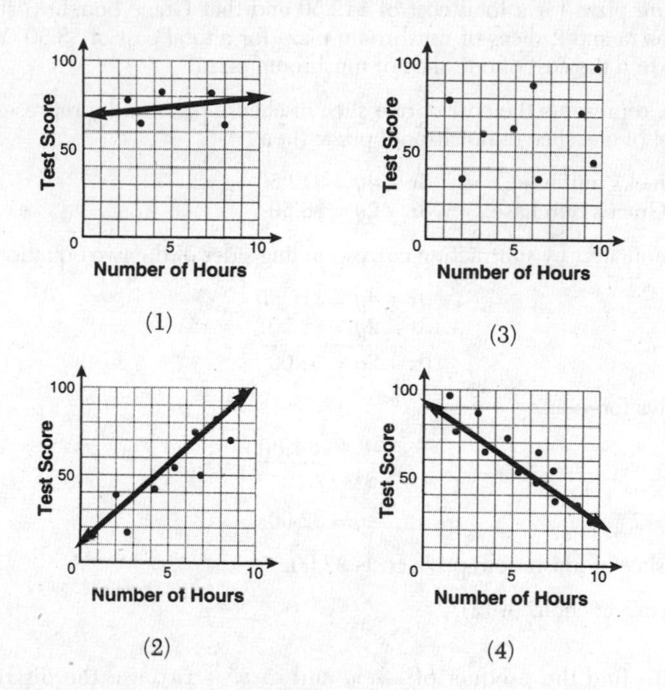

Then examine each answer choice in turn:

- Choice (1): The approximate line of best fit has a slope close to 0. This indicates that the student's test score does not depend on the number of hours of television watched. ✗
- Choice (2): The approximate line of best fit has a positive slope. This indicates that the student's test score increases when the number of hours of television watched increases. Hence, this scatter plot displays a positive correlation between the two quantities. ✗
- Choice (3): The scatter plot indicates that no correlation exists between the two quantities. ✗

- Choice (4): The approximate line of best fit has a *negative* slope. This indicates that the student's test score decreases when the number of hours of television watched increases. Thus, this scatter plot displays a negative correlation between the two quantities. ✔

The correct choice is **(4)**.

6. It is given that Jack bought 3 slices of cheese pizza and 4 slices of mushroom pizza for a total cost of \$12.50 and that Grace bought 3 slices of cheese pizza and 2 slices of mushroom pizza for a total cost of \$8.50. You are asked to find the cost of one slice of mushroom pizza.

- If x represents the cost of one slice of cheese pizza and y represents the cost of one slice of mushroom pizza, then

Jack's purchase: $3x + 4y = \$12.50$
Grace's purchase: $3x + 2y = \$8.50$

- Eliminate x by subtracting corresponding sides of the two equations:

$$\begin{array}{r} 3x + 4y = \$12.50 \\ -\ (3x + 2y = \$8.50) \\ \hline 0x + 2y = \$4.00 \end{array}$$

- Solve for y:

$$\frac{2y}{2} = \frac{\$4.00}{2}$$

$$y = \$2.00$$

One slice of mushroom pizza costs \$2.00.

The correct choice is **(2)**.

7. To find the product of $-3x^2y$ and $(5xy^2 + xy)$, use the distributive property.

- Remove the parentheses by multiplying each term inside of the parentheses by $-3x^2y$:

$$-3x^2y(5xy^2 + xy) = (-3x^2y)(5xy^2) + (-3x^2y)(xy)$$

- Multiply powers of the same base by adding their exponents:

$$= (-3x^2y^1)(5x^1y^2) + (-3x^2y^1) + (x^1y^1)$$

$$= (-3)(5)x^{2+1}y^{1+2} + (-3)(x^{2+1})(y^{1+1})$$

$$= -15x^3y^3 - 3x^3y^2$$

The correct choice is **(1)**.

8. It is given that from 10 members on the bowling team, a president, vice president, and secretary will be chosen. You are asked to represent the number of different ways in which these three officer positions could be filled.

- Any one of the 10 members on the bowling team can be selected to fill the position of president:

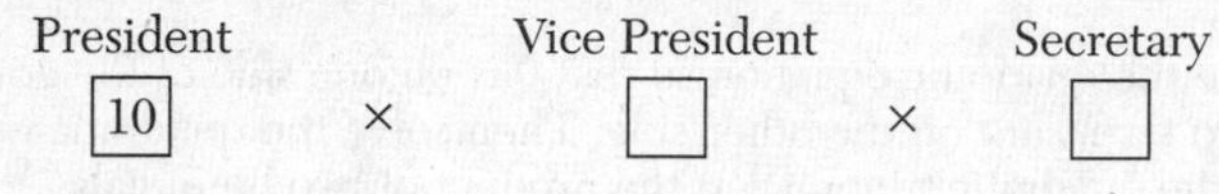

- Once the position of president is filled, any one of the remaining 9 members of the bowling team can be selected to fill the position of vice president:

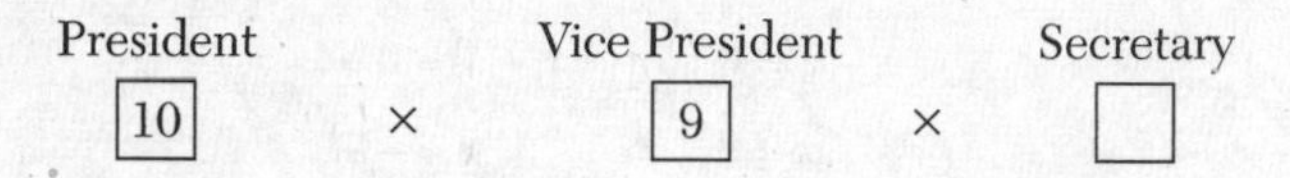

- After the positions of president and vice president are filled, any one of the remaining 8 members of the bowling team can fill the position of secretary:

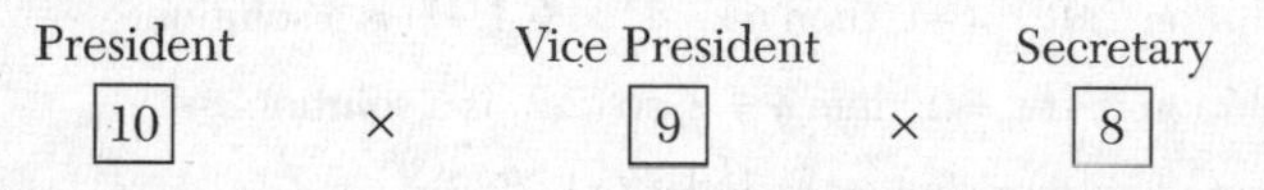

According to the multiplication principle of counting, the three officer positions can be filled in $10 \times 9 \times 8$ ways. The notation ${}_nP_r$ represents the permutation of n objects taken r at a time and is numerically equal to the product of the r greatest factors of n! Thus,

$$10 \times 9 \times 8 = {}_{10}P_3$$

The correct choice is **(3)**.

9. The volume, V, of a cube with an edge length of e is e^3:

$$V = e^3$$

$$= (1.5)^3$$

Use your calculator: $= 3.375$

The correct choice is **(2)**.

10. You are required to find the ordered pair that is a solution to the system of equations $y = x$ and $y = x^2 - 2$.

Method 1: Solve the linear-quadratic system of equations algebraically by replacing y with x in the quadratic equation:

$$x = x^2 - 2$$

Rewrite the quadratic equation so that 0 is on one side of the equation and all nonzero terms are on the other side. Then solve the quadratic equation by factoring the quadratic trinomial as the product of two binomials:

$$0 = x^2 - x - 2$$

$$0 = (x - 2)(x + 1)$$

$$x - 2 = 0 \quad \text{or} \quad x + 1 = 0$$

$$x = 2 \qquad\qquad x = -1$$

To find the corresponding values of y, use the equation $y = x$:

- Solution 1: If $x = -1$, then $y = -1$, so $(-1,-1)$ is a solution.
- Solution 2: If $x = 2$, then $y = 2$, so $(2,2)$ is a solution.

The set of answer choices includes only (2,2).

Method 2: Test each ordered pair in both equations until you find the ordered pair that satisfies the two equations at the same time:

- Choice (1): Test $(-2,-2)$.

$$y = x^2 - 2$$

$$-2 \;\boxed{?}\; (-2)^2 - 2$$

$$-2 \neq 4 - 2 \quad ✗$$

- Choice (2): Test $(-1,-1)$.

$$y = x$$

$$-1 \neq 1 \quad ✗$$

- Choice (3): Test $(0,0)$.

$$y = x^2 - 2$$

$$0 \;\boxed{?}\; (0)^2 - 2$$

$$0 \neq -2 \quad ✗$$

- Choice (4): Test (2,2).

$y = x$ $\qquad\qquad$ $y = x^2 - 2$

$2 = 2$ ✔ $\qquad\qquad$ $2 \overset{?}{=} (2)^2 - 2$

$2 = 4 - 2$ ✔

The correct choice is **(4)**.

11. The vertex of a parabola is its maximum or minimum point. The axis of symmetry is the line through the vertex that divides the parabola into two mirror image parts, as shown in the accompanying figure.

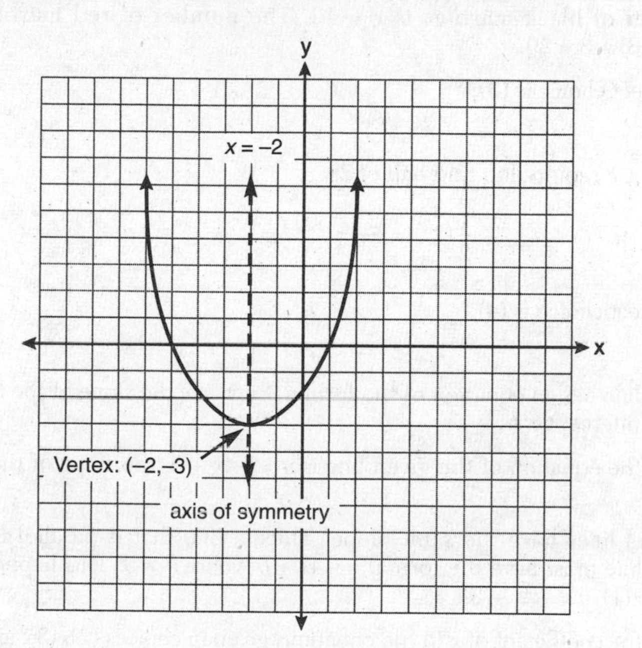

The vertex of the parabola is (–2,–3). The axis of symmetry is the vertical line through the vertex, an equation of which is $x = -2$.

The correct choice is **(1)**.

12. It is given that the number of red marbles Pam has is three more than twice the number of black marbles she has. If x represents the number of black marbles, then $2x + 3$ represents the number of red marbles. Since it is also given that she has 42 marbles in all:

$$x + (2x + 3) = 42$$

$$3x + 3 = 42$$

$$3x = 42 - 3$$

$$\frac{3x}{3} = \frac{39}{3}$$

The number of black marbles is $x = 13$. The number of red marbles is $2x + 3 = 2(13) + 3 = 29$.

The correct choice is **(3)**.

13. You are required to find half of 2^6:

$$\frac{1}{2}(2^6) = \frac{2^6}{2^1} = 2^{6-1} = 2^5$$

The correct choice is **(4)**.

14. If a line has an equation of the form $y = mx + b$, the slope of the line is m and its y-intercept is b.

- Since the equation of the given line is $y = -4x + 5$, the slope of the line is -4.
- Parallel lines have the same slope. Thus, a line that is parallel to the given line must have the form $y = -4x + b$ where $b \neq 5$. This happens in choice (1), $y = -4x + 3$.
- Since the coefficient of x in the equations given in choices (2), (3), and (4) are different than -4, their graphs cannot be parallel to the given line.

The correct choice is **(1)**.

15. To find the product of $\frac{x^2-1}{x+1}$ and $\frac{x+3}{3x-3}$, first factor where possible.

Factor the first numerator as the difference of two squares, and factor out the greatest common factor of 3 from the last denominator:

$$\left(\frac{x^2-1}{x+1}\right)\left(\frac{x+3}{3x-3}\right) = \frac{(x-1)(x+1)}{x+1} \cdot \frac{x+3}{3(x-1)}$$

Divide out any factor that appears in both a numerator and a denominator as their quotient is 1:

$$= \frac{\overset{1}{\cancel{(x-1)}}\overset{1}{\cancel{(x+1)}}}{\underset{1}{\cancel{x+1}}} \cdot \frac{x+3}{3\underset{1}{\cancel{(x-1)}}}$$

Multiply the numerators together, and multiply the denominators together:

$$= \frac{x+3}{3}$$

The correct choice is **(4)**.

16. It is given that the center pole of a tent is 8 feet long and a side of the tent is 12 feet long as shown in the accompanying figure:

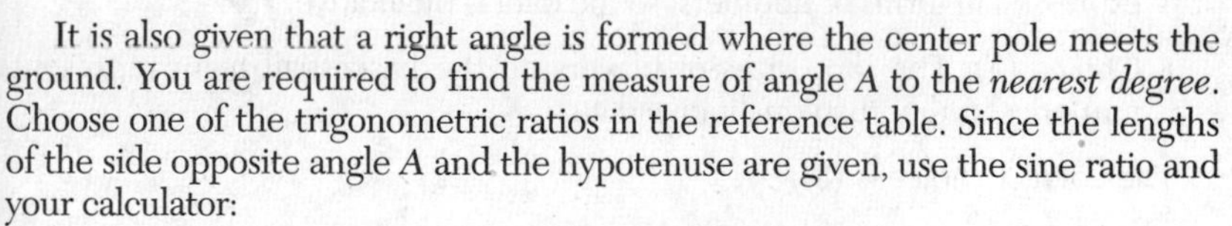

It is also given that a right angle is formed where the center pole meets the ground. You are required to find the measure of angle A to the *nearest degree*. Choose one of the trigonometric ratios in the reference table. Since the lengths of the side opposite angle A and the hypotenuse are given, use the sine ratio and your calculator:

$$\sin A = \frac{\text{Leg opposite } \angle A}{\text{Hypotenuse}}$$

$$= \frac{8}{12}$$

$$A = \sin^{-1}(8 \div 12)$$

$$= 41.8103149°$$

$$\approx 42°$$

The correct choice is **(2)**.

17. A fraction is undefined when its denominator equals 0. The given fraction, $\frac{x+4}{x-3}$, is undefined when $x - 3 = 0$. This happens when $x = 3$.

The correct choice is **(3)**.

18. Represent the given information using set notation.

- If A represents the set of integers greater than –2 and less than 6:

$$A = \{-1, 0, 1, 2, 3, 4, 5\}$$

- If B is the subset of A that consists of the positive factors of 5:

$$B = \{1, 5\}$$

- The complement of subset B consists of those elements in A but not in B:

$$\text{Complement of subset } B = \{-1, 0, \cancel{1}, 2, 3, 4, \cancel{5}\} = \{-1, 0, 2, 3, 4\}$$

The correct choice is **(2)**.

19. Qualitative data refers to nonnumerical data. Consider each of the answer choices.

- Choice (1): Elevations of the five highest mountains in the world represent numerical rather than qualitative data. ✗
- Choice (2): The ages of presidents at the time of their inauguration represent numerical rather than qualitative data. ✗
- Choice (3): The opinions of students regarding school lunches are not expressed in terms of numbers, so the data is qualitative. ✔
- Choice (4): The shoe sizes of players on the basketball team represent numerical rather than qualitative data. ✗

The correct choice is **(3)**.

20. To find the slope, m, of the line that passes through the points (–6,1) and (4,–4), use the slope formula found on your formula sheet.

$$m = \frac{y_2 - y_1}{x_2 - x_1}$$

If $(x_1,y_1) = (-6,1)$ and $(x_2,y_2) = (4,-4)$:

$$m = \frac{-4-1}{4-(-6)}$$

$$= \frac{-5}{4+6}$$

$$= \frac{-5}{10}$$

$$= -\frac{1}{2}$$

The correct choice is **(3)**.

21. It is given that students in a ninth grade class measured their heights, h, in centimeters. If the height of the shortest student was 155 cm and the height of the tallest student was 190 cm, then

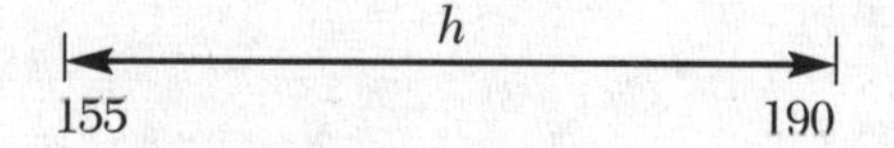

Thus, $155 \leq h$ and $h \leq 190$. As a result, h is sandwiched between 155 and 190. The two inequalities can be written in an equivalent form as the compound inequality $155 \leq h \leq 190$.

The correct choice is **(2)**.

22. To find the number of runners who are in their forties, use the data in the given cumulative frequency table to construct the corresponding frequency table. Then find the number of runners in the 40–49 age group.

Cumulative Frequency Table

Age Group (Cumulative)	Total (Cumulative Frequency)
20–29	8
20–39	18
20–49	25
20–59	31
20–69	35

Frequency Table

Age Group	Frequency
20–29	8
30–39	18 – 8 =10
40–49	**25 – 18 = 7**
50–59	31 – 25 = 6
60–69	35 – 31 = 4

In the accompanying table, the frequency for each age group interval after the first is obtained by taking the difference between the cumulative frequencies on the same line and the line immediately before it. Since 7 runners are in the interval 40–49, 7 runners are in their forties.

The correct choice is **(3)**.

23. It is given that Mr. Turner bought x boxes of pencils. If he left 3 boxes at home and took the rest to school, he took $x - 3$ boxes to school. Since it is also given that each box holds 25 pencils, the total number of pencils he took to school is represented by

$$25(x - 3) = 25x - 75$$

The correct choice is **(4)**.

24. To simplify the given expression, $\frac{2x^2 - 12x}{x - 6}$, factor the numerator. Then divide out any factor that is common to both the numerator and the denominator:

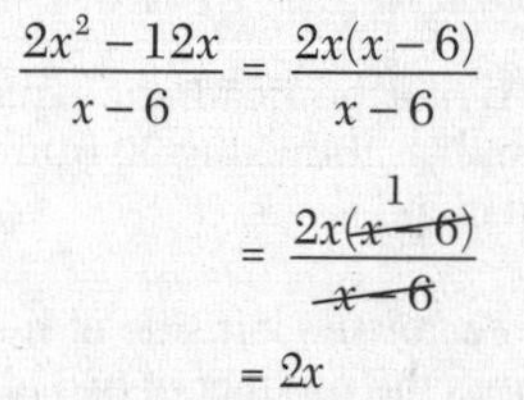

Factor out the greatest common factor (GCF) of $2x$ from the numerator: $\frac{2x^2 - 12x}{x - 6} = \frac{2x(x-6)}{x-6}$

Divide out the common factor of $x - 6$: $= \frac{2x\overset{1}{\cancel{(x-6)}}}{\cancel{x-6}}$

$= 2x$

The correct choice is **(2)**.

25. It is given that Don placed a ladder against the side of his house as shown in the accompanying diagram.

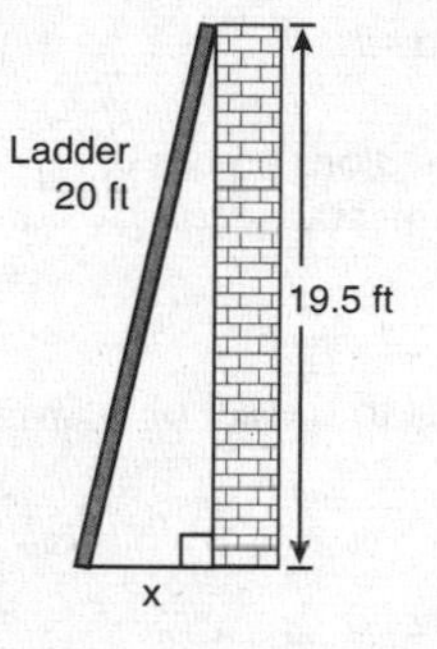

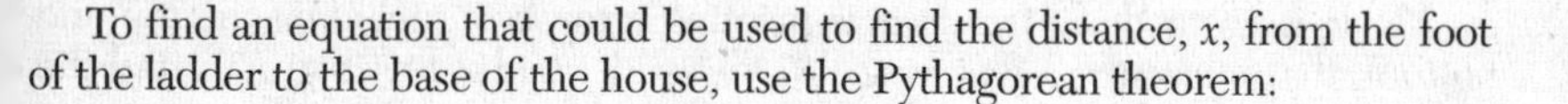

To find an equation that could be used to find the distance, x, from the foot of the ladder to the base of the house, use the Pythagorean theorem:

$$x^2 + (19.5)^2 = (20)^2$$

Solve for x^2: $x^2 = 20^2 - 19.5^2$

Solve for x: $\sqrt{x^2} = \sqrt{20^2 - 19.5^2}$

$$x = \sqrt{20^2 - 19.5^2}$$

The correct choice is **(3)**.

26. The given equation is $\frac{5}{x} = \frac{x+13}{6}$. In a proportion, the product of the means is equal to the product of the extremes:

$$x(x+13) = (5)(6)$$
$$x^2 + 13x = 30$$

Write the quadratic equation in standard form by subtracting 30 from both sides of the equation:

$$x^2 + 13x - 30 = 0$$

Factor the left side of the quadratic equation as the product of two binomials:

$$(x + ?)\,(x + ?) = 0$$

The missing terms in the binomial factors are the two integers whose product is −30, the last term of $x^2 + 13x - 30$, and whose sum is +13, the coefficient of the x-term. Since (+15)(−2) = −30 and (+15) + (−2) = +13, the missing terms are +15 and −2:

$$(x + 15)\,(x - 2) = 0$$

If the product of two expressions equals 0, then either expression may equal 0:

$$x + 15 = 0 \quad \text{or} \quad x - 2 = 0$$
$$x = -15 \quad \text{or} \quad x = 2$$

Compare the two solutions to the set of answer choices. The answer choices include −15 but not 2.

The correct choice is **(4)**.

27. It is given that Mrs. Ayer is painting a toy box, including the top and bottom, that measures 3 feet long, 1.5 feet wide, and 2 feet high. To find the total surface she will paint, use the formula for surface area, SA, of a rectangular prism found on your formula sheet:

$$\text{SA} = 2lw + 2hw + 2lh$$

where $l = 3$, $w = 1.5$, and $h = 2$. Thus:

$$\text{SA} = 2(3)(1.5) + 2(2)(1.5) + 2(3)(2)$$
$$= 9.0 + 6.0 + 12.0$$
$$= 27.0 \text{ ft}^2$$

The correct choice is **(4)**.

28. The given expression is: $\frac{\sqrt{32}}{4}$

To simplify the radical, factor the radicand into two positive integers such that one of these is the greatest perfect square factor of 32: $\frac{\sqrt{32}}{4} = \frac{\sqrt{16 \cdot 2}}{4}$

Write the radical over each factor of the radicand: $= \frac{\sqrt{16} \cdot \sqrt{2}}{4}$

Evaluate the square root of the perfect square factor: $= \frac{4\sqrt{2}}{4}$

Simplify: $= \frac{\overset{1}{\cancel{4}}\sqrt{2}}{\cancel{4}}$

$= \sqrt{2}$

The correct choice is **(1)**.

29. The graph of the equation $y = ax^2 + bx + c$ is a parabola provided $a \neq 0$. Multiplying a, the coefficient of the x^2-term, by 3 makes the new graph rise more rapidly over a given interval of x. The graph is steeper and, as a result, narrower than the original parabola.

You can easily verify this fact by graphing $y = x^2$ and $y = 3x^2$ on the same set of axes using your graphing calculator, as shown in the accompanying figure.

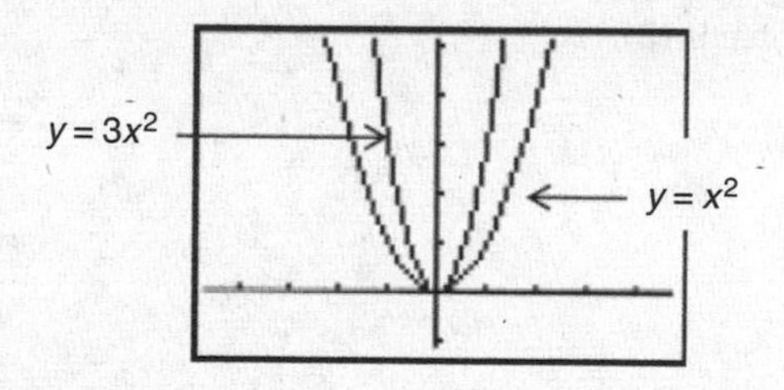

The correct choice is **(4)**.

30. It is given that Kathy plans to purchase a car that depreciates (loses value) at a rate of 14% per year. The initial cost of the car is $21,000. You are required to find an equation that represents the value, v, of the car after 3 years.

- At the end of the first year, the car depreciates or loses 14% of its original cost. Hence, at the end of the first year:

$$\begin{aligned} v &= 21{,}000 - (14\% \times 21{,}000) \\ &= 21{,}000 - (0.14 \times 21{,}000) \end{aligned}$$

- Factor out the GCF of 21,000:

$$\begin{aligned} &= 21{,}000(1 - 0.14) \\ &= \overbrace{21{,}000}^{\text{current value}} \times \underbrace{(0.86)}_{\text{depreciation factor}} \end{aligned}$$

- At the end of the second year, the value of the car is 0.86 times its value at the end of the first year:

$$\begin{aligned} v &= \overbrace{[21{,}000(0.86)]}^{\text{previous value of } v} \times 0.86 \\ &= 21{,}000(0.86)^2 \end{aligned}$$

- At the end of the third year, the pattern continues. The value of the car is again 0.86 times its previous value:

$$\begin{aligned} v &= [21{,}000(0.86)^2] \times 0.86 \\ &= 21{,}000(0.86)^3 \end{aligned}$$

The correct choice is **(2)**.

PART II

31. The greater the number of miles per gallon a vehicle can travel, the better the gas mileage of the vehicle.

- It is given that Tom drove 290 miles using 23.2 gallons of gasoline. Tom's vehicle's gas mileage:

$$\frac{290 \text{ miles}}{23.2 \text{ gallons}} = 12.5 \frac{\text{miles}}{\text{gallon}}$$

- It is also given that Ann drove 225 miles using 15 gallons of gasoline. Ann's vehicle's gas mileage:

$$\frac{225 \text{ miles}}{15 \text{ gallons}} = 15.0 \frac{\text{miles}}{\text{gallon}}$$

Ann's vehicle has better gas mileage since it gets more miles per gallon of gasoline.

32. It is given that a designer created the logo in the accompanying figure. The logo consists of a square and four quarter-circles of equal size.

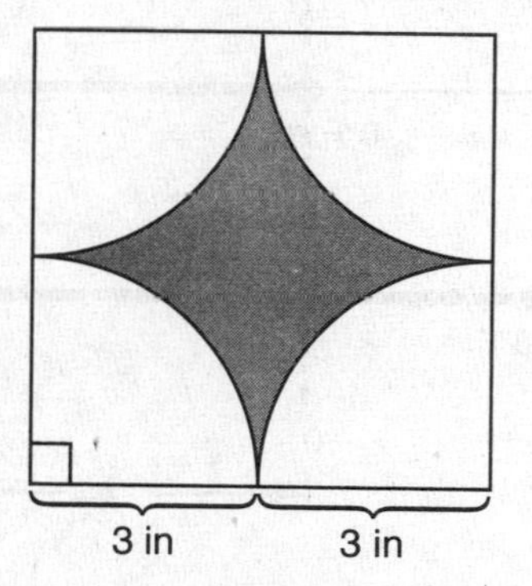

- Since a side of the square measures 3 in + 3 in = 6 in, first find the area of the square:

$$6 \text{ in} \times 6 \text{ in} = 36 \text{ in}^2$$

- Because each of the four quarter-circles have the same radius of 3 in, the sum of their areas is equivalent to the area of a whole circle with radius of 3 in. The area of a circle is $\pi(\text{radius})^2$. Find the sum of the areas of the four quarter-circles:

$$\pi 3^2 = 9\pi \text{ in}^2$$

- The area of the shaded region is the difference between the area of the square and the sum of the areas of the four quarter-circles:

$$\text{Area shaded region} = (36 - 9\pi) \text{ in}^2.$$

The area of the shaded region, expressed in terms of π, is **$36 - 9\pi$** square inches.

33. It is given that Maureen tracks the range of outdoor temperatures over three days and records the information in the three graphs that are given. The intersection of the three sets of temperatures is the set of temperatures that are common to all three graphs, as shown in the accompanying figure.

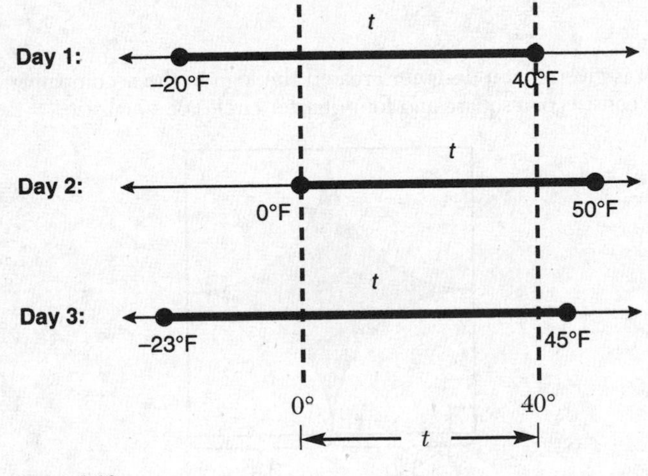

Thus, **$0 \leq t \leq 40$**.

PART III

34. It is given that Peter begins his kindergarten year able to spell 10 words and is going to learn to spell 2 new words every day.

- After d days, Peter will be able to spell a total of $10 + 2d$ words. The number of days until Peter is able to spell *at least* (greater than or equal to) 75 words is represented by the inequality:

$$\mathbf{10 + 2d \geq 75}$$

- To determine the minimum number of whole days until Peter is able to spell at least 75 words, solve $10 + 2d \geq 75$ for the least-possible integer value of d:

$$\begin{array}{rcl} 10 + 2d & \geq & 75 \\ -10 & & -10 \\ \hline 2d & \geq & 65 \end{array}$$

$$\frac{2d}{2} \geq \frac{65}{2}$$

$$d \geq 32.5$$

Since d must be a positive integer, round 32.5 up to the nearest integer.

Peter will take a minimum of **33** days to be able to spell at least 75 words.

35. It is given that the Hudson Record Store is having a going-out-of-business sale. CDs that normally sell for $18.00 will sell for $15.00 during the first week of the sale.

- The rate of discount is equal to the change in price compared with the original price:

$$\text{Rate of discount} = \frac{\text{Original price} - \text{Discounted price}}{\text{Original price}}$$

$$= \frac{18 - 15}{18}$$

$$= \frac{3}{18}$$

The rate of discount, written as a fraction, is $\mathbf{\frac{3}{18}}$.

- To express the rate of discount as a percent, use your calculator to divide 3 by 18 and then multiply the result by 100%:

$$\text{Percent of discount} = (3 \div 18) \times 100\%$$
$$= 16.666666667\%$$
$$\approx 16.67\%$$

The percent of discount is **16.67%** to the *nearest hundredth of a percent*.

- It is given that during the second week of the sale, the same CDs will be on sale for 25% off the *original* price. Find the amount of the discount:

$$0.25 \times \$18.00 = \$4.50$$

So the price of a CD will be \$18.00 – \$4.50 = \$13.50.

The price of a CD during the second week of the sale is **\$13.50**.

36. The graph of the quadratic equation $y = x^2 - 2x - 3$ is a parabola. The equation of the axis of symmetry of this parabola has the form $x = -\frac{b}{2a}$ where $a = 1$ and $b = -2$:

$$x = -\frac{(-2)}{2(1)}$$
$$= \frac{2}{2}$$
$$= 1$$

Hence, the x-coordinate of the vertex of the parabola is 1. To graph $y = x^2 - 2x - 3$:

- Prepare a table of values that includes three consecutive-integer x-values on either side of $x = 1$.

x	$x^2 - 2x - 3 = y$	(x,y)
–2	$(-2)^2 - 2(-2) - 3 = 5$	(–2,5)
–1	$(-1)^2 - 2(-1) - 3 = 0$	(–1,0)
0	$(0)^2 - 2(0) - 3 = -3$	(0,–3)
1	$(1)^2 - 2(1) - 3 = -4$	(1,–4)
2	$(2)^2 - 2(2) - 3 = -3$	(2,–3)
3	$(3)^2 - 2(3) - 3 = 0$	(3,0)
4	$(4)^2 - 2(4) - 3 = 5$	(4,5)

TIP: Use the table feature of your graphing calculator to create a table of values.

- Check for symmetry in the y-values listed in the table. Corresponding pairs of points on either side of $x = 1$ have matching y-coordinates. This confirms that (1,–4) is the vertex of the parabola.
- Plot (–2,5), (–1,0), (0,–3), (1,–4), (2,–3), (3,0), and (4,5) on the grid provided. Then connect these points with a smooth, U-shaped curve, as shown in the accompanying figure. Label the graph with its equation.

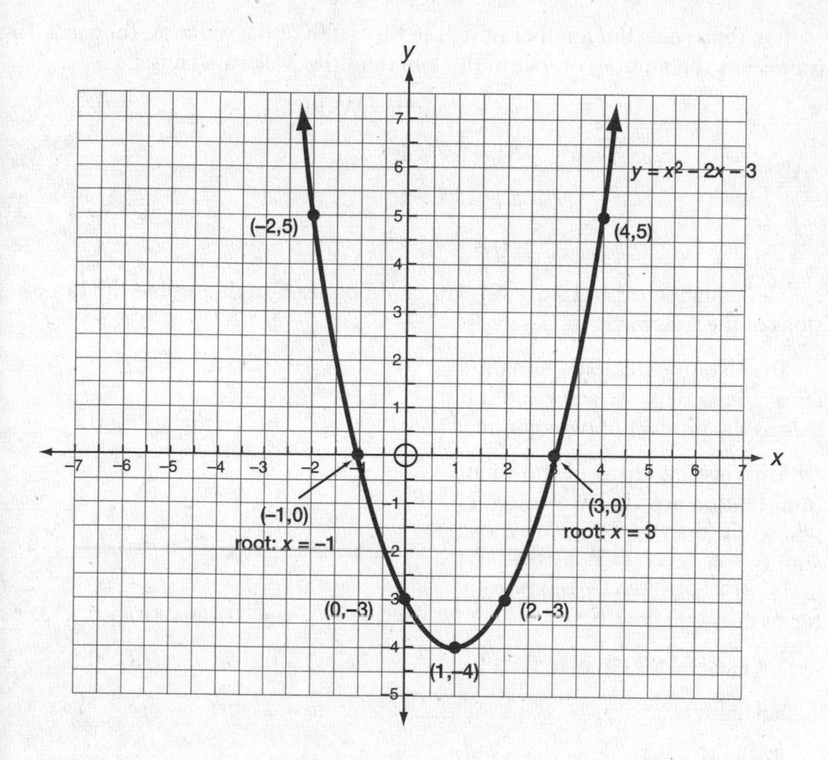

The roots of the equation $x^2 - 2x - 3 = 0$ are the x-coordinates of the points at which the graph of $y = x^2 - 2x - 3$ intersects the x-axis since at each of these intercepts, $y - 0$. Thus, the roots of $x^2 - 2x - 3 = 0$ arc **–1** and **3**.

PART IV

37. It is given that a contractor needs 54 square feet of brick to construct a rectangular walkway and that the length of the walkway is 15 feet more than the width.

If w represents the number of feet in the width of the walkway, then $w + 15$ represents the number of feet in the length of the walkway. Thus:

$$\text{Area} = \text{Length} \times \text{Width}$$
$$54 = (w + 15) \times w$$
$$54 = w^2 + 15w$$
$$0 = w^2 + 15w - 54$$

The equation $\mathbf{w^2 + 15w - 54 = 0}$ could be used to determine the dimensions of the walkway.

To solve the quadratic equation $w^2 + 15w - 54 = 0$, factor the left side as the product of two binomials: $(w + ?)(w + ?) = 0$

The missing terms of the binomial factors are the two integers whose product is -54 and whose sum is $+15$. Since $(-3)(+18) = -54$ and $(-3) + (+18) = +15$, the missing terms are -3 and $+18$: $(w - 3)(w + 18) = 0$

Set each factor equal to 0: $w - 3 = 0$ or $w + 18 = 0$

Solve for w: $w = 3$ or $w = -18$

Reject $w = -18$ since the width must be positive. Now determine the length of the walkway:

$$l = w + 15$$
$$= 3 + 15$$
$$= 18$$

The length of the walkway is **18 feet**, and the width of the walkway is **3 feet**.

38. It is given that Sophie measured a piece of paper to be 21.7 cm by 28.5 cm and that the piece of paper is actually 21.6 cm by 28.4 cm.

- Using Sophie's measurements and assuming the piece of paper is rectangular, find the area of the piece of paper Sophie measured:

$$21.7 \text{ cm} \times 28.5 \text{ cm} = \mathbf{618.45} \text{ cm}^2$$

- Find the actual area of the piece of paper:

$$21.6 \text{ cm} \times 28.4 \text{ cm} = \mathbf{613.44} \text{ cm}^2$$

- The relative error in calculating the area of the piece of paper is the difference in the measured and actual areas divided by the actual area:

$$\text{Relative error} = \frac{\text{Measured area} - \text{Actual area}}{\text{Actual area}}$$

$$= \frac{618.45 - 613.44}{613.44}$$

$$= \frac{5.01}{613.44}$$

$$= 0.0081670579$$

$$\approx \mathbf{0.008}$$

Sophie is correct in thinking that **the amount of error is *not* significant since the error is less than 1%**. This error in calculating the area of a notebook-sized piece of paper should be acceptable for most practical purposes. On the other hand, an error of this magnitude might not be acceptable in critical engineering applications, such as when designing computer circuits.

39. The prices of seven race cars sold last week are listed in the accompanying table:

Price per Race Car	Number of Race Cars
$126,000	1
$140,000	2
$180,000	1
$400,000	2
$819,000	1

- The mean or average value of these race cars is the sum of their values divided by the number of values in the sum:

$$\text{Mean} = \frac{(1\times 126{,}000)+(2\times 140{,}000)+(1\times 180{,}000)+(2\times 400{,}000)+(1\times 819{,}000)}{1+2+1+2+1}$$

$$= \frac{2{,}205{,}000}{7}$$

$$= 315{,}000$$

The mean value of the seven race cars is **$315,000**.

- The median value is the middle value when the values in a list are arranged in size order. In an ordered list of 7 values, as contained in the given table, the mean is the fourth value—$180,000.

The median race car value is **$180,000**.

- One of the race car values, $819,000, differs greatly from each of the other race car values. Such a value is called an outlier. When a relatively small set of data values, 7 in this case, includes an outlier, the mean becomes distorted. Thus, the mean race car value does not accurately represent the value of the seven race cars. On the other hand, the median is not affected by an outlier since the median depends on the relative positions of the data values rather than on their actual sizes.

The median is the best measure of central tendency because the outlier car price of $819,000 distorts the mean but does not affect the median.

Topic	Question Numbers	Number of Points	Your Points	Your Percentage
1. Sets and Numbers; Intersection and Complements of Sets; Interval Notation; Properties of Real Numbers	17, 18, 33	2 + 2 + 2 = 6		
2. Operations on Rat'l. Numbers & Monomials	—	—		
3. Laws of Exponents for Integer Exponents; Scientific Notation	13	2		
4. Operations on Polynomials	7, 15	2 + 2 = 4		
5. Square Root; Operations with Radicals	28	2		
6. Evaluating Formulas & Algebraic Expressions	23	2		
7. Solving Linear Eqs. & Inequalities	12, 34	2 + 3 = 5		
8. Solving Literal Eqs. & Formulas for a Given Letter	—	—		
9. Alg. Operations (including factoring)	4, 24	2 + 2 = 4		
10. Quadratic Equations (incl. alg. and graphical solutions; parabolas)	11, 29, 36, 37	2 + 2 + 3 + 4 = 11		
11. Coordinate Geometry (eq. of a line; graphs of linear eqs.; slope)	14, 20	2 + 2 = 4		
12. Systems of Linear Eqs. & Inequalities (algebraic & graphical solutions)	6	2		
13. Mathematical Modeling (using eqs.; tables; graphs)	21	2		
14. Linear-Quadratic Systems	10	2		
15. Perimeter; Circumference; Area of Common Figures	—	—		
16. Volume and Surface Area; Area of Overlapping Figures; Relative Error in Measurement	9, 27, 32, 38	2 + 2 + 2 + 4 = 10		
17. Fractions and Percent	35	3		
18. Ratio & Proportion (incl. similar polygons, scale drawings, & rates)	26, 31	2 + 2 = 4		
19. Pythagorean Theorem	25	2		
20. Right Triangle Trigonometry	16	2		
21. Functions (def.; domain and range; vertical line test; absolute value)	1	2		

Topic	Question Numbers	Number of Points	Your Points	Your Percentage
22. Exponential Functions (properties; growth and decay)	30	2		
23. Probability (incl. tree diagrams & sample spaces)	2	2		
24. Permutations and Counting Methods (incl. Venn diagrams)	8	2		
25. Statistics (mean, median, percentiles, quartiles; freq. dist., histograms; box-and-whisker plots; causality, bivariate data; qualitative vs. quantitative data; unbiased vs. biased samples; circle graphs)	3, 19, 22, 39	2 + 2 + 2 +4 = 10		
26. Line of Best Fit (including linear regression, scatter plots, and linear correlation)	5	2		
27. Nonroutine Word Problems Requiring Arith. or Alg. Reasoning	—	—		

MAP TO LEARNING STANDARDS

The table below shows which content strand each item is aligned to. The numbers in the table represent the question numbers on the test.

Key Ideas	Item Numbers
Number Sense and Operations	8, 28, 35
Algebra	4, 6, 7, 10, 12, 13, 14, 15, 16, 17, 18, 20, 21, 23, 24, 25, 26, 30, 33, 34, 37
Geometry	1, 9, 11, 27, 29, 32, 36
Measurement	31, 38
Probability and Statistics	2, 3, 5, 19, 22, 39

HOW TO CONVERT YOUR RAW SCORE TO YOUR INTEGRATED ALGEBRA REGENTS EXAMINATION SCORE

Below is the conversion chart that must be used to determine your final score on the June 2008 Regents Examination in Integrated Algebra. To find your final exam score, locate in the column labeled "Raw Score" the total number of points you scored out of a possible 87 points. Since partial credit is allowed in Parts II, III, and IV of the test, you may need to approximate the credit you would receive for a solution that is not completely correct. Then locate in the adjacent column to the right the scale score that corresponds to your raw score. The scale score is your final Integrated Algebra Regents Examination score.

Raw Score	Scaled Score	Raw Score	Scaled Score	Raw Score	Scaled Score
87	100	57	82	27	61
86	99	56	82	26	60
85	98	55	81	25	59
84	97	54	81	24	57
83	96	53	80	23	56
82	95	52	80	22	54
81	94	51	80	21	53
80	93	50	79	20	51
79	93	49	79	19	49
78	92	48	78	18	47
77	91	47	78	17	45
76	91	46	77	16	44
75	90	45	77	15	41
74	89	44	76	14	39
73	89	43	76	13	37
72	88	42	75	12	35
71	88	41	74	11	32
70	87	40	74	10	30
69	87	39	73	9	27
68	86	38	72	8	25
67	86	37	72	7	22
66	86	36	71	6	19
65	85	35	70	5	16
64	84	34	69	4	13
63	84	33	68	3	10
62	84	32	67	2	7
61	83	31	66	1	3
60	83	30	65	0	0
59	83	29	64		
58	82	28	63		

Examination August 2008

Integrated Algebra

FORMULAS

Trigonometric ratio $\sin A = \frac{\text{opposite}}{\text{hypotenuse}}$

$\cos A = \frac{\text{adjacent}}{\text{hypotenuse}}$

$\tan A = \frac{\text{opposite}}{\text{adjacent}}$

Area Trapezoid $A = \frac{1}{2}h(b_1 + b_2)$

Volume Cylinder $V = \pi r^2 h$

Surface area Rectangular prism $SA = 2lw + 2hw + 2lh$

Cylinder $SA = 2\pi r^2 + 2\pi rh$

Coordinate geometry $m = \frac{\Delta y}{\Delta x} = \frac{y_2 - y_1}{x_2 - x_1}$

PART I

Answer all questions in this part. Each correct answer will receive 2 credits. No partial credit will be allowed. For each question, write in the space provided the numeral preceding the word or expression that best completes the statement or answers the question. [60]

1 Which value of p is the solution of $5p - 1 = 2p + 20$?

(1) $\frac{19}{7}$
(2) $\frac{19}{3}$
(3) 3
(4) 7

1 ____

2 The statement $2 + 0 = 2$ is an example of the use of which property of real numbers?

(1) associative
(2) additive identity
(3) additive inverse
(4) distributive

2 ____

3 Mrs. Smith wrote “Eight less than three times a number is greater than fifteen” on the board. If x represents the number, which inequality is a correct translation of this statement?

(1) $3x - 8 > 15$
(2) $3x - 8 < 15$
(3) $8 - 3x > 15$
(4) $8 - 3x < 15$

3 ____

4 Which statement is true about the data set 3, 4, 5, 6, 7, 7, 10?

(1) mean = mode
(2) mean > mode
(3) mean = median
(4) mean < median

4 _____

5 Which value of x is in the solution set of the inequality $-4x + 2 > 10$?

(1) -2
(2) 2
(3) 3
(4) -4

5 _____

6 Factored completely, the expression $2x^2 + 10x - 12$ is equivalent to

(1) $2(x - 6)(x + 1)$
(2) $2(x + 6)(x - 1)$
(3) $2(x + 2)(x + 3)$
(4) $2(x - 2)(x - 3)$

6 _____

7 The gas tank in a car holds a total of 16 gallons of gas. The car travels 75 miles on 4 gallons of gas. If the gas tank is full at the beginning of a trip, which graph represents the rate of change in the amount of gas in the tank?

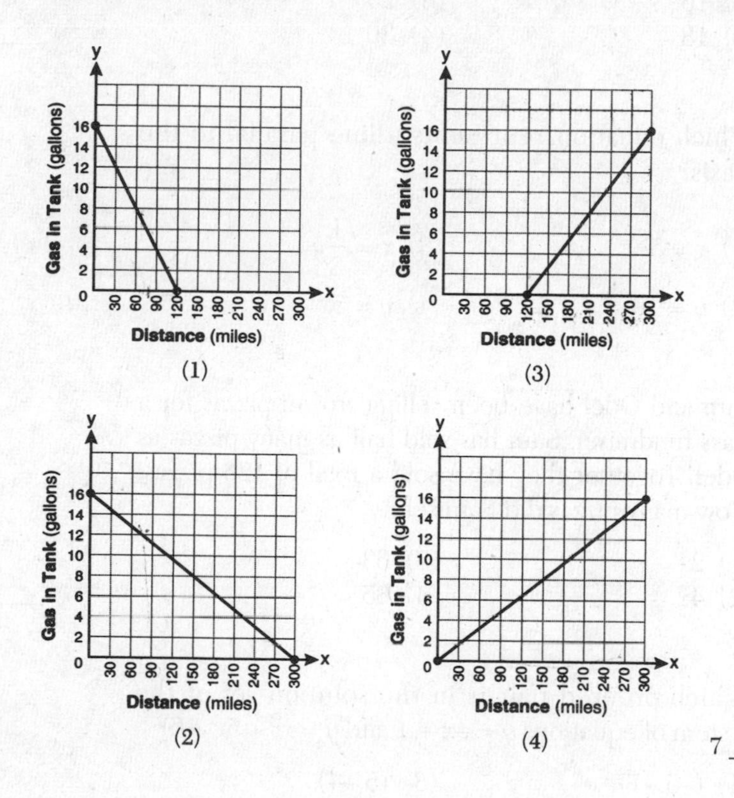

7 ______

8 If $3ax + b = c$, then x equals

(1) $c - b + 3a$

(2) $c + b - 3a$

(3) $\frac{c-b}{3a}$

(4) $\frac{b-c}{3a}$

8 ______

9 The length of the hypotenuse of a right triangle is 34 inches and the length of one of its legs is 16 inches. What is the length, in inches, of the other leg of this right triangle?

(1) 16
(2) 18
(3) 25
(4) 30

9 _____

10 Which equation represents a line parallel to the x-axis?

(1) $x = 5$
(2) $y = 10$
(3) $x = \frac{1}{3}y$
(4) $y = 5x + 17$

10 _____

11 Sam and Odel have been selling frozen pizzas for a class fundraiser. Sam has sold half as many pizzas as Odel. Together they have sold a total of 126 pizzas. How many pizzas did Sam sell?

(1) 21
(2) 42
(3) 63
(4) 85

11 _____

12 Which ordered pair is in the solution set of the system of equations $y = -x + 1$ and $y = x^2 + 5x + 6$?

(1) $(-5,-1)$
(2) $(-5,6)$
(3) $(5,-4)$
(4) $(5,2)$

12 _____

13 A swim team member performs a dive from a 14-foot-high springboard. The parabola below shows the path of her dive.

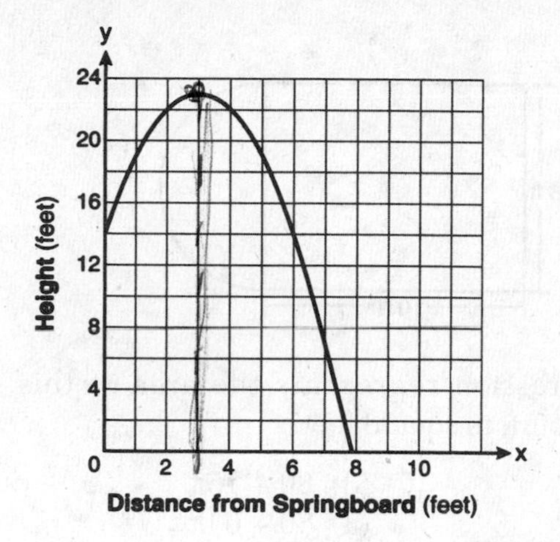

Which equation represents the axis of symmetry?

(1) $x = 3$ (3) $x = 23$
(2) $y = 3$ (4) $y = 23$ 13 ____

14 Nicole's aerobics class exercises to fast-paced music. If the rate of the music is 120 beats per minute, how many beats would there be in a class that is 0.75 hour long?

(1) 90 (3) 5,400
(2) 160 (4) 7,200 14 ____

15 Luis is going to paint a basketball court on his driveway, as shown in the diagram below. This basketball court consists of a rectangle and a semicircle.

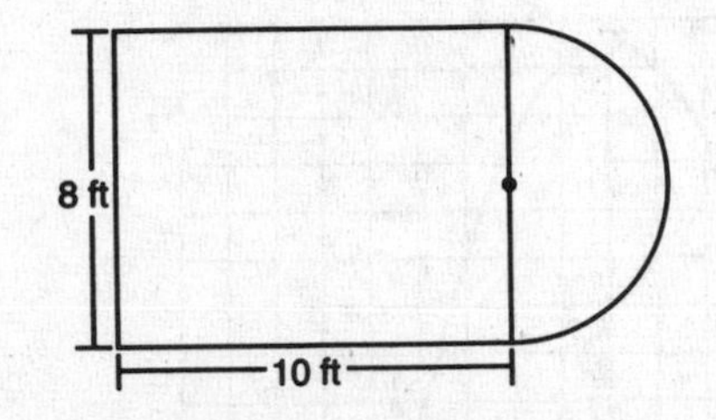

Which expression represents the area of this basketball court, in square feet?

(1) 80
(2) $80 + 8\pi$
(3) $80 + 16\pi$
(4) $80 + 64\pi$

15 _____

16 John is going to line up his four golf trophies on a shelf in his bedroom. How many different possible arrangements can he make?

(1) 24
(2) 16
(3) 10
(4) 4

16 _____

17 A rectangle has an area of 24 square units. The width is 5 units less than the length. What is the length, in units, of the rectangle?

(1) 6
(2) 8
(3) 3
(4) 19

17 _____

18 What is the value of the third quartile shown on the box-and-whisker plot below?

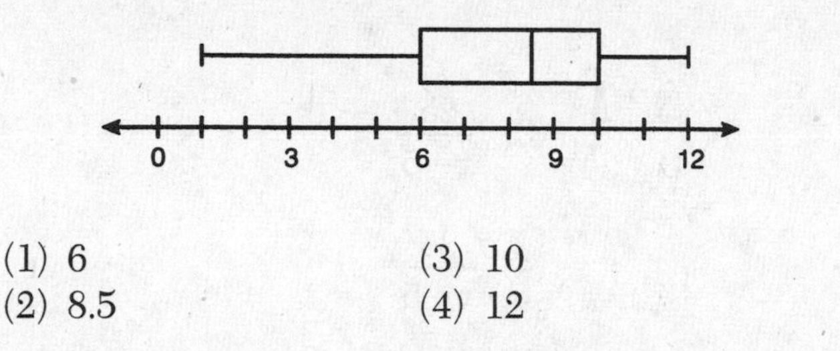

(1) 6
(2) 8.5
(3) 10
(4) 12

18 ______

19 When $3y^2 - 4g + 2$ is subtracted from $7g^2 + 5g - 1$, the difference is

(1) $-4g^2 - 9g + 3$
(2) $4g^2 + g + 1$
(3) $4g^2 + 9g - 3$
(4) $10g^2 + g + 1$

19 ______

20 Which value of x is the solution of $\frac{2x}{5} + \frac{1}{3} = \frac{7x-2}{15}$?

(1) $\frac{3}{5}$
(2) $\frac{31}{26}$
(3) 3
(4) 7

20 ______

21 Which expression represents $\frac{25x-125}{x^2-25}$ in simplest form?

(1) $\frac{5}{x}$

(2) $\frac{-5}{x}$

(3) $\frac{25}{x-5}$

(4) $\frac{25}{x+5}$

21 _____

22 The table below shows a cumulative frequency distribution of babysitting wages.

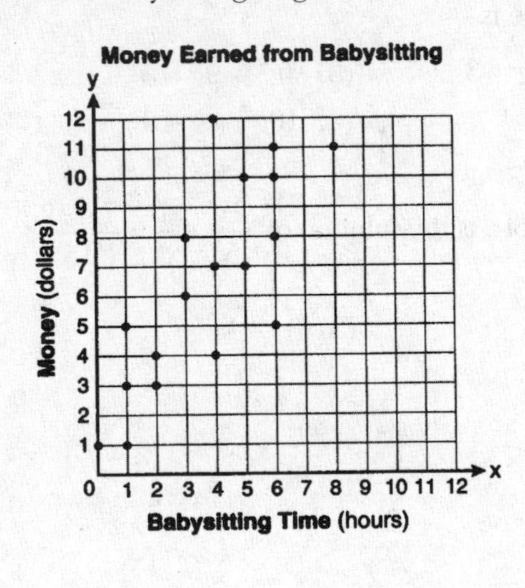

(1) $y = x$

(2) $y = \frac{2}{3}x + 1$

(3) $y = \frac{3}{2}x + 4$

(4) $y = \frac{3}{2}x + 1$

22 _____

23 In a linear equation, the independent variable increases at a constant rate while the dependent variable decreases at a constant rate. The slope of this line is

(1) zero
(2) negative
(3) positive
(4) undefined

23 _____

24 Which equation could be used to find the measure of one acute angle in the right triangle shown below?

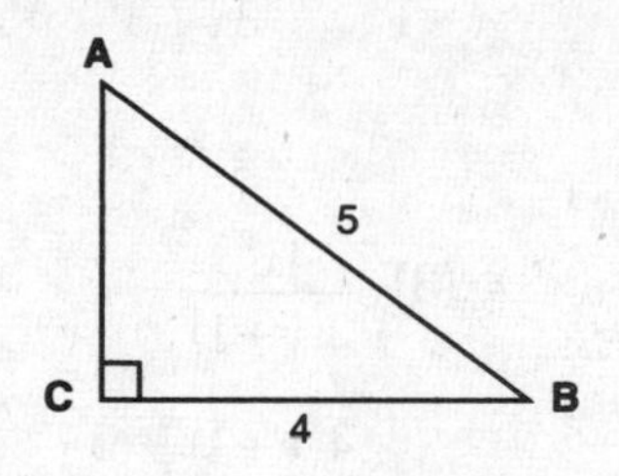

(1) $\sin A = \frac{4}{5}$
(2) $\tan A = \frac{5}{4}$
(3) $\cos B = \frac{5}{4}$
(4) $\tan B = \frac{4}{5}$

24 _____

25 Which ordered pair is in the solution set of the following system of inequalities?

$$y < \frac{1}{2}x + 4$$

$$y \geq -x + 1$$

(1) (–5,3)
(2) (0,4)
(3) (3,–5)
(4) (4,0)

25 ______

26 What is the product of $\frac{4x}{x-1}$ and $\frac{x^2-1}{3x+3}$ expressed in simplest form?

(1) $\frac{4x}{3}$
(2) $\frac{4x^2}{3}$
(3) $\frac{4x^2}{3(x+1)}$
(4) $\frac{4(x+1)}{3}$

26 ______

27 Which expression is equivalent to $(3x^2)^3$?

(1) $9x^5$
(2) $9x^6$
(3) $27x^5$
(4) $27x^6$

27 ______

28 Ryan estimates the measurement of the volume of a popcorn container to be 282 cubic inches. The actual volume of the popcorn container is 289 cubic inches. What is the relative error of Ryan's measurement to the *nearest thousandth*?

(1) 0.024
(2) 0.025
(3) 0.096
(4) 1.025

28 ______

29 In the diagram of $\triangle ABC$ shown below, $BC = 10$ and $AB = 16$.

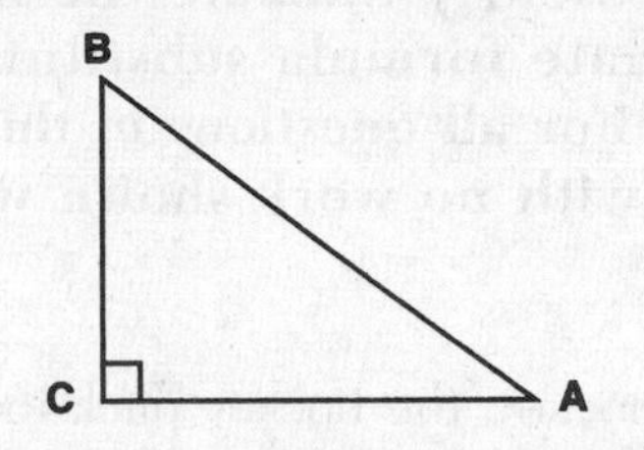

To the *nearest tenth of a degree*, what is the measure of the largest acute angle in the triangle?

(1) 32.0
(2) 38.7
(3) 51.3
(4) 90.0

29 ____

30 The faces of a cube are numbered from 1 to 6. If the cube is tossed once, what is the probability that a prime number or a number divisible by 2 is obtained?

(1) $\frac{6}{6}$
(2) $\frac{5}{6}$
(3) $\frac{4}{6}$
(4) $\frac{1}{6}$

30 ____

PART II

Answer all questions in this part. Each correct answer will receive 2 credits. Clearly indicate the necessary steps, including appropriate formula substitutions, diagrams, graphs, charts, etc. For all questions in this part, a correct numerical answer with no work shown will receive only 1 credit. [6]

31 In a game of ice hockey, the hockey puck took 0.8 second to travel 89 feet to the goal line. Determine the average speed of the puck in feet per second.

32 Brianna is using the two spinners shown below to play her new board game. She spins the arrow on each spinner once. Brianna uses the first spinner to determine how many spaces to move. She uses the second spinner to determine whether her move from the first spinner will be forward or backward.

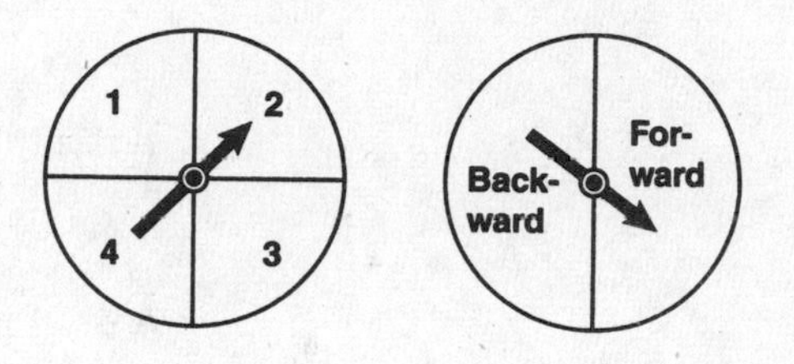

Find the probability that Brianna will move *fewer than* four spaces and *backward*.

33 Twelve players make up a high school basketball team. The team jerseys are numbered 1 through 12. The players wearing the jerseys numbered 3, 6, 7, 8, and 11 are the only players who start a game. Using set notation, list the complement of this subset.

PART III

Answer all questions in this part. Each correct answer will receive 3 credits. Clearly indicate the necessary steps, including appropriate formula substitutions, diagrams, graphs, charts, etc. For all questions in this part, a correct numerical answer with no work shown will receive only 1 credit. [9]

34 Express the product of $3\sqrt{20}\,(2\sqrt{5} - 7)$ in simplest radical form.

35 On the set of axes below, draw the graph of $y = 2^x$ over the interval $-1 \leq x \leq 3$. Will this graph ever intersect the x-axis? Justify your answer.

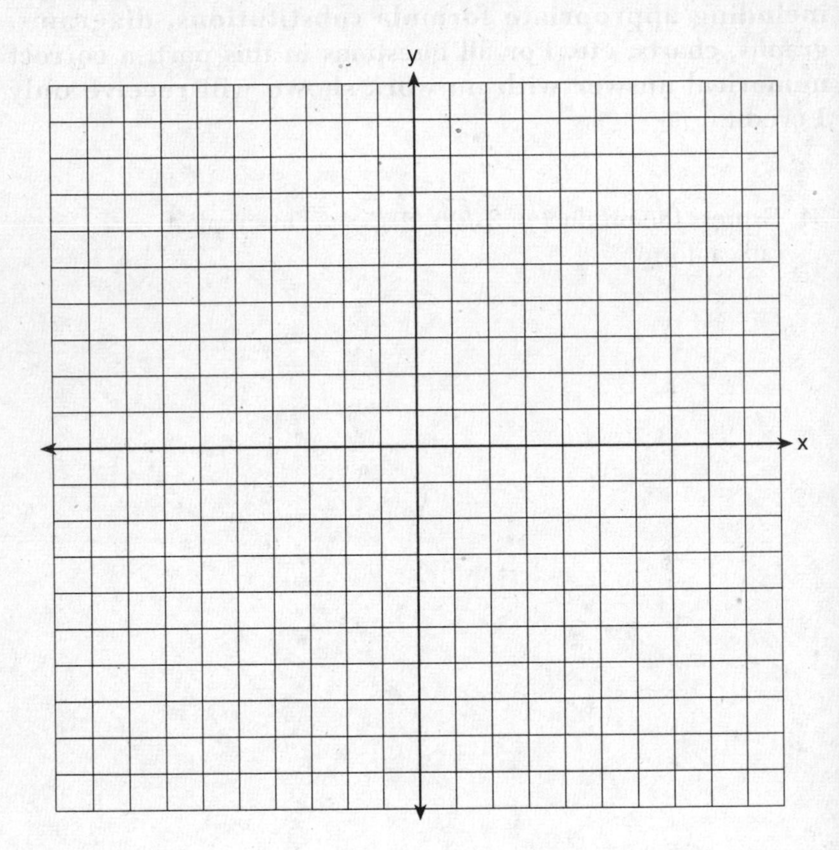

36 Write an equation that represents the line that passes through the points (5,4) and (–5,0).

PART IV

Answer all questions in this part. Each correct answer will receive 4 credits. Clearly indicate the necessary steps, including appropriate formula substitutions, diagrams, graphs, charts, etc. For all questions in this part, a correct numerical answer with no work shown will receive only 1 credit. [12]

37 The cost of 3 markers and 2 pencils is $1.80. The cost of 4 markers and 6 pencils is $2.90. What is the cost of *each* item? Include appropriate units in your answer.

38 Twenty students were surveyed about the number of days they played outside in one week. The results of this survey are shown below.

{6, 5, 4, 3, 0, 7, 1, 5, 4, 4, 3, 2, 2, 3, 2, 4, 3, 4, 0, 7}

Complete the frequency table below for these data.

Number of Days Outside

Interval	Tally	Frequency
0–1		
2–3		
4–5		
6–7		

Complete the cumulative frequency table below using these data.

Number of Days Outside

Interval	Cumulative Frequency
0–1	
0–3	
0–5	
0–7	

This question continued on the next page.

Question 38 continued

On the grid below, create a cumulative frequency histogram based on the table you made on the previous page.

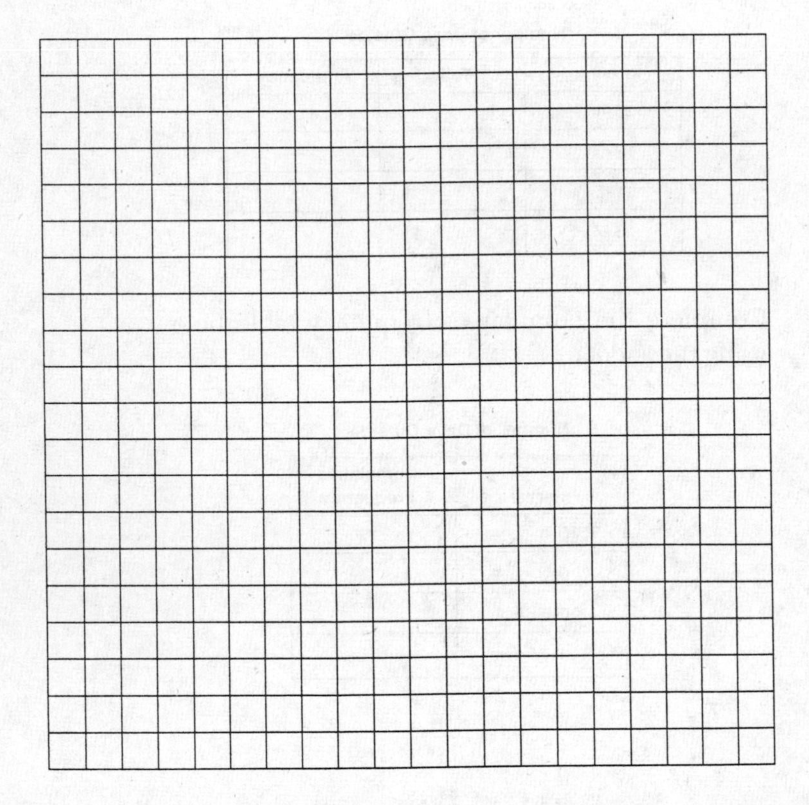

39 On the set of axes below, solve the following system of equations graphically and state the coordinates of all points in the solution set.

$$y = x^2 + 4x - 5$$

$$y = x - 1$$

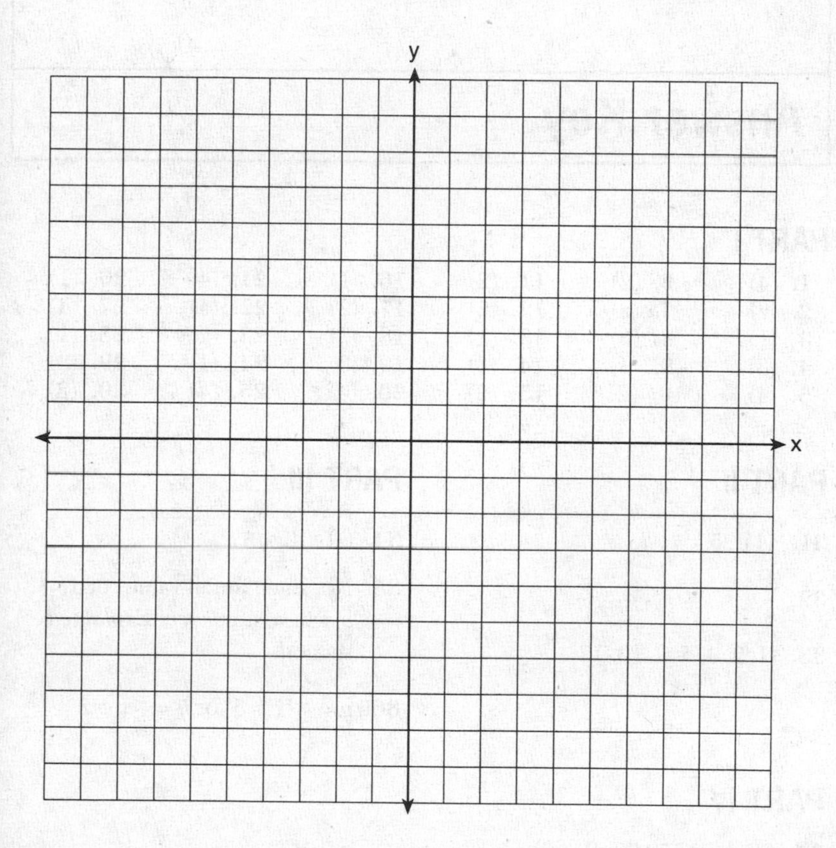

Answers August 2008

Integrated Algebra

Answer Key

PART I

1. (4)	**6.** (2)	**11.** (2)	**16.** (1)	**21.** (4)	**26.** (1)
2. (2)	**7.** (2)	**12.** (2)	**17.** (2)	**22.** (4)	**27.** (4)
3. (1)	**8.** (3)	**13.** (1)	**18.** (3)	**23.** (2)	**28.** (1)
4. (3)	**9.** (4)	**14.** (3)	**19.** (3)	**24.** (1)	**29.** (3)
5. (4)	**10.** (2)	**15.** (2)	**20.** (4)	**25.** (4)	**30.** (2)

PART II

31. 111.25

32. $\frac{3}{8}$

33. {1, 2, 4, 5, 9, 10, 12}

PART III

34. $60 - 42\sqrt{5}$

35. No, since for all values of x, $y > 0$; see *Answers Explained* section.

36. $y = \frac{2}{5}(x + 5)$ or $y = \frac{2}{5}x + 2$

PART IV

37. Marker = $0.50; Pencil = $0.15

38. See *Answers Explained* section.

39. (–4,–5) and (1,0); see *Answers Explained* section.

In **PARTS II–IV** you are required to show how you arrived at your answers. For sample methods of solutions, see the *Answers Explained* section.

Answers Explained

PART I

1. The given equation is: $5p - 1 = 2p + 20$

Add 1 to each side of the equation:

$$\begin{array}{r} +1 = \quad +1 \\ \hline 5p + 0 = 2p + 21 \end{array}$$

Subtract $2p$ from both sides of the equation:

$$\begin{array}{r} -2p = -2p \\ \hline 3p = 0 + 21 \end{array}$$

Divide each side of the equation by 3:

$$\frac{\overset{1}{\cancel{3}}p}{\cancel{3}} = \frac{21}{3}$$

$$p = 7$$

The correct choice is **(4)**.

2. The number 0 is the additive identity element for the set of real numbers since the sum of any real number and 0 is that same real number. Thus, the given statement, $2 + 0 = 2$, is an example of the additive identity property for real numbers.

The correct choice is **(2)**.

3. It is given that Mrs. Smith wrote "Eight less than three times a number is greater than fifteen" on the board. If x represents the number, then

$$\overbrace{\text{Eight less than three times a number}}^{3x-8}\ \underbrace{\text{is greater than}}_{>}\ \overbrace{\text{fifteen}}^{15}$$

$$3x - 8 > 15$$

The correct choice is **(1)**.

4. The given data set consists of 3, 4, 5, 6, 7, 7, and 10. Find the mean, median, and mode.

- Mean $= \dfrac{\text{Sum of data values}}{\text{Number of data values}}$

$$= \frac{3+4+5+6+7+7+10}{7}$$

$$= \frac{42}{7}$$

$$= 6$$

- The median is the middle value in the list when the numbers are arranged in size order:

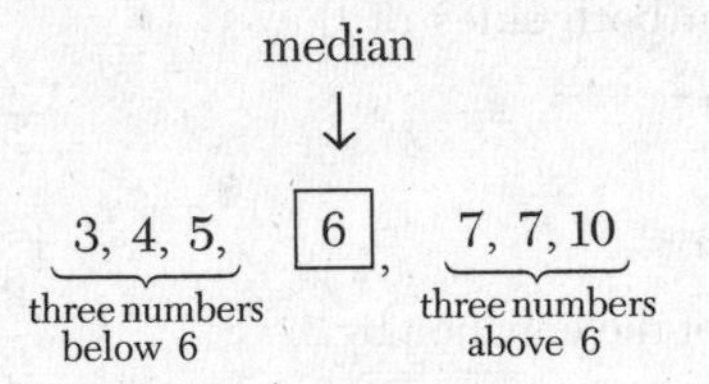

- The mode is the value that occurs the most often in the data set. For the given data set, the mode is 7.

Examine each of the answer choices in turn:

- Choice (1): mean = mode? Since 6 does not equal 7, this statement is false. ✗
- Choice (2): mean > mode? Since 6 is not greater than 7, this statement is false. ✗
- Choice (3): mean = median? Because the mean and median are each 6, this statement is true. ✔
- Choice (4): mean < median? Since 6 is not less than 6, this statement is false. ✗

The correct choice is **(3)**.

5. The given inequality is: $-4x + 2 > 10$

Subtract 2 from each side of the inequality:

$$\begin{array}{r} -4x + 2 > 10 \\ \underline{\quad -2 = -2} \\ -4x + 0 > 8 \end{array}$$

Divide each side of the inequality by −2, and reverse the direction of the inequality:

$$\frac{-4x}{-4} \boxed{<} \frac{8}{-4}$$

$$x < -2$$

Examine each of the answer choices in turn until you find the value that is less than −2. In this case, that answer choice is −4.

The correct choice is **(4)**.

6. The given expression is: $2x^2 + 10x - 12$

Factor out the greatest common factor (GCF) of 2 for each term: $2(x^2 + 5x - 6)$

Factor the quadratic trinomial as the product of two binomials: $2(x + ?)(x + ?)$

The missing terms of the binomial factors are the two integers whose product is −6, the constant term of $x^2 + 5x - 6$, and whose sum is +5, the coefficient of the x-term of $x^2 + 5x - 6$. Since $(+6) + (-1) = +5$ and $(+6)(-1) = -6$, the missing terms are +6 and −1: $2(x + 6)(x - 1)$

The correct choice is **(2)**.

7. It is given that the gas tank holds a total of 16 gallons of gas and that the car travels 75 miles on 4 gallons of gas. Assuming that the rate is constant and the gas tank is full, you are required to identify the graph that represents the rate of change in the amount of gas in the tank.

- For each graph, the rate of change in the amount of gas is the slope of the line. Since the distance traveled increases as the number of gallons of gas left in the tank decreases, the slope of the line must be negative. Thus, you can eliminate the graphs in choices (3) and (4) because these lines have a positive slope.
- The rate of change in the amount of gas in the tank is given as $-\frac{4}{75}$. Consider the graph in answer choice (1), which includes the points (0,16) and (120,0):

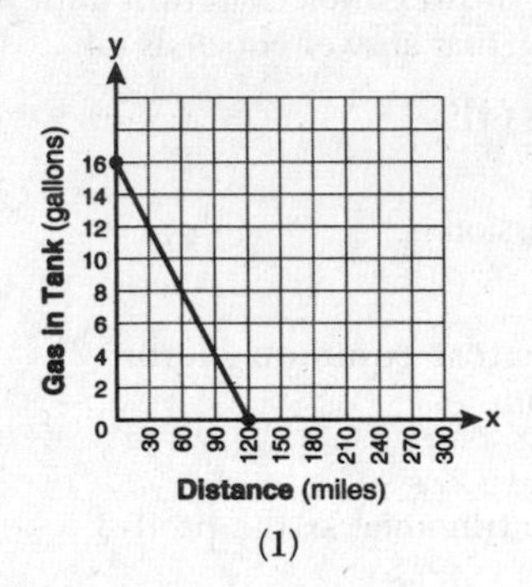

(1)

The slope of this line is:

$$\frac{\Delta y}{\Delta x} = \frac{0-16}{120-0} = -\frac{16 \div 8}{120 \div 8} = -\frac{2}{15} \text{ ✗}$$

- Next, consider the graph in choice (2), which includes the points (0,16) and (300,0):

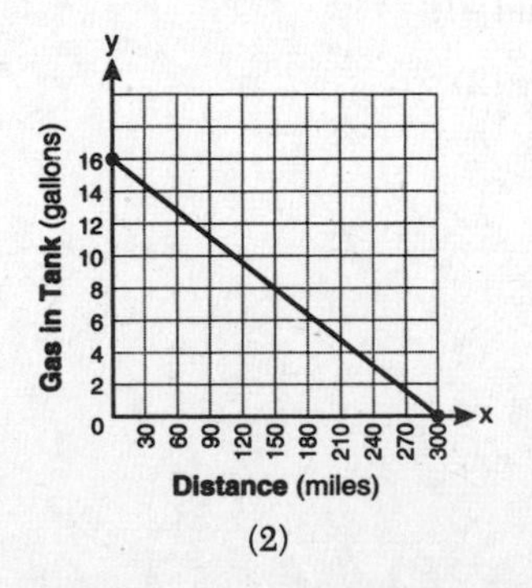

(2)

The slope of this graph is:

$$\frac{\Delta y}{\Delta x} = \frac{0-16}{300-0} = -\frac{16 \div 4}{300 \div 4} = -\frac{4}{75} \checkmark$$

The correct choice is **(2)**.

8. The given equation is $3ax + b = c$.

- Isolate x in the usual way by first subtracting b from both sides of the equation:

$$3ax = c - b$$

- Divide each side of the equation by $3a$:

$$\frac{\overset{1}{\cancel{3}}ax}{\cancel{3}a} = \frac{c-b}{3a}$$

$$x = \frac{c-b}{3a}$$

The correct choice is **(3)**.

9. It is given that the length of the hypotenuse of a right triangle is 34 inches and the length of one of its legs is 16 inches. To find the number of inches in the length of the other leg of this right triangle, use the Pythagorean theorem:

$$\begin{aligned}
(\text{leg } 1)^2 + (\text{leg } 2)^2 &= (\text{hyp})^2 \\
(16)^2 + x^2 &= (34)^2 \\
256 + x^2 &= 1156 \\
x^2 &= 1156 - 256 \\
x^2 &= 900 \\
\sqrt{x^2} &= \sqrt{900} \\
x &= 30
\end{aligned}$$

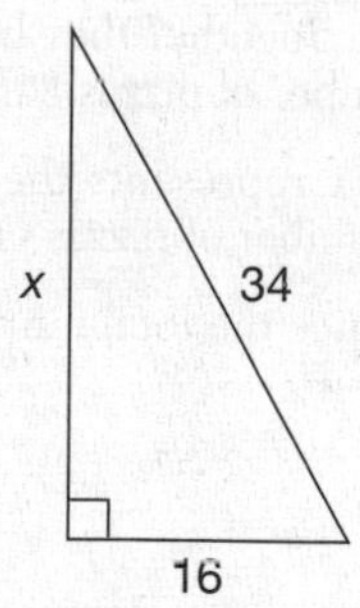

x

34

16

The correct choice is **(4)**.

10. A line parallel to the x-axis is a horizontal line whose equation has the form $y = b$ where b represents the y-intercept of the line, as in $y = 10$:

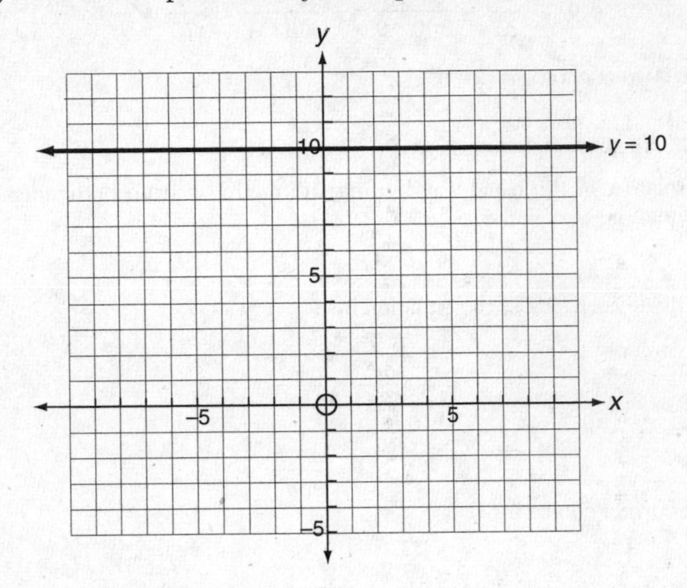

The correct choice is **(2)**.

11. It is given that Sam has sold half as many pizzas for a class fundraiser as Odel. Together they have sold a total of 126 pizzas. You are required to find the number of pizzas Sam sold.

- If x represents the number of pizzas Sam sold, then $2x$ represents the number of pizzas Odel sold.
- Since the total number of pizzas they sold is 126:

$$x + 2x = 126$$

$$3x = 126$$

$$\frac{\overset{1}{\cancel{3}}x}{\cancel{3}} = \frac{126}{3}$$

$$x = 42$$

- Sam sold 42 pizzas.

The correct choice is **(2)**.

12. You are required to find the ordered pair that is in the solution set of the system of equations $y = -x + 1$ and $y = x^2 + 5x + 6$.

Method 1: Solve the linear-quadratic system algebraically by substituting $-x + 1$ for y in the quadratic equation:

$$-x + 1 = x^2 + 5x + 6$$

$$0 = x^2 + 5x + x + 6 - 1$$

$$0 = x^2 + 6x + 5$$

Solve the quadratic equation by factoring:

$$(x + ?)(x + ?) = 0$$

The missing terms of the binomial factors are the two integers whose product is +5, the constant term of $x^2 + 6x + 5$, and whose sum is +6, the coefficient of the x-term of $x^2 + 6x + 5$. As $(+5)(+1) = +5$ and $(+5) + (+1) = +6$, the missing terms are +5 and +1:

$$(x + 5)(x + 1) = 0$$

If the product of two terms is 0, then at least one of the terms equals 0:

$$x + 5 = 0 \quad \text{or} \quad x + 1 = 0$$

Solve for x:

$$x = -5 \quad \Big| \quad x = -1$$

Find the corresponding values of y by substituting the solution values for x into the linear equation:

- If $x = -5$, then $y = -(-5) + 1 = 5 + 1 = 6$. One solution is $(-5,6)$.
- If $x = -1$, then $y = -(-1) + 1 = 1 + 1 = 2$. The other solution is $(-1,2)$.

Look at the answer choices for either $(-5,6)$ or $(-1,2)$. Answer choice (2) gives $(-5,6)$.

Method 2: Test each ordered pair in both equations until you find the ordered pair that satisfies the two equations at the same time:

- Choice (1): Test (–5,–1).

$$y = -x + 1$$
$$-1 \boxed{?} -(-5) + 1$$
$$-1 \boxed{?} 5 + 1$$
$$-1 \neq 6 \quad ✗$$

- Choice (2): Test (–5,6).

$$y = -x + 1 \qquad\qquad y = x^2 + 5x + 6$$
$$6 \boxed{?} -(-5) + 1 \qquad\qquad 6 \boxed{?} (-5)^2 + 5(-5) + 6$$
$$6 \boxed{?} 5 + 1 \qquad\qquad 6 \boxed{?} 25 - 25 + 6$$
$$6 = 6 \ ✔ \qquad\qquad 6 = 6 \ ✔$$

The correct choice is **(2)**.

13. The vertex of a vertical parabola is its highest or lowest point. The axis of symmetry is the vertical line that contains the vertex, as shown in the accompanying figure.

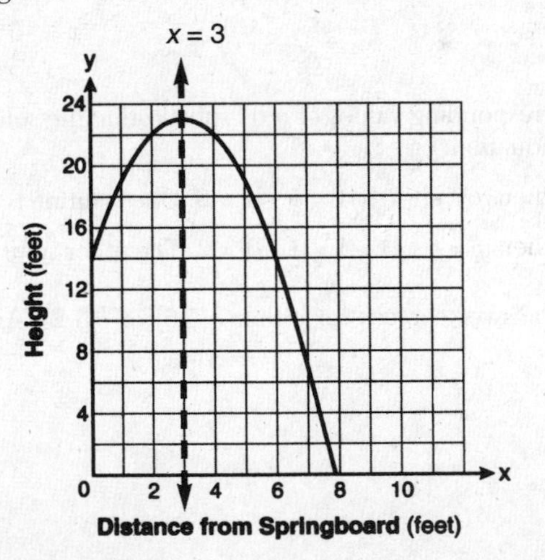

The correct choice is **(1)**.

14. It is given that Nicole's aerobics class exercises to music with a rate of 120 beats per minute. To find the number of beats in a class that is 0.75 hour long, multiply the rate by the number of minutes in the class.

- Since 60 minutes are in 1 hour:

$$0.75 \text{ hour} = 0.75 \text{ hour} \times 60 \frac{\text{min}}{\text{hour}} = 45 \text{ min}$$

- The number of beats in class:

$$120 \frac{\text{beats}}{\text{min}} \times 45 \text{ min} = 5{,}400 \text{ beats}$$

The correct choice is **(3)**.

15. It is given that Luis is going to paint a basketball court on his driveway. This basketball court consists of a rectangle and a semicircle, as shown in the accompanying diagram.

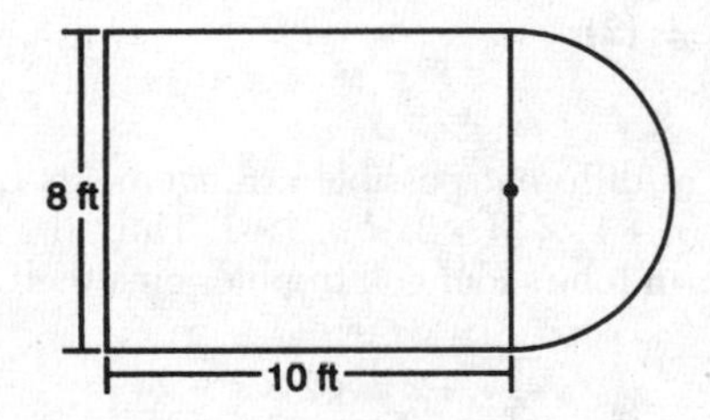

To determine the area of the basketball court, find the sum of the areas of the rectangle and semicircle.

- The area, A, of the rectangle is length $\times$ width:

$$A = 10 \text{ ft} \times 8 \text{ ft} = 80 \text{ ft}^2$$

- As opposite sides of a rectangle are equal in length, the diameter of the semicircle is 8 ft, so the radius of the semicircle is 4 ft:

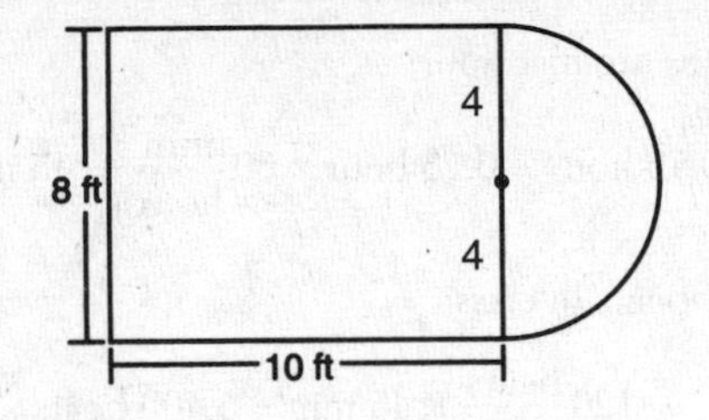

Since the area, A, of a circle with radius r is πr^2,

$$A_{\text{circle}} = \pi \times (4 \text{ ft})^2 = 16\pi \text{ ft}^2$$

The area of the semicircle is half that area, or 8π ft^2.

- The number of square feet in the area of this basketball court is $80 + 8\pi$.

The correct choice is **(2)**.

16. The number of different possible arrangements of n objects is given by $n!$ where $n! = n \times (n - 1) \times (n - 2) \times \ldots \times 1$. Thus, the number of different possible arrangements of John's four golf trophies on a shelf is:

$$4! = 4 \times 3 \times 2 \times 1 = 24$$

The correct choice is **(1)**.

17. It is given that a rectangle has an area of 24 square units and that the width is 5 units less than the length. You are required to find the length of the rectangle. If x represents the length of the rectangle, then $x - 5$ represents the width of the rectangle:

Length × Width = Area

$$x(x - 5) = 24$$

$$x^2 - 5x - 24 = 0$$

$$(x - 8)(x + 3) = 0$$

$x - 8 = 0$ or $x + 3 = 0$

$x = 8$ | $x = -3$ ← reject since x cannot be negative

$x - 5$

Area = 24

x

The correct choice is **(2)**.

18. You are required to determine the value of the third quartile from the box-and-whisker plot in the accompanying diagram:

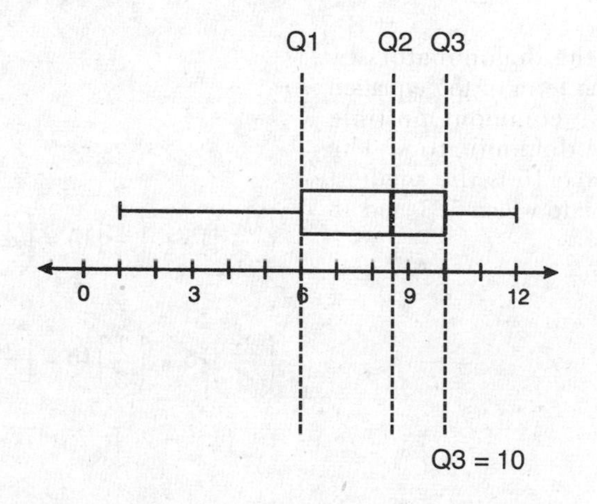

In a box-and-whisker plot, the vertical ends of the box represent the first (Q1) and third (Q3) quartiles. The vertical bar in the interior of the box corresponds to the second (Q2) quartile, which is also the median.

The correct choice is **(3)**.

19. To subtract $3g^2 - 4g + 2$ from $7g^2 + 5g - 1$, write the first polynomial under the second polynomial, aligning like terms in the same column:

$$\begin{array}{r} 7g^2 + 5g - 1 \\ - \quad 3g^2 - 4g + 2 \\ \hline \end{array}$$

Change to an equivalent additional example by taking the opposite of each term of the polynomial that is being subtracted. Then combine like terms:

$$\begin{array}{r} 7g^2 + 5g - 1 \\ + \quad -3g^2 + 4g - 2 \\ \hline 4g^2 + 9g - 3 \end{array}$$

The correct choice is **(3)**.

20. The given equation is:

$$\frac{2x}{5} + \frac{1}{3} = \frac{7x - 2}{15}$$

Eliminate the denominators by multiplying each term of the equation by the lowest common multiple (LCM) of the denominators. The LCM is 15 since 15 is the smallest whole number into which 5, 3, and 15 divide evenly:

$$\left(\frac{2x}{5}\right)15 + \left(\frac{1}{3}\right)15 = \left(\frac{7x - 2}{15}\right)15$$

$$\left(\frac{2x}{\cancel{5}}\right)\overset{3}{\cancel{15}} + \left(\frac{1}{\cancel{3}}\right)\overset{5}{\cancel{15}} = \left(\frac{7x - 2}{\cancel{15}}\right)\overset{1}{\cancel{15}}$$

$$6x \quad + \quad 5 \quad = 7x - 2$$

Collect like terms on the same side of the equation by transposing $7x$ and $+5$ to the opposite sides of the equation:

$$6x - 7x = -2 - 5$$

$$-x = -7$$

$$x = 7$$

The correct choice is **(4)**.

21. The given expression is:

$$\frac{25x - 125}{x^2 - 25}$$

Factor the greatest common factor of 25 from each term of the numerator:

$$\frac{25(x-5)}{x^2 - 25}$$

The denominator represents the difference of two squares, so it can be factored as the sum and difference of the terms that are being squared:

$$\frac{25(x-5)}{(x+5)(x-5)}$$

Divide out any factor that appears in both the numerator and the denominator since their quotient is 1:

$$\frac{25\overset{1}{\cancel{(x-5)}}}{(x+5)\cancel{(x-5)}}$$

Simplify:

$$\frac{25}{x+5}$$

The correct choice is (4).

22. You are required to identify from among the answer choices the equation that most closely represents the line of best fit for the scatter plot in the accompanying figure.

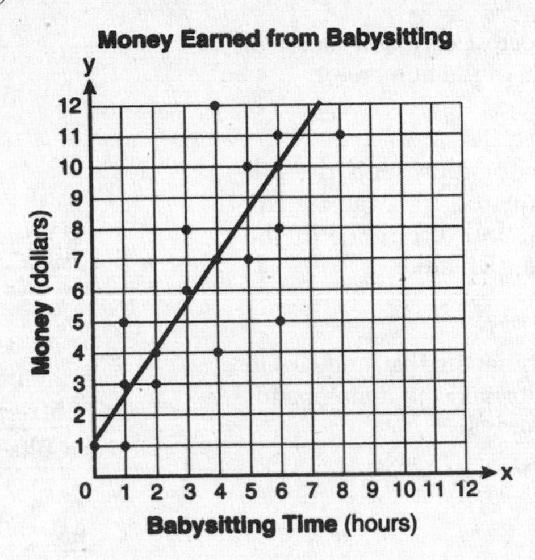

A linear equation has the form $y = mx + b$, where m is the slope of the line and b is its y-intercept. Sketch the line of best fit by drawing the line that has approximately the same number of data points on either side of it while also having a y-intercept that matches the y-intercept of one of the equations in the answer choices, as shown in the figure above.

Answer Choice	y-intercept
(1) $y = x$	0
(2) $y = \frac{2}{3}x + 1$	1
(3) $y = \frac{3}{2}x + 4$	4
(4) $y = \frac{3}{2}x + 1$	1

Since the line of best fit contains the points (0,1) and (4,7), its y-intercept is 1 and its slope is:

$$m = \frac{\Delta y}{\Delta x} = \frac{7-1}{4-0} = \frac{6}{4} = \frac{3}{2}$$

Because $m = \frac{3}{2}$ and $b = 1$, an equation of the line that most closely represents the line of best fit is $y = \frac{3}{2}x + 1$.

The correct choice is **(4)**.

23. The sign of the slope of a line depends on whether the independent (x) variable and the dependent (y) variable move in the same or in opposite directions.

- If x and y move in the same direction, the slope of the line is positive. Thus, if the dependent variable increases at a constant rate as the independent variable also increases at a constant rate, the slope of the line is positive.
- If x and y move in opposite directions, the slope of the line is negative. Thus, if the dependent variable decreases at a constant rate as the independent variable increases at a constant rate, as is given, the slope of the line is negative.

The correct choice is **(2)**.

24. You are required to identify from among the answer choices the equation that could be used to find the measure of one of the acute angles of the right triangle in the accompanying figure.

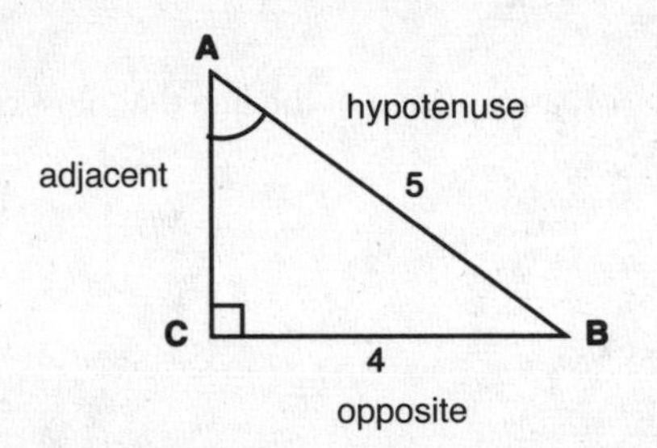

Since the lengths of the leg opposite angle A and the hypotenuse are given, use the sine ratio with respect to angle A:

$$\sin A = \frac{\text{leg opposite}\angle A}{\text{hypotenuse}} = \frac{4}{5}$$

The correct choice is **(1)**.

25. You are required to determine from among the answer choices the ordered pair that is in the solution set of the system of inequalities $y < \frac{1}{2}x + 4$ and $y \geq -x + 1$. Test each answer choice in turn until you find the ordered pair that satisfies both inequalities.

- Choice (1): Test (–5,3).

$$y \geq -x + 1$$
$$3 \geq -(-5) + 1$$
$$3 \ngeq 6 \quad ✗$$

- Choice (2): Test (0,4).

$$y < \frac{1}{2}x + 4$$
$$4 < \frac{1}{2}(0) + 4$$
$$4 \nless 4 \quad ✗$$

- Choice (3): Test (3,–5).

$$y \geq -x + 1$$
$$-5 \geq -(3) + 1$$
$$-5 \ngeq -2 \quad ✗$$

- Choice (4): Test (4,0).

$$y < \frac{1}{2}x + 4 \qquad y \geq -x + 1$$
$$0 < \frac{1}{2}(4) + 4 \qquad 0 \geq -4 + 1$$
$$0 < 6 \; ✔ \qquad 0 \geq -3 \; ✔$$

The correct choice is **(4)**.

26. To find the product of $\frac{4x}{x-1}$ and $\frac{x^2-1}{3x+3}$, first factor where possible.

In the second fraction, factor the numerator as the difference of two squares. Factor out the greatest common factor of 3 from the denominator:

$$\left(\frac{4x}{x-1}\right)\left(\frac{x^2-1}{3x+3}\right) = \frac{4x}{x-1} \cdot \frac{(x-1)(x+1)}{3(x+1)}$$

Divide out any factor that appears in both a numerator and a denominator since their quotient is 1:

$$= \frac{4x}{\cancel{x-1}} \cdot \frac{\overset{1}{\cancel{(x-1)}}\overset{1}{\cancel{(x+1)}}}{3\cancel{(x+1)}}$$

Multiply the numerators together, and multiply the denominators together:

$$= \frac{4x}{3}$$

The correct choice is **(1)**.

27. The given expression is: $(3x^2)^3$

Remove the parentheses by raising each factor inside the parentheses to the third power: $3^3(x^2)^3$

Raise a power to a power by multiplying exponents: $3^3x^{2\times 3}$

Evaluate 3^3 as $3 \times 3 \times 3 = 27$: $27x^6$

The correct choice is **(4)**.

28. It is given that Ryan estimates the measurement of the volume of a popcorn container to be 282 cubic inches while the actual volume is 289 cubic inches.

$$\text{Relative error in measurement} = \frac{\text{Actual volume} - \text{Measured volume}}{\text{Actual volume}}$$

$$= \frac{289 - 282}{289}$$

$$= \frac{7}{289}$$

$$\approx 0.024$$

The correct choice is **(1)**.

29. In the accompanying diagram of $\triangle ABC$, it is given that $BC = 10$ and $AB = 16$.

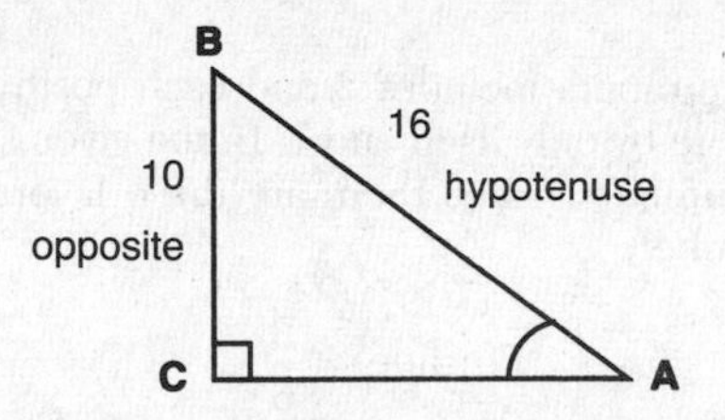

You are required to find the measure of the largest acute angle of the triangle to the *nearest tenth* of a degree.

- Since the lengths of the leg opposite angle A and the hypotenuse are given, use the sine ratio with respect to angle A:

$$\sin A = \frac{\text{leg opposite}\angle A}{\text{hypotenuse}}$$

$$= \frac{BC}{AB}$$

$$= \frac{10}{16}$$

$$A = \sin^{-1}\left(\frac{10}{16}\right)$$

Use your calculator: $m\angle A \approx 38.7$

- Find $m\angle B$. Because the measures of the acute angles of a right triangle add up to 90:

$$38.7 + m\angle B = 90$$

$$m\angle B = 90 - 38.7$$

$$= 51.3$$

- Angle B is, therefore, the largest acute angle in the triangle.

The correct choice is **(3)**.

30. It is given that the faces of a cube are numbered from 1 to 6 and that the cube is tossed once. Thus, the sample space is {1, 2, 3, 4, 5, 6}. You are required to find the probability of tossing a prime number or a number divisible by 2.

- The set of prime numbers includes 2 and each positive integer greater than 2 that is divisible by only itself and 1. In the given sample space, 2, 3, and 5 are prime numbers. Since there are three favorable outcomes for tossing a prime number:

$$P(\text{prime}) = \frac{3}{6}$$

- In the given sample space, 2, 4, and 6 are each divisible by 2. Because there are three favorable outcomes for tossing a number divisible by 2:

$$P(\text{even}) = \frac{3}{6}$$

- These two events, however, are not mutually exclusive as they have one favorable outcome in common, namely, tossing a 2. Thus,

$$P(\text{prime and even}) = \frac{1}{6}$$

To find the probability tossing a prime number or a number divisible by 2, add the probabilities for each event and then subtract the probability of both events occurring at the same time, thereby avoiding counting the same outcome twice.

$$P(\text{prime or even}) = P(\text{prime}) + P(\text{even}) - P(\text{prime and even})$$

$$= \frac{3}{6} + \frac{3}{6} - \frac{1}{6}$$

$$= \frac{5}{6}$$

The correct choice is **(2)**.

PART II

31. It is given that in a game of ice hockey, the hockey puck took 0.8 second to travel 89 feet to the goal line. To find the average speed, r, of the puck, use the formula $r \times t = d$ where $t = 0.8$ sec and $d = 89$ ft:

$$r \times t = d$$

$$r \times 0.8 \text{ sec} = 89 \text{ ft}$$

$$r = \frac{89 \text{ ft}}{0.8 \text{ sec}}$$

$$= 111.25 \frac{\text{ft}}{\text{sec}}$$

The average rate of speed of the puck in feet per second is **111.25**.

32. It is given that in her new board game, Brianna uses one spinner to determine how many spaces to move and a second spinner to determine whether her move from the first spinner will be forward or backward.

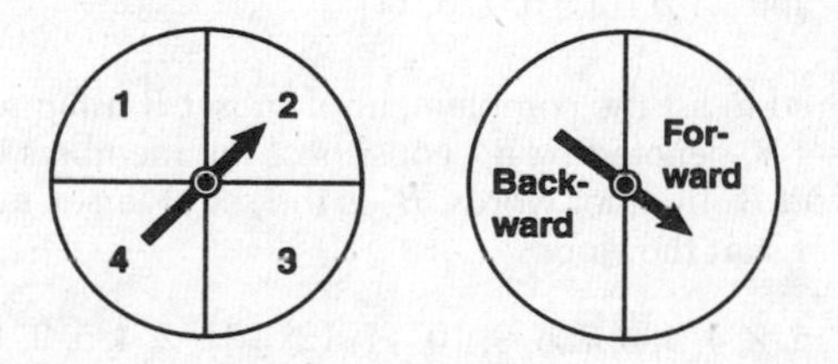

To find the probability that Brianna will move fewer than four spaces and backward, multiply the probability she will move fewer than four spaces by the probability she will move backward.

- The first spinner is divided into four equal regions with three of the four regions representing a move of fewer than 4 spaces:

$$P(\text{move fewer than 4 spaces}) = \frac{3}{4}$$

- Since the second spinner is divided into two equal regions:

$$P(\text{move backward}) = \frac{1}{2}$$

- Thus:

$$P(\text{move fewer than 4 spaces and backward}) = \frac{3}{4} \times \frac{1}{2} = \frac{3}{8}$$

The probability that Brianna will move fewer than four spaces and backward is $\mathbf{\frac{3}{8}}$.

33. It is given that twelve players make up a high school basketball team and that the team jerseys are numbered 1 through 12. It is also given that the players wearing the jerseys numbered 3, 6, 7, 8, and 11 are the only players who start a game.

Let A represent the set of team jersey numbers and B the subset of team jersey numbers that are worn by the players who start a game. Thus:

$$A = \{1, 2, 3, 4, 5, 6, 7, 8, 9, 10, 11, 12\}$$

$$\text{and} \quad B = \{3, 6, 7, 8, 11\}$$

You are required to list the complement of subset B using set notation. The complement of set B, denoted by B', consists of the members of set A that are not members of set B. In other words, B' is the set of jersey numbers of those players who do *not* start the game:

$$B' = \{1, 2, \not{3}, 4, 5, \not{6}, \not{7}, \not{8}, 9, 10, \not{11}, 12\} = \{1, 2, 4, 5, 9, 10, 12\}$$

The complement of the subset of jersey numbers worn by the players who start a game is **{1, 2, 4, 5, 9, 10, 12}**.

PART III

34. To express $3\sqrt{20}(2\sqrt{5}-7)$ in simplest radical form, use the distributive property. Multiply each term of the expression inside the parentheses by the radical expression in front of the parentheses:

$$3\sqrt{20}(2\sqrt{5}-7) = (3\sqrt{20})(2\sqrt{5}) + (3\sqrt{20})(-7)$$
$$= (3\times 2)(\sqrt{20}\times\sqrt{5}) + (3)(-7)\sqrt{20}$$
$$= 6\sqrt{100} - 21\sqrt{20}$$
$$= (6\times 10) - 21\sqrt{20}$$
$$= 60 - 21\sqrt{20}$$

Rewrite the number underneath the radical sign as the product of two positive integers, one of which is the greatest perfect square factor of 20: $= (60) - 21\sqrt{4\cdot 5}$

Write the radical over each factor: $= 60 - 21\sqrt{4}\cdot\sqrt{5}$

Evaluate the square root of the perfect square: $= 60 - 21\cdot 2\sqrt{5}$

Simplify: $= 60 - 42\sqrt{5}$

In simplest radical form, the given product is $\mathbf{60 - 42\sqrt{5}}$.

35. You are required to draw the graph of $y = 2^x$ over the interval $-1 \leq x \leq 3$. Graph a few representative points by first making a table of values.

x	$y = 2^x$	Ordered Pairs
-1	$y = 2^{-1} = \frac{1}{2}$	$\left(-1, \frac{1}{2}\right)$
0	$y = 2^0 = 1$	(0,1)
1	$y = 2^1 = 2$	(1,2)
2	$y = 2^2 = 4$	(2,4)
3	$y = 2^3 = 8$	(3,8)

See the accompanying graph. The graph will never intersect the x-axis since for all values of x, $y = 2^x$ is greater than 0.

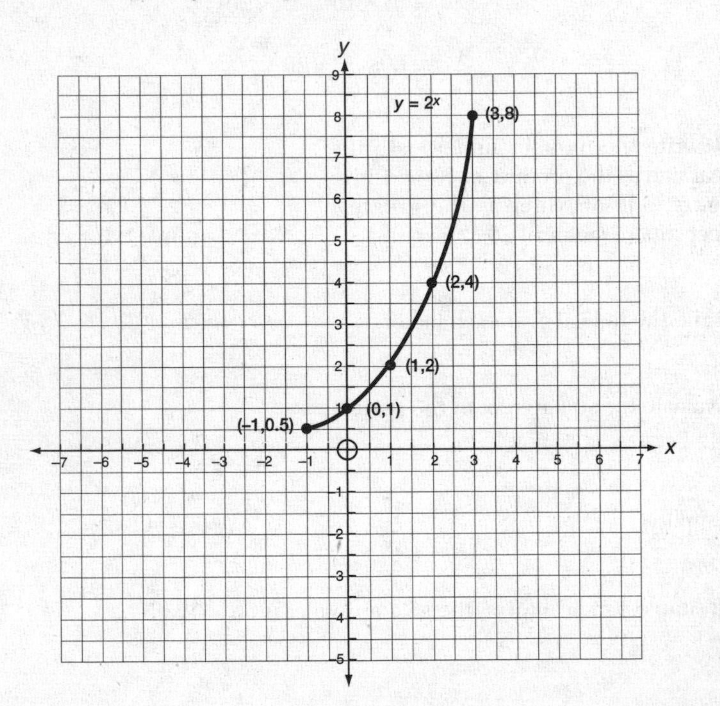

36. To write an equation that represents the line that passes through (5,4) and (–5,0), first find its slope.

- Substitute the coordinates of the given points into the slope formula:

$$m = \frac{\Delta y}{\Delta x} = \frac{0-4}{-5-5}$$

$$= \frac{-4}{-10}$$

$$= \frac{2}{5}$$

- If (a,b) is a point on a line with a slope of m, then an equation of this line has the point-slope form $y - b = m(x - a)$, where $m = \frac{2}{5}$. Replace (a,b) with the coordinates of either of the two given points. If $(a,b) = (-5,0)$, then

$$y - 0 = \frac{2}{5}(x - (-5))$$

$$y = \frac{2}{5}(x + 5)$$

Although not required, you can express the equation in slope-intercept form by simplifying the right side of the equation:

$$= \frac{2}{5}x + \frac{2}{\cancel{5}}(\overset{1}{\cancel{+5}})$$

$$= \frac{2}{5}x + 2$$

An equation of the required line is $\boldsymbol{y = \frac{2}{5}(x + 5)}$ or, equivalently, $\boldsymbol{y = \frac{2}{5}x + 2}$.

PART IV

37. It is given that the cost of 3 markers and 2 pencils is \$1.80 and that the cost of 4 markers and 6 pencils is \$2.90. If m represents the price of a marker and p represents the price of a pencil:

$$\begin{aligned} 3m + 2p &= 1.80 \\ 4m + 6p &= 2.90 \end{aligned}$$

Eliminate the variable p by multiplying the first equation by –3 and then adding corresponding sides of the two equations:

$$\begin{aligned} -3(3m + 2p) &= -3(1.80) \\ 4m + 6p &= 2.90 \end{aligned} \quad \Rightarrow \quad \begin{aligned} -9m - 6p &= -5.40 \\ 4m + 6p &= 2.90 \\ \hline -5m &= -2.50 \\ \frac{-5m}{-5} &= \frac{-2.50}{-5} \\ m &= \$0.50 \end{aligned}$$

Substitute $m = 0.50$ into either of the original equations to find the corresponding value of p:

$$\begin{aligned} 3m + 2p &= 1.80 \\ 3(0.50) + 2p &= 1.80 \\ 1.50 + 2p &= 1.80 \\ 2p &= 1.80 - 1.50 \\ \frac{2p}{2} &= \frac{0.30}{2} \\ p &= \$0.15 \end{aligned}$$

The cost of a marker is **\$0.50**, and the cost of a pencil is **\$0.15**.

38. It is given that twenty students were surveyed about the number of days they played outside in one week with the following results:

{6, 5, 4, 3, 0, 7, 1, 5, 4, 4, 3, 2, 2, 3, 2, 4, 3, 4, 0, 7}

- Complete the frequency table for these data:

Number of Days Outside

Interval	Tally	Frequency
0–1	///	3
2–3	~~////~~ //	7
4–5	~~////~~ //	7
6–7	///	3

- Complete the cumulative frequency table. On the first line of the cumulative frequency table, copy the frequency from the first line of the frequency table.

  On each line after the first, add the entry from the corresponding line of the frequency table to the previous line of the cumulative frequency table as shown:

Number of Days Outside

Interval	Cumulative Frequency
0–1	3
0–3	3 + 7 = 10
0–5	10 + 7 = 17
0–7	17 + 3 = 20

- Based on the cumulative frequency table, draw a cumulative frequency histogram:

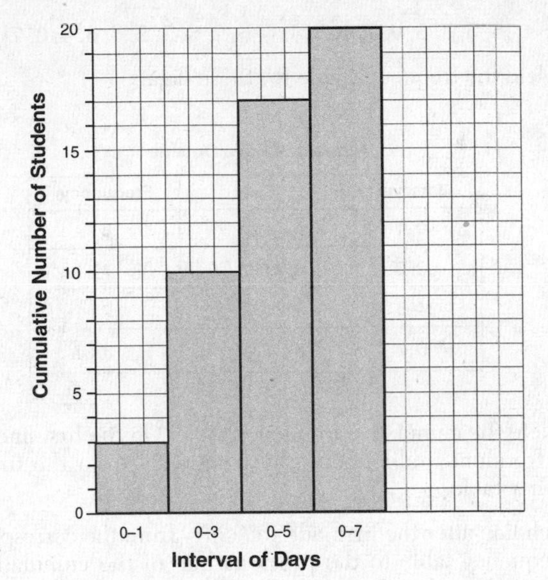

39. To solve the system of equations $y = x^2 + 4x - 5$ and $y = x - 1$ graphically, graph each equation on the same set of axes. Then determine the coordinates of the points of intersection of the two graphs.

- Graph the parabola $y = x^2 + 4x - 5$ by following these steps:

Step 1: Determine an equation of the axis of symmetry. The equation of the axis of symmetry of this parabola has the form $x = -\frac{b}{2a}$, where $a = 1$ and $b = 4$:

$$\begin{aligned} x &= -\frac{4}{2(1)} \\ &= -\frac{4}{2} \\ &= -2 \end{aligned}$$

Hence, the x-coordinate of the vertex of the parabola is -2.

Step 2: Prepare a table of values that includes three consecutive integer x-values on either side of $x = -2$.

x	$x^2 + 4x - 5 = y$	(x,y)
−5	$(-5)^2 + 4(-5) - 5 = 0$	(−5,0)
−4	$(-4)^2 + 4(-4) - 5 = -5$	(−4,−5)
−3	$(-3)^2 + 4(-3) - 5 = -8$	(−3,−8)
−2	$(-2)^2 + 4(-2) - 5 = -9$	(−2,−9)
−1	$(-1)^2 + 4(-1) - 5 = -8$	(−1,−8)
0	$(0)^2 + 4(0) - 5 = -5$	(0,−5)
1	$(1)^2 + 4(1) - 5 = 0$	(1,0)

TIP: Use the table feature of your graphing calculator to create a table of values.

Step 3: Check for symmetry in the y-values listed in the table. Corresponding pairs of points on either side of $x = -2$ have matching y-coordinates. This confirms that $(-2,-9)$ is the vertex of the parabola.

Step 4: Plot (–5,0), (–4,–5), (–3,–8), (–2,–9), (–1,–8), (0,–5), and (1,0) on the grid provided. Then connect these points with a smooth, U-shaped curve, as shown in the accompanying figure. Label the graph with its equation.

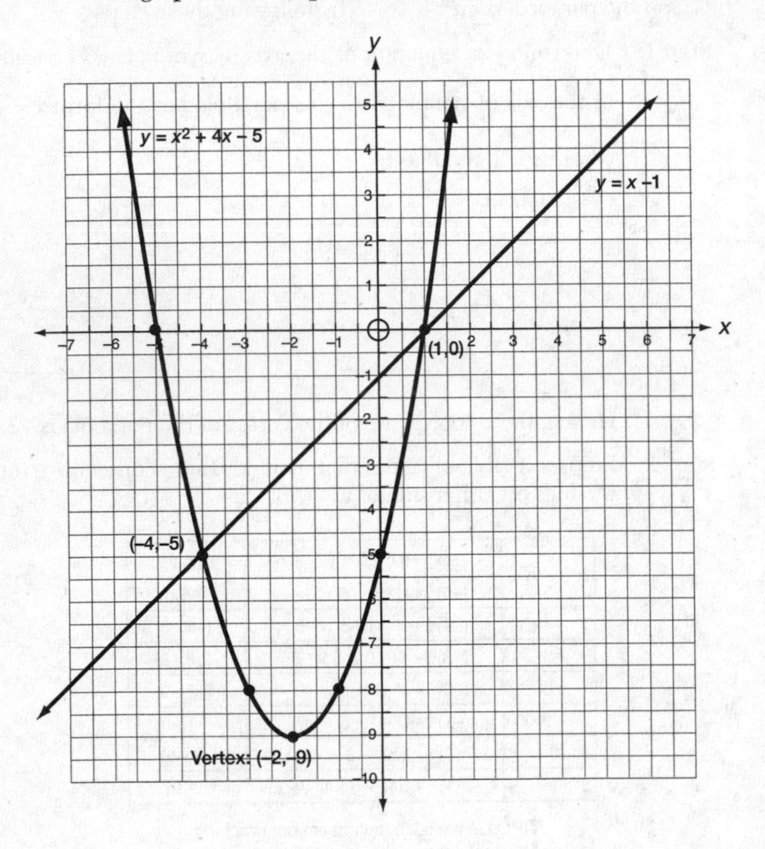

- Graph $y = x - 1$ by plotting any two convenient points that satisfy the equation, such as (0,–1) and (2,1). Then connect these points with a straight line as shown in the accompanying figure. Label the graph with its equation.

- Read the coordinates of the points of intersection from the graph. The two graphs intersect at (–4,–5) and (1,0).

The points in the solution set are **(–4,–5)** and **(1,0)**.

Topic	Question Numbers	Number of Points	Your Points	Your Percentage
1. Sets and Numbers; Intersection and Complements of Sets; Interval Notation; Properties of Real Numbers	2, 33	2 + 2 = 4		
2. Operations on Rat'l. Numbers & Monomials	—	—		
3. Laws of Exponents for Integer Exponents; Scientific Notation	27	2		
4. Operations on Polynomials	19	2		
5. Square Root; Operations with Radicals	34	3		
6. Evaluating Formulas & Algebraic Expressions	—	—		
7. Solving Linear Eqs. & Inequalities	1, 5, 11, 20	2 + 2 + 2 + 2 = 8		
8. Solving Literal Eqs. & Formulas for a Given Letter	8	2		
9. Alg. Operations (including factoring)	6, 21, 26	2 + 2 + 2 = 6		
10. Quadratic Equations (incl. alg. and graphical solutions; parabolas)	13, 17	2 + 2 = 4		
11. Coordinate Geometry (eq. of a line; graphs of linear eqs.; slope)	10, 23, 36	2 + 2 + 3 = 7		
12. Systems of Linear Eqs. & Inequalities (algebraic & graphical solutions)	25, 37	2 + 4 = 6		
13. Mathematical Modeling (using eqs.; tables; graphs)	3, 7	2 + 2 = 4		
14. Linear-Quadratic Systems	12, 39	2 + 4 = 6		
15. Perimeter; Circumference; Area of Common Figures	—	—		
16. Volume and Surface Area; Area of Overlapping Figures; Relative Error in Measurement	15, 28	2 + 2 = 4		
17. Fractions and Percent	—	—		
18. Ratio & Proportion (incl. similar polygons, scale drawings, & rates)	14, 31	2 + 2 = 4		
19. Pythagorean Theorem	9	2		
20. Right Triangle Trigonometry	24, 29	2 + 2 = 4		
21. Functions (def.; domain and range; vertical line test; absolute value)	—	—		

Topic	Question Numbers	Number of Points	Your Points	Your Percentage
22. Exponential Functions (properties; growth and decay)	35	3		
23. Probability (incl. tree diagrams & sample spaces)	30, 32	2 + 2 = 4		
24. Permutations and Counting Methods (incl. Venn diagrams)	16	2		
25. Statistics (mean, median, percentiles, quartiles; freq. dist., histograms; box-and-whisker plots; causality, bivariate data; qualitative vs. quantitative data; unbiased vs. biased samples; circle graphs)	4, 18, 38	2 + 2 + 4 = 8		
26. Line of Best Fit (including linear regression, scatter plots, and linear correlation)	22	2		
27. Nonroutine Word Problems Requiring Arith. or Alg. Reasoning	—	—		

MAP TO LEARNING STANDARDS

The table below shows which content strand each item is aligned to. The numbers in the table represent the question numbers on the test.

Key Ideas	Item Numbers
Number Sense and Operations	2, 16, 34
Algebra	1, 3, 5, 6, 8, 9, 10, 11, 12, 17, 19, 20, 21, 23, 24, 25, 26, 27, 29, 33, 36, 37
Geometry	7, 13, 15, 35, 39
Measurement	14, 28, 31
Probability and Statistics	4, 18, 22, 30, 32, 38

HOW TO CONVERT YOUR RAW SCORE TO YOUR INTEGRATED ALGEBRA REGENTS EXAMINATION SCORE

Below is the conversion chart that must be used to determine your final score on the August 2008 Regents Examination in Integrated Algebra. To find your final exam score, locate in the column labeled "Raw Score" the total number of points you scored out of a possible 87 points. Since partial credit is allowed in Parts II, III, and IV of the test, you may need to approximate the credit you would receive for a solution that is not completely correct. Then locate in the adjacent column to the right the scale score that corresponds to your raw score. The scale score is your final Integrated Algebra Regents Examination score.

Raw Score	Scaled Score	Raw Score	Scaled Score	Raw Score	Scaled Score
87	100	57	82	27	61
86	98	56	81	26	60
85	97	55	81	25	59
84	96	54	81	24	57
83	95	53	80	23	56
82	94	52	80	22	55
81	93	51	79	21	53
80	92	50	79	20	52
79	91	49	79	19	50
78	91	48	78	18	48
77	90	47	78	17	46
76	89	46	77	16	45
75	89	45	77	15	43
74	88	44	76	14	41
73	88	43	76	13	39
72	87	42	75	12	37
71	87	41	74	11	34
70	86	40	74	10	32
69	86	39	73	9	29
68	86	38	72	8	27
67	86	37	71	7	24
66	85	36	71	6	21
65	84	35	70	5	18
64	84	34	69	4	15
63	84	33	68	3	12
62	83	32	67	2	8
61	83	31	66	1	4
60	83	30	65	0	0
59	82	29	64		
58	82	28	63		

Examination June 2009

Integrated Algebra

FORMULAS

Trigonometric ratios		$\sin A = \frac{\text{opposite}}{\text{hypotenuse}}$
		$\cos A = \frac{\text{adjacent}}{\text{hypotenuse}}$
		$\tan A = \frac{\text{opposite}}{\text{adjacent}}$
Area	Trapezoid	$A = \frac{1}{2}h(b_1 + b_2)$
Volume	Cylinder	$V = \pi r^2 h$
Surface area	Rectangular prism	$SA = 2lw + 2hw + 2lh$
	Cylinder	$SA = 2\pi r^2 + 2\pi rh$
Coordinate geometry		$m = \frac{\Delta y}{\Delta x} = \frac{y_2 - y_1}{x_2 - x_1}$

PART I

Answer all questions in this part. Each correct answer will receive 2 credits. No partial credit will be allowed. For each question, write in the space provided the numeral preceding the word or expression that best completes the statement or answers the question. [60]

1 It takes Tammy 45 minutes to ride her bike 5 miles. At this rate, how long will it take her to ride 8 miles?

(1) 0.89 hour
(2) 1.125 hours
(3) 48 minutes
(4) 72 minutes

1 ____

2 What are the roots of the equation $x^2 - 7x + 6 = 0$?

(1) 1 and 7
(2) −1 and 7
(3) −1 and −6
(4) 1 and 6

2 ____

3 Which expression represents $\frac{27x^{18}y^5}{9x^6y}$ in simplest form?

(1) $3x^{12}y^4$
(2) $3x^3y^5$
(3) $18x^{12}y^4$
(4) $18x^3y^5$

3 ____

4 Marie currently has a collection of 58 stamps. If she buys s stamps each week for w weeks, which expression represents the total number of stamps she will have?

(1) $58sw$
(2) $58 + sw$
(3) $58s + w$
(4) $58 + s + w$

4 ____

5 Which data set describes a situation that could be classified as qualitative?

(1) the ages of the students in Ms. Marshall's Spanish class
(2) the test scores of the students in Ms. Fitzgerald's class
(3) the favorite ice cream flavor of each of Mr. Hayden's students
(4) the heights of the players on the East High School basketball team

5 ____

6 The sign shown below is posted in front of a roller coaster ride at the Wadsworth County Fairgrounds.

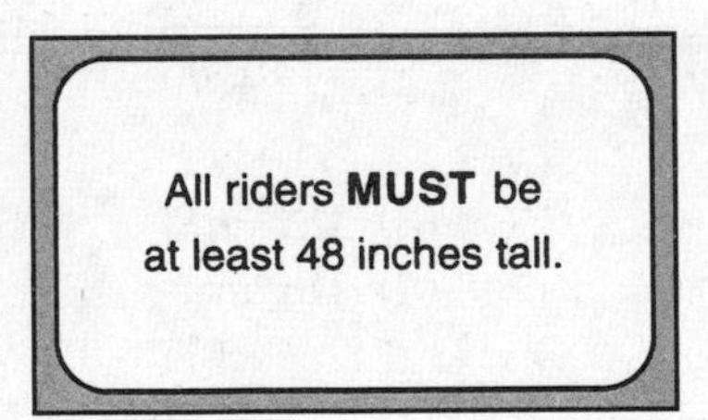

If h represents the height of a rider in inches, what is a correct translation of the statement on this sign?

(1) $h < 48$
(2) $h > 48$
(3) $h \leq 48$
(4) $h \geq 48$

6 ____

7 Which value of x is the solution of the equation $\frac{2x}{3}+\frac{x}{6}=5$?

(1) 6
(2) 10
(3) 15
(4) 30

7 ______

8 Students in Ms. Nazzeer's mathematics class tossed a six-sided number cube whose faces are numbered 1 to 6. The results are recorded in the table below.

Result	Frequency
1	3
2	6
3	4
4	6
5	4
6	7

Based on these data, what is the empirical probability of tossing a 4?

(1) $\frac{8}{30}$
(2) $\frac{6}{30}$
(3) $\frac{5}{30}$
(4) $\frac{1}{30}$

8 ______

9 What is the value of x, in inches, in the right triangle below?

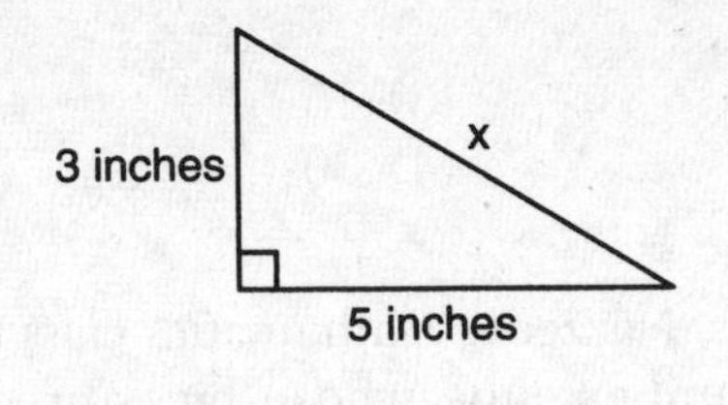

(1) $\sqrt{15}$ (3) $\sqrt{34}$
(2) 8 (4) 4

9 ____

10 What is $\sqrt{32}$ expressed in simplest radical form?

(1) $16\sqrt{2}$ (3) $4\sqrt{8}$
(2) $4\sqrt{2}$ (4) $2\sqrt{8}$

10 ____

11 If the speed of sound is 344 meters per second, what is the approximate speed of sound, in meters per hour?

60 seconds = 1 minute
60 minutes = 1 hour

(1) 20,640 (3) 123,840
(2) 41,280 (4) 1,238,400

11 ____

12 The sum of two numbers is 47, and their difference is 15. What is the larger number?

(1) 16
(2) 31
(3) 32
(4) 36

12 _____

13 If $a + ar = b + r$, the value of a in terms of b and r can be expressed as

(1) $\frac{b}{r}+1$
(2) $\frac{1+b}{r}$
(3) $\frac{b+r}{1+r}$
(4) $\frac{1+b}{r+b}$

13 _____

14 Which value of x is in the solution set of $\frac{4}{3}x+5<17$?

(1) 8
(2) 9
(3) 12
(4) 16

14 _____

15 The box-and-whisker plot below represents students' scores on a recent English test.

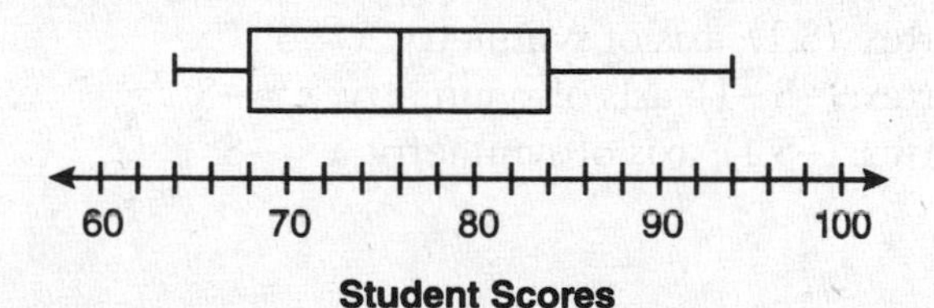

What is the value of the upper quartile?

(1) 68
(2) 76
(3) 84
(4) 94

15 _____

16 Which value of n makes the expression $\frac{5n}{2n-1}$ undefined?

(1) 1
(2) 0
(3) $-\frac{1}{2}$
(4) $\frac{1}{2}$

16 _____

17 At Genesee High School, the sophomore class has 60 more students than the freshman class. The junior class has 50 fewer students than twice the students in the freshman class. The senior class is three times as large as the freshman class. If there are a total of 1,424 students at Genesee High School, how many students are in the freshman class?

(1) 202
(2) 205
(3) 235
(4) 236

17 _____

18 What are the vertex and axis of symmetry of the parabola $y = x^2 - 16x + 63$?

(1) vertex: (8,–1); axis of symmetry: $x = 8$
(2) vertex: (8,1); axis of symmetry: $x = 8$
(3) vertex: (–8,–1); axis of symmetry: $x = -8$
(4) vertex: (–8,1); axis of symmetry: $x = -8$

18 _____

19 Which statement is true about the relation shown on the graph below?

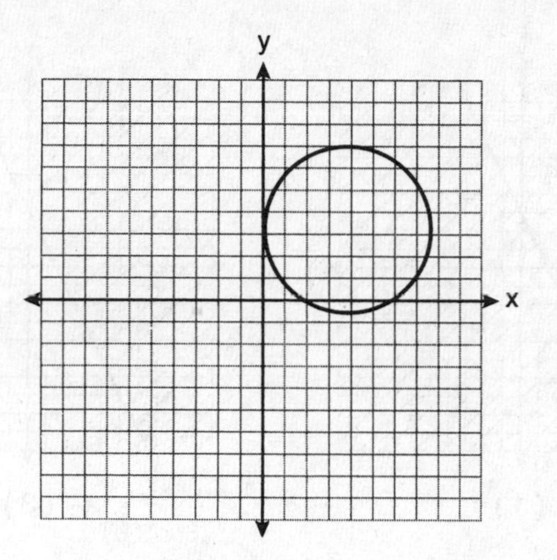

(1) It is a function because there exists one x-coordinate for each y-coordinate.
(2) It is a function because there exists one y-coordinate for each x-coordinate.
(3) It is *not* a function because there are multiple y-values for a given x-value.
(4) It is *not* a function because there are multiple x-values for a given y-value.

19 ______

20 Which graph represents the solution of $3y - 9 \leq 6x$?

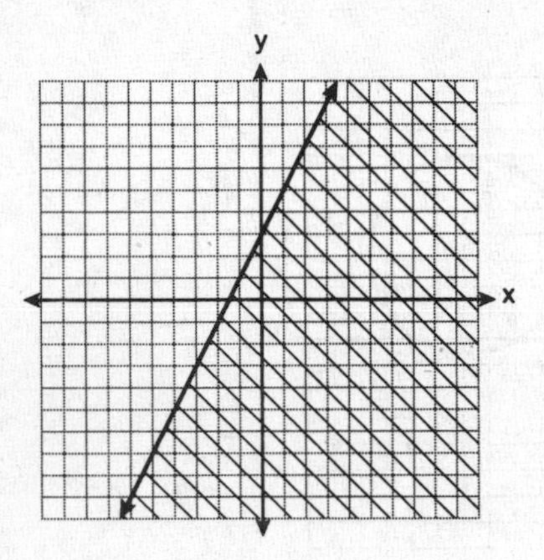

(1)

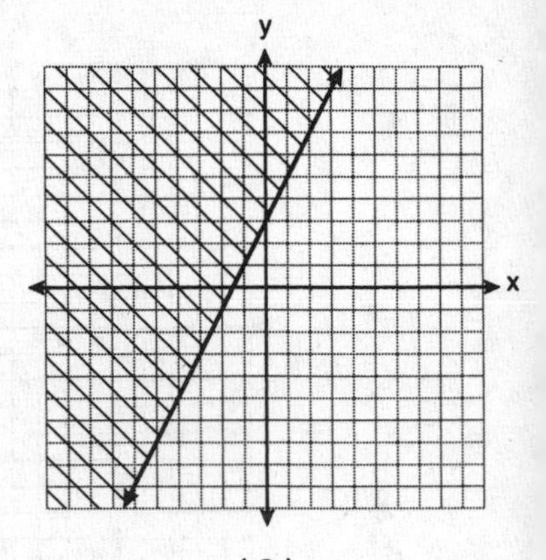

(3)

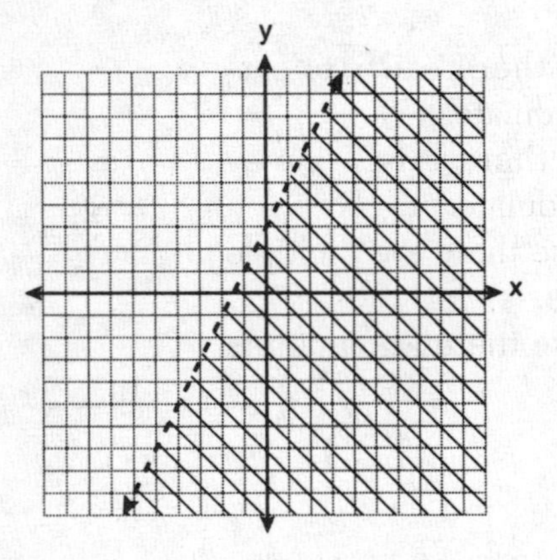

(2)

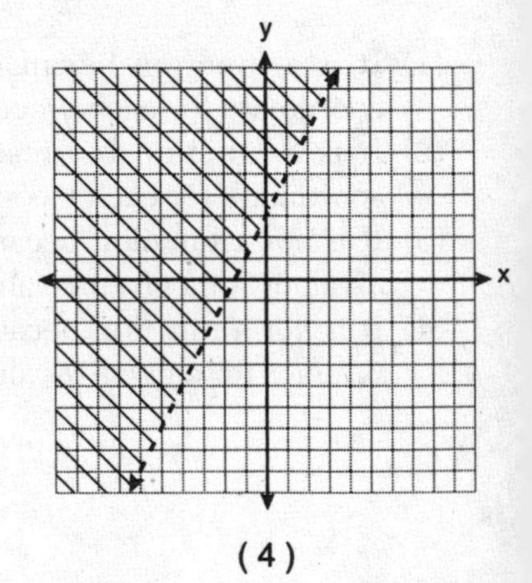

(4)

20 ______

21 Which expression represents $\frac{x^2 - 2x - 15}{x^2 + 3x}$ in simplest form?

(1) -5

(2) $\frac{x-5}{x}$

(3) $\frac{-2x-5}{x}$

(4) $\frac{-2x-15}{3x}$

21 ____

22 What is an equation of the line that passes through the point (4,–6) and has a slope of –3?

(1) $y = -3x + 6$

(2) $y = -3x - 6$

(3) $y = -3x + 10$

(4) $y = -3x + 14$

22 ____

23 When $4x^2 + 7x - 5$ is subtracted from $9x^2 - 2x + 3$, the result is

(1) $5x^2 + 5x - 2$

(2) $5x^2 - 9x + 8$

(3) $-5x^2 + 5x - 2$

(4) $-5x^2 + 9x - 8$

23 ____

24 The equation $y = x^2 + 3x - 18$ is graphed on the set of axes below.

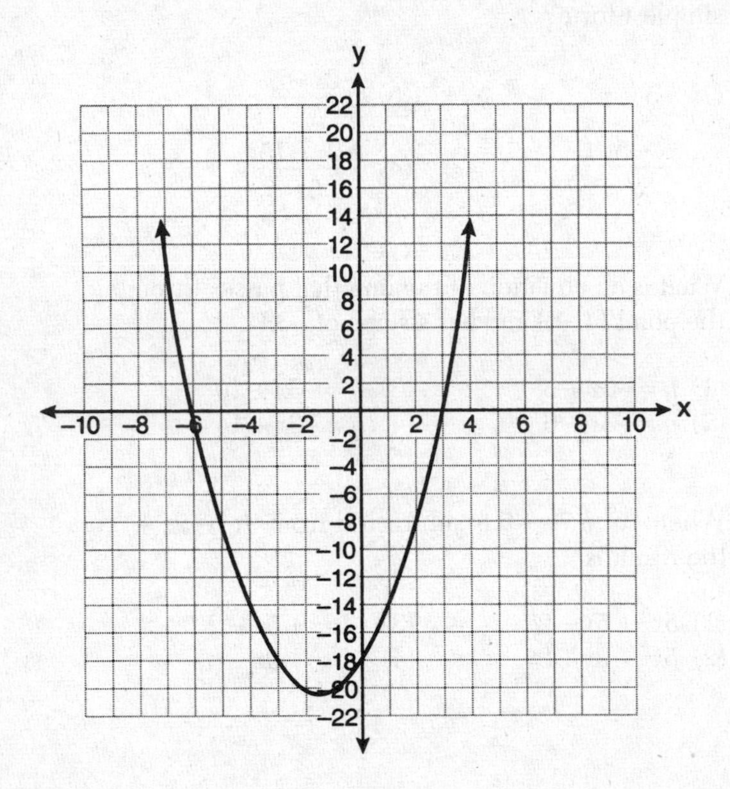

Based on this graph, what are the roots of the equation $x^2 + 3x - 18 = 0$?

(1) −3 and 6
(2) 0 and −18
(3) 3 and −6
(4) 3 and −18

24 ______

25 What is the value of the y-coordinate of the solution to the system of equations $x + 2y = 9$ and $x - y = 3$?

(1) 6
(2) 2
(3) 3
(4) 5

25 ____

26 What is the additive inverse of the expression $a - b$?

(1) $a + b$
(2) $a - b$
(3) $-a + b$
(4) $-a - b$

26 ____

27 What is the product of 12 and 4.2×10^6 expressed in scientific notation?

(1) 50.4×10^6
(2) 50.4×10^7
(3) 5.04×10^6
(4) 5.04×10^7

27 ____

28 To calculate the volume of a small wooden cube, Ezra measured an edge of the cube as 2 cm. The actual length of the edge of Ezra's cube is 2.1 cm. What is the relative error in his volume calculation to the *nearest hundredth*?

(1) 0.13
(2) 0.14
(3) 0.15
(4) 0.16

28 ____

29 What is $\frac{6}{4a} - \frac{2}{3a}$ expressed in simplest form?

(1) $\frac{4}{a}$

(2) $\frac{5}{6a}$

(3) $\frac{8}{7a}$

(4) $\frac{10}{12a}$

29 ____

30 The set $\{11,12\}$ is equivalent to

(1) $\{x|11 < x < 12$, where x is an integer$\}$

(2) $\{x|11 < x \leq 12$, where x is an integer$\}$

(3) $\{x|10 \leq x < 12$, where x is an integer$\}$

(4) $\{x|10 < x \leq 12$, where x is an integer$\}$

30 ____

PART II

Answer all questions in this part. Each correct answer will receive 2 credits. Clearly indicate the necessary steps, including appropriate formula substitutions, diagrams, graphs, charts, etc. For all questions in this part, a correct numerical answer with no work shown will receive only 1 credit. [6]

31 Determine how many three-letter arrangements are possible with the letters *A*, *N*, *G*, *L*, and *E* if no letter may be repeated.

32 Factor completely: $4x^3 - 36x$

33 Some books are laid on a desk. Two are English, three are mathematics, one is French, and four are social studies. Theresa selects an English book and Isabelle then selects a social studies book. Both girls take their selections to the library to read. If Truman then selects a book at random, what is the probability that he selects an English book?

PART III

Answer all questions in this part. Each correct answer will receive 3 credits. Clearly indicate the necessary steps, including appropriate formula substitutions, diagrams, graphs, charts, etc. For all questions in this part, a correct numerical answer with no work shown will receive only 1 credit. [9]

34 In the diagram below, the circumference of circle O is 16π inches. The length of $\overline{BC}$ is three-quarters of the length of diameter $\overline{AD}$ and $CE = 4$ inches. Calculate the area, in square inches, of trapezoid $ABCD$.

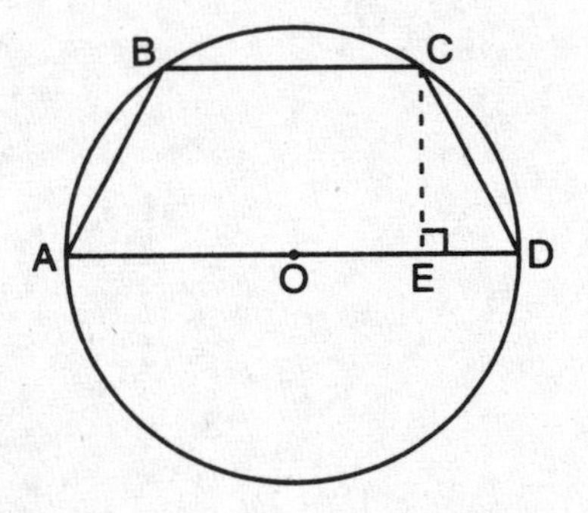

35 A bank is advertising that new customers can open a savings account with a $3\frac{3}{4}\%$ interest rate compounded annually. Robert invests \$5,000 in an account at this rate. If he makes no additional deposits or withdrawals on his account, find the amount of money he will have, to the *nearest cent*, after three years.

36 The table below shows the number of prom tickets sold over a ten-day period.

Prom Ticket Sales

Day (x)	1	2	5	7	10
Number of Prom Tickets Sold (y)	30	35	55	60	70

Plot these data points on the coordinate grid below. Use a consistent and appropriate scale. Draw a reasonable line of best fit and write its equation.

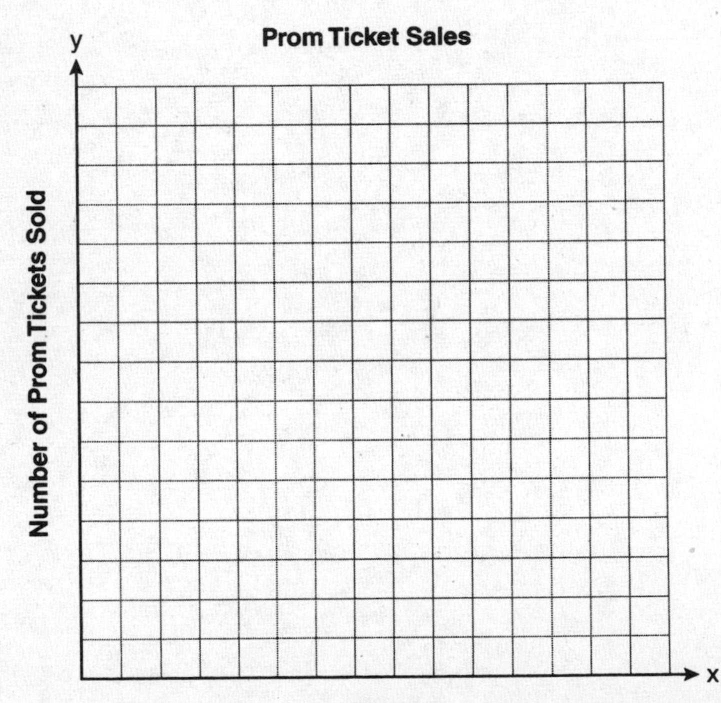

PART IV

Answer all questions in this part. Each correct answer will receive 4 credits. Clearly indicate the necessary steps, including appropriate formula substitutions, diagrams, graphs, charts, etc. For all questions in this part, a correct numerical answer with no work shown will receive only 1 credit. [12]

37 A stake is to be driven into the ground away from the base of a 50-foot pole, as shown in the diagram below. A wire from the stake on the ground to the top of the pole is to be installed at an angle of elevation of 52°.

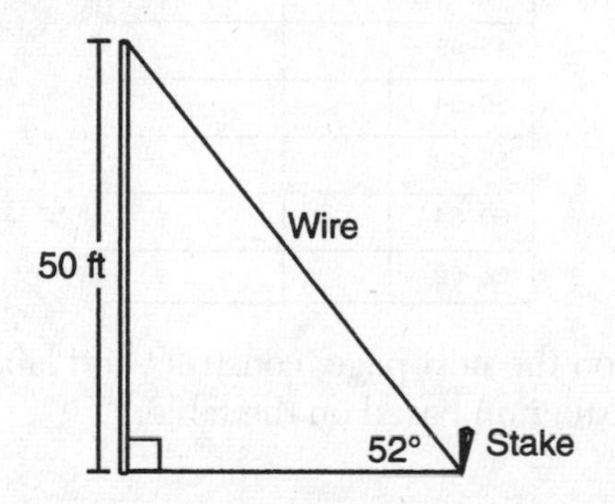

How far away from the base of the pole should the stake be driven in, to the *nearest foot*?

What will be the length of the wire from the stake to the top of the pole, to the *nearest foot*?

38 The Fahrenheit temperature readings on 30 April mornings in Stormville, New York, are shown below.

41°, 58°, 61°, 54°, 49°, 46°, 52°, 58°, 67°, 43°,
47°, 60°, 52°, 58°, 48°, 44°, 59°, 66°, 62°, 55°,
44°, 49°, 62°, 61°, 59°, 54°, 57°, 58°, 63°, 60°

Using the data, complete the frequency table below.

Interval	Tally	Frequency
40–44		
45–49		
50–54		
55–59		
60–64		
65–69		

On the grid on the next page, construct and label a frequency histogram based on the table.

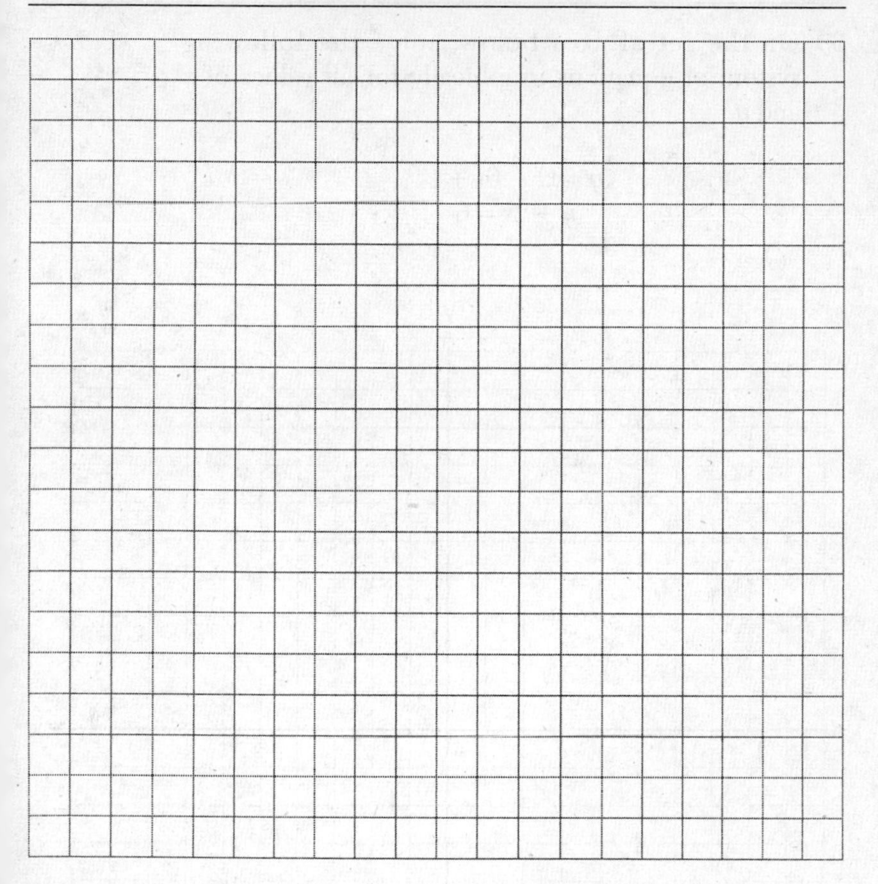

39 On the set of axes below, solve the following system of equations graphically for all values of x and y.

$$y = x^2 - 6x + 1$$
$$y + 2x = 6$$

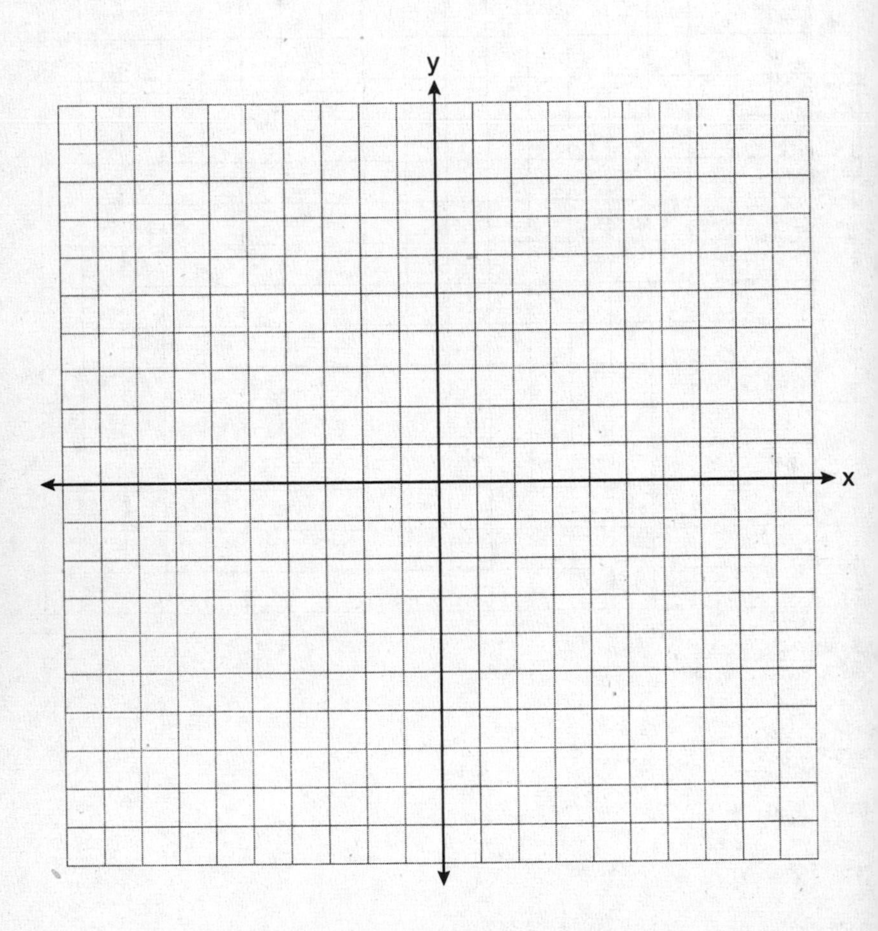

Answers June 2009

Integrated Algebra

Answer Key

PART I

1. (4)	**7.** (1)	**13.** (3)	**19.** (3)	**25.** (2)
2. (4)	**8.** (2)	**14.** (1)	**20.** (1)	**26.** (3)
3. (1)	**9.** (3)	**15.** (3)	**21.** (2)	**27.** (4)
4. (2)	**10.** (2)	**16.** (4)	**22.** (1)	**28.** (2)
5. (3)	**11.** (4)	**17.** (1)	**23.** (2)	**29.** (2)
6. (4)	**12.** (2)	**18.** (1)	**24.** (3)	**30.** (4)

PART II

31. 60

32. $4x(x-3)(x+3)$

33. $\frac{1}{8}$

PART III

34. 56 in^2

35. 5,583.86

36. $y = 5x + 25$; See the graph and best-fit line in the *Answers Explained* section.

PART IV

37. 39 feet; 63 feet

38. See the histogram in the *Answers Explained* section.

39. (–1,8) and (5,–4); See the graph in the *Answers Explained* section.

In **PARTS II–IV** you are required to show how you arrived at your answers. For sample methods of solutions, see the *Answers Explained* section.

Answers Explained

PART I

1. If Tammy takes 45 minutes to ride her bike 5 miles, then her rate is $\frac{5 \text{ miles}}{45 \text{ min}}$. If x represents the number of minutes it will take her to ride 8 miles, then her rate for this distance is $\frac{8 \text{ miles}}{x \text{ min}}$. Since the rates are the same:

$$\frac{5 \text{ miles}}{45 \text{ min}} = \frac{8 \text{ miles}}{x \text{ min}}$$
$$5x = 8 \cdot 45$$
$$\frac{5x}{5} = \frac{360}{5}$$
$$x = 72 \text{ min}$$

The correct choice is **(4)**.

2. The given equation is: $x^2 - 7x + 6 = 0$

Factor the quadratic trinomial as the product of two binomials: $(x + ?)(x + ?) = 0$

The missing terms are the two integers whose sum is −7, the coefficient of the x-term of $x^2 - 7x + 6$, and whose product is +6, the constant term of $x^2 - 7x + 6$. Since $(-1) + (-6) = -7$ and $(-1)(-6) = +6$: $(x - 1)(x - 6) = 0$

Set each factor equal to 0: $x - 1 = 0$ or $x - 6 = 0$

Solve each linear equation: $x = 1$ or $x = 6$

The roots of the given equation are 1 and 6.

The correct choice is **(4)**.

3. The given expression is: $\dfrac{27x^{18}y^5}{9x^6y}$

Divide the numerical coefficients, and then divide powers of the same base by subtracting their exponents: $\dfrac{27x^{18}y^5}{9x^6y^1} = \dfrac{27}{9} \cdot x^{18-6} \cdot y^{5-1}$

Simplify: $= 3x^{12}y^4$

The correct choice is **(1)**.

4. Marie currently has 58 stamps and buys s stamps each week. After 1 week, she has a total of $58 + s$ stamps. After 2 weeks, she has a total of $58 + s + s$ or $58 + 2s$ stamps. After 3 weeks, she has a total of $58 + s + s + s$ or $58 + 3s$ stamps, and so forth. After w weeks, she will have a total of $58 + sw$ stamps.

The correct choice is **(2)**.

5. *Qualitative* data refer to nonnumerical data. To identify the situation that could be classified as qualitative, examine each answer choice. Reject those answer choices that involve numerical data.

Choice (1): The ages of students are numerical data. ✗

Choice (2): Test scores are numerical data. ✗

Choice (3): Ice cream flavors are not numerical data. ✓

Choice (4): Heights of basketball players are numerical data. ✗

The correct choice is **(3)**.

6. It is given that a sign states "All riders **MUST** be at least 48 inches tall." If h represents the height of a rider in inches, then the statement on the sign may be translated as $h \geq 48$ since a rider's height may be equal to 48 inches or be greater than 48 inches.

The correct choice is **(4)**.

7. The given equation is: $\frac{2x}{3}+\frac{x}{6}=5$

Clear the equation of its fractions by multiplying each member of the equation by 6, the lowest common multiple of its denominators: $6\left(\frac{2x}{3}\right)+6\left(\frac{x}{6}\right)=6(5)$

Simplify: $\overset{2}{\cancel{6}}\left(\frac{2x}{\cancel{3}}\right)+\overset{1}{\cancel{6}}\left(\frac{x}{\cancel{6}}\right)=30$

$4x+x=30$

Combine like terms: $5x=30$

Divide both sides of the equation by 5: $\frac{5x}{5}=\frac{30}{5}$

$x=6$

The correct choice is **(1)**.

8. It is given that the accompanying table represents the results of students tossing a six-sided cube whose faces are numbered 1 to 6:

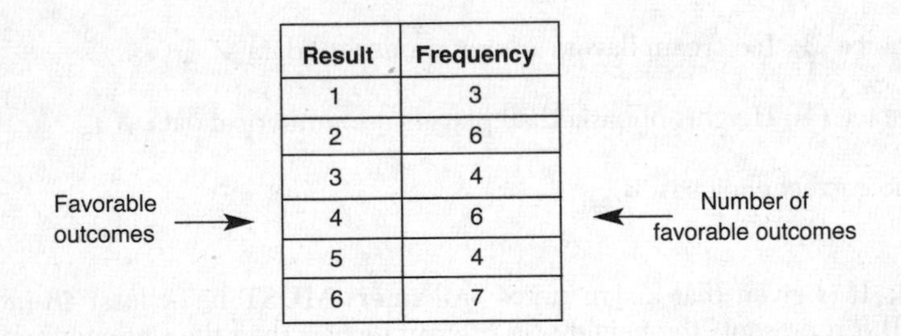

Result	Frequency
1	3
2	6
3	4
4	6
5	4
6	7

Empirical probability refers to calculating probability based on experimental results rather than theory. To find the empirical probability of tossing a 4 using the data in the table, form the ratio of the number of favorable outcomes to the total number of trials:

$$P(\text{tossing a } 4) = \frac{\text{Number of favorable outcomes}}{\text{Total number of trials}}$$
$$= \frac{6}{3+6+4+6+4+7}$$
$$= \frac{6}{30}$$

The correct choice is **(2)**.

9. You are asked to find the value of x, in inches, for the accompanying right triangle:

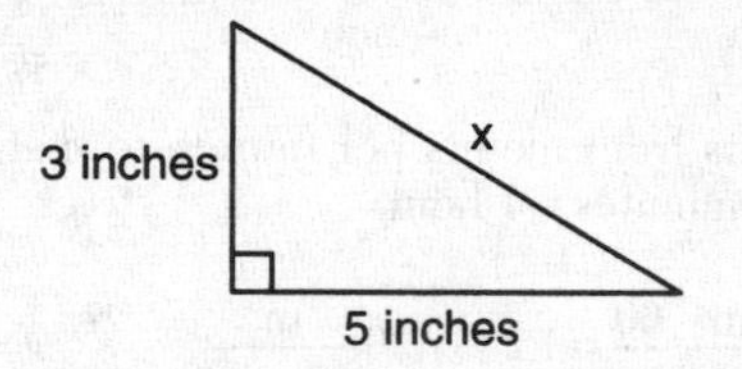

Since the side labeled x is opposite the right angle, it is the hypotenuse of the right triangle. Find x using the Pythagorean theorem:

$$(\text{Leg } 1)^2 + (\text{Leg } 2)^2 = (\text{Hypotenuse})^2$$
$$3^2 + 5^2 = x^2$$
$$9 + 25 = x^2$$
$$34 = x^2$$
$$\sqrt{34} = x$$

The correct choice is **(3)**.

10. The given radical is: $\sqrt{32}$

Factor the radicand as the product of two whole numbers such that one of these numbers is the greatest perfect square factor of 32: $\sqrt{16 \cdot 2}$

Write the radical over each factor: $\sqrt{16} \cdot \sqrt{2}$

Evaluate the square root of the perfect square factor: $4\sqrt{2}$

The correct choice is **(2)**.

11. You are asked to find the approximate speed of sound in meters per hour given that the speed of sound is 344 meters per second.

- Change the units from meters per second to meters per minute by using the fact that 60 seconds = 1 minute:

$$344\frac{\text{m}}{\text{sec}}\cdot\frac{60}{60}=20640\frac{\text{m}}{60\,\text{sec}}$$

$$=20640\frac{\text{m}}{\text{min}} \quad \leftarrow 60\,\text{sec}=1\text{ min}$$

- Change the units from meters per minute to meters per hour by using the fact that 60 minutes = 1 hour:

$$20640\frac{\text{m}}{\text{min}}\cdot\frac{60}{60}=1{,}238{,}400\frac{\text{m}}{60\text{ min}}$$

$$=1{,}238{,}400\frac{\text{m}}{\text{hour}} \quad \leftarrow 60\text{ min}=1\text{ hour}$$

The correct choice is **(4)**.

12. It is given that the sum of two numbers is 47 and their difference is 15. If x is the larger of the two numbers and y is the smaller:

$$x-y=15$$
$$x+y=47$$

Add the two equations:

$$2x+0\cdot y=62$$

$$\frac{2x}{2}=\frac{62}{2}$$

$$x=31$$

The correct choice is **(2)**.

13. The given equation is: $a + ar = b + r$

Isolate a by factoring it out from the left-side of the equation: $a(1 + r) = b + r$

Divide both sides of the equation by $1 + r$: $\frac{a\cancel{(1+r)}}{\cancel{1+r}} = \frac{b+r}{1+r}$

$$a = \frac{b+r}{1+r}$$

The correct choice is **(3)**.

14. If $\frac{4}{3}x + 5 < 17$, then

$$\frac{4}{3}x + 5 - 5 < 17 - 5$$

$$\frac{4}{3}x < 12$$

$$\frac{3}{4}\left(\frac{4}{3}x\right) < \frac{3}{\cancel{4}}(\cancel{12}^{3})$$

$$x < 9$$

The only number less than 9 in the set of answer choices is 8.

The correct choice is **(1)**.

15. In a box-and-whisker plot, the right-most vertical side of the box represents the upper or third quartile, as shown in the accompanying figure.

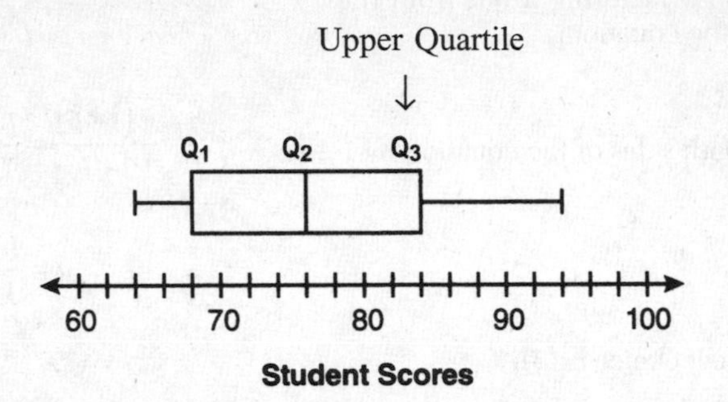

According to the scale used for student scores, each tick mark between 60 and 100 represents 2 units. The value of Q_3, the upper quartile, is 84.

The correct choice is **(3)**.

16. Since division by 0 is not defined, a fraction with a variable in the denominator is undefined for any value of the variable that makes the denominator equal 0. The given fraction, $\frac{5n}{2n-1}$, is undefined when $2n - 1 = 0$. Solving for n gives $2n = 1$, so $n = \frac{1}{2}$.

The correct choice is **(4)**.

17. At Genesee High School, the sophomore class has 60 more students than the freshman class. If x represents the number of students in the freshman class, then $x + 60$ is the number of students in the sophomore class.

The junior class has 50 fewer students less than twice the students in the freshman class. In terms of x, the junior class has $2x - 50$ students.

The senior class is three times as large as the freshman class. In terms of x, the senior class has $3x$ students.

There is a total of 1,424 students at Genesee High School. To find the number of students in the freshman class, solve for x:

$$\underbrace{x}_{\text{Freshmen}} + \underbrace{(x+60)}_{\text{Sophomores}} + \underbrace{(2x-50)}_{\text{Juniors}} + \underbrace{3x}_{\text{Seniors}} = \underbrace{1{,}424}_{\text{Total}}$$

$$7x + 10 = 1{,}424$$

$$7x = 1{,}424 - 10$$

$$\frac{7x}{7} = \frac{1{,}414}{7}$$

$$x = 202$$

There are 202 students in the freshman class.

The correct choice is **(1)**.

18. For a parabola whose equation is $y = ax^2 + bx + c$, an equation of the axis of symmetry is given by the formula $x = -\frac{b}{2a}$. If $y = x^2 - 16x + 63$, then $a = 1$ and $b = -16$:

$$x = -\frac{b}{2a}$$

$$= -\frac{(-16)}{2(1)}$$

$$= \frac{16}{2}$$

$$= 8$$

See the accompanying figure.

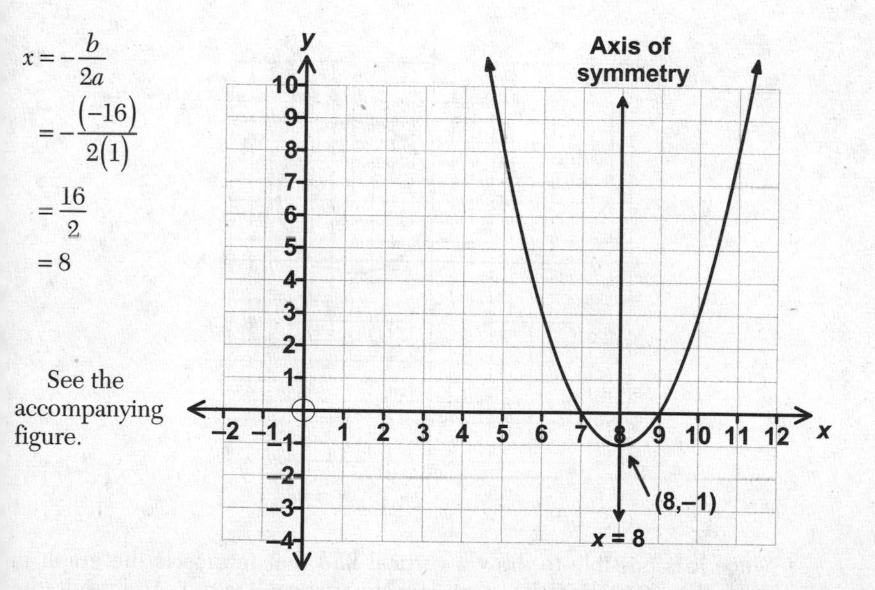

Since the axis of symmetry is a vertical line that contains the vertex, the x-coordinate of the vertex is 8.

Find the y-coordinate of the vertex by substituting 8 for x in the equation of the parabola:

$$\begin{aligned} y &= x^2 - 16x + 63 \\ &= 8^2 - 16(8) + 63 \\ &= 64 - 128 + 63 \\ &= -1 \end{aligned}$$

Hence, the vertex is at (8,–1), and the equation of the axis of symmetry is $x = 8$.

The correct choice is **(1)**.

19. You are asked to identify from the answer choices the statement that is true about the relation shown in the accompanying figure.

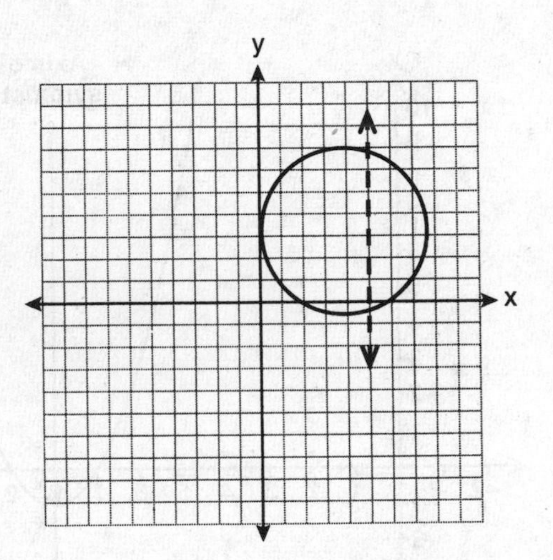

- Since it is possible to draw a vertical line that intersects the graph in more than one point, the graph fails the vertical line test. As a result, the graph does *not* represent a function. You can therefore eliminate answer choices (1) and (2) as each state the relation is a function.

- At the two points that the vertical line intersects the given graph, the x-coordinates are the same but the y-coordinates are different. Thus, there are two y-values for the same value of x. This corresponds to choice (3), which states the relation is not a function because there are multiple y-values for a given x-value.

The correct choice is **(3)**.

20. You are asked to identify the graph that represents the solution of $3y - 9 \leq 6x$.

- Since the inequality symbol includes a bar, it represents "is less than or equal to." Hence, the boundary line of its graph must be a solid rather than a broken line. This eliminates choices (2) and (4).
- Because the boundary line is the same in choices (1) and (3), you need to decide which side of the line includes the solution points. Solving the inequality for y gives $3y \leq 6x + 9$, so $y \leq \frac{6}{3}x + \frac{9}{3}$. Because the inequality relation is "less than or equal to," the solution set must consist of all points on or below the line, as shown by the shaded region in the graph in choice (1):

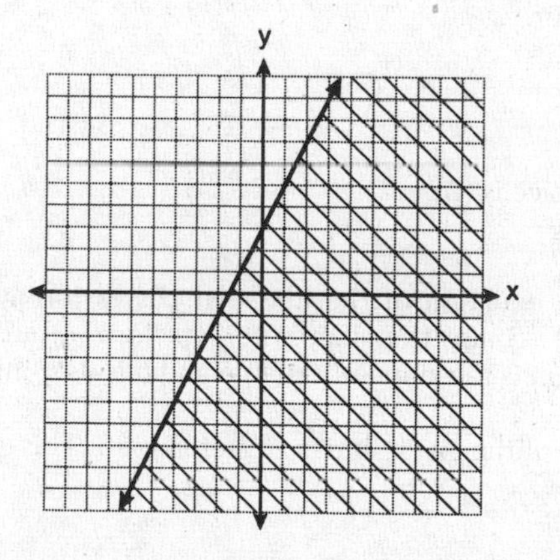

(1)

- Confirm your answer by choosing any convenient point in the solution set of the graph in choice (1), such as (0,0). Verify that it satisfies the given inequality:

$$3y - 9 \le 6x$$
$$3(0) - 9 \le 6(0)$$
$$-9 \le 0 \checkmark$$

The correct choice is **(1)**.

21. The given expression is: $\dfrac{x^2 - 2x - 15}{x^2 + 3x}$

Factor the quadratic trinomial in the numerator as the product of two binomials, and factor out the greatest common factor of x from the denominator: $\dfrac{(x+3)(x-5)}{x(x+3)}$

Divide out any factors common to both the numerator and the denominator as their quotient is 1: $\dfrac{\overset{1}{\cancel{(x+3)}}(x-5)}{x\cancel{(x+3)}}$

$$\frac{x-5}{x}$$

The correct choice is **(2)**.

22. To find an equation of the line that passes through the point (4,–6) and has a slope of –3, use the slope-intercept form of an equation of a line, $y = mx + b$, where m is the slope of the line and b is its y-intercept.

Since the slope of the given line is –3, set $m = -3$. This gives $y = -3x + b$.

To find b, substitute 4 for x and –6 for y into the equation $y = -3x + b$:

$$y = -3x + b$$

$$-6 = -3(4) + b$$

$$-6 = -12 + b$$

$$-6 + 12 = b$$

$$6 = b$$

Because $m = -3$ and $b = 6$, the equation of the required line is $y = -3x + 6$.

The correct choice is **(1)**.

23. To subtract $4x^2 + 7x - 5$ from $9x^2 - 2x + 3$, write the first polynomial underneath the second one. Then change this to an addition problem by taking the opposite of each term of the polynomial that is being subtracted:

$$\begin{array}{rr} & 9x^2 - 2x + 3 \\ - & \\ & \underline{4x^2 + 7x - 5} \end{array} \Rightarrow \begin{array}{rr} & 9x^2 - 2x + 3 \\ + & \\ & \underline{-4x^2 - 7x + 5} \\ & 5x^2 - 9x + 8 \end{array}$$

The difference is $5x^2 - 9x + 8$.

The correct choice is **(2)**.

24. The real roots of a quadratic equation of the form $ax^2 + bx + c = 0$ correspond to the x-coordinates of the points at which the parabola $y = ax^2 + bx + c$ crosses the x-axis.

The graph of $y = x^2 + 3x - 18$ is shown in the accompanying figure:

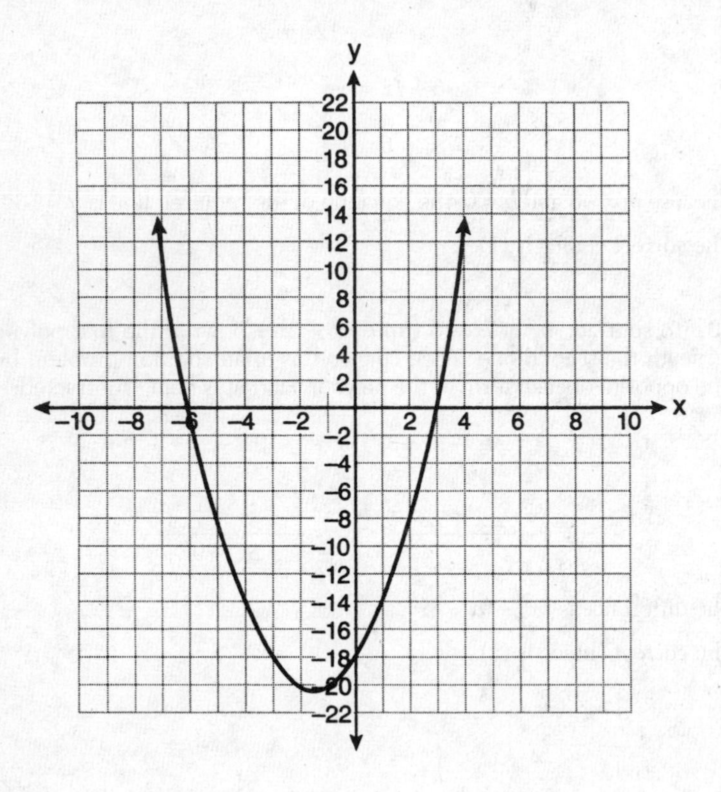

Based on the graph, you are asked to find the roots of $x^2 + 3x - 18 = 0$. Since the x-intercepts of the graph are at (3,0) and (–6,0), 3 and –6, are the solutions of the equation $x^2 + 3x - 18 = 0$.

The correct choice is **(3)**.

25. To find the y-coordinate of the solution of the system of equations $x + 2y = 9$ and $x - y = 3$, solve the system algebraically for y by subtracting corresponding sides of the two equations:

$$
\begin{array}{r}
x = 2y = 9 \\
- \qquad\qquad \\
x - y = 3 \\
\hline
0 \cdot x + 3y = 6 \\
\frac{3y}{3} = \frac{6}{3} \\
y = 2
\end{array}
$$

The correct choice is **(2)**.

26. The additive inverse of a real number is its opposite since their sum is 0. To find the additive inverse of $a - b$, find its opposite:

$$-(a - b) = -a + b.$$

The correct choice is **(3)**.

27. To write the product of 12 and 4.2×10^6 in scientific notation, multiply the numbers together and then write the product in the form $a \times 10^N$, where a is between 1 and 10 and N is an integer:

$$
\begin{aligned}
12 \times (4.2 \times 10^6) &= 50.4 \times 10^6 \\
&= (5.04 \times 10) \times 10^6 \\
&= 5.04 \times 10^7
\end{aligned}
$$

The correct choice is **(4)**.

28. To determine the relative error in a measurement, evaluate the following ratio:

$$\frac{\text{Actual value} - \text{Measured value}}{\text{Actual value}}$$

You are asked to find the relative error in the measurement of the volume of a cube to the *nearest hundredth*, given the measurements of an edge.

- The actual length of an edge of a cube is given as 2.1 cm. The actual volume of the cube is $2.1 \text{ cm} \times 2.1 \text{ cm} \times 2.1 \text{ cm} = 9.261 \text{ cm}^3$.

- You are also told that Ezra measured an edge of the same cube as 2 cm. Based on his measurement, the measured volume of the cube is 2 cm × 2 cm × 2 cm = 8 cm^3.
- The relative error is

$$\frac{9.261-8}{9.261}=\frac{1.261}{9.261}\approx 0.1361624015.$$

The relative error in Ezra's volume calculation, to the *nearest hundredth*, is 0.14.

The correct choice is **(2)**.

29. To subtract unlike fractions, rewrite each fraction with a common denominator. Then write the difference of their numerators over the common denominator. The given expression is:

$$\frac{6}{4a}-\frac{2}{3a}$$

The lowest common denominator (LCD) is $12a$ since $12a$ is the smallest expression into which $4a$ and $3a$ both divide evenly. Change the first fraction into an equivalent fraction with the LCD as its denominator by multiplying it by 1 in the form of $\frac{3}{3}$. Change the second fraction into an equivalent fraction with the LCD as its denominator by multiplying it by 1 in the form of $\frac{4}{4}$:

$$\frac{3}{3}\left(\frac{6}{4a}\right)-\frac{4}{4}\left(\frac{2}{3a}\right)=\frac{18}{12a}-\frac{8}{12a}$$

Write the difference of the numerators over the common denominator:

$$=\frac{18-8}{12a}$$

Simplify:

$$=\frac{10}{12a}$$

$$=\frac{10\div 2}{12a\div 2}$$

$$=\frac{5}{6a}$$

The correct choice is **(2)**.

30. To find the set that is equivalent to the set {11,12}, determine the specific numbers that are contained in each of the sets in the answer choices:

- Choice (1): The notation $\{x|11 < x < 12$, where x is an integer$\}$ represents the set of all integers between 11 and 12, which is the empty set. ✗
- Choice (2): The notation $\{x|11 < x \leq 12$, where x is an integer$\}$ represents the set of all integers greater than 11 and less than or equal to 12. This set is equivalent to {12}. ✗
- Choice (3): The notation $\{x|10 \leq x < 12$, where x is an integer$\}$ represents the set of all integers greater than or equal to 10 and less than 12. This set is equivalent to {10,11}. ✗
- Choice (4): The notation $\{x|10 < x \leq 12$, where x is an integer$\}$ represents the set of all integers greater than 10 and less than or equal to 12. This set is equivalent to {11,12}. ✓

The correct choice is **(4)**.

PART II

31. To determine the number of three-letter arrangements that are possible with the letters A, N, G, L, and E if no letter may be repeated, draw three boxes to indicate the positions of the letters in a three-letter arrangement:

Any one of the 5 letters can be used to fill the first position:

$$\begin{array}{ccccc} \text{First} & & \text{Second} & & \text{Third} \\ \boxed{5} & \times & \boxed{} & \times & \boxed{} \end{array}$$

Once the first position is filled, any one of the remaining 4 letters can be selected to fill the second position:

$$\begin{array}{ccccc} \text{First} & & \text{Second} & & \text{Third} \\ \boxed{5} & \times & \boxed{4} & \times & \boxed{} \end{array}$$

After the first two positions are filled, any one of the remaining 3 letters can fill the third position:

$$\begin{array}{ccccc} \text{First} & & \text{Second} & & \text{Third} \\ \boxed{5} & \times & \boxed{4} & \times & \boxed{3} \end{array}$$

According to the multiplication principle of counting, the three positions can be filled in $5 \times 4 \times 3 = 60$ ways.

There are **60** three-letter arrangements that are possible.

32. To factor $4x^3 - 36x$ completely, first factor out the greatest common factor, which is $4x$:

$$4x(x^2 - 9)$$

Factor the binomial next. Since $x^2 - 9$ represents the difference of two squares, it can be factored as the product of the sum and difference of the same two terms:

$$4x(x - 3)(x + 3)$$

The answer is $\mathbf{4x(x - 3)(x + 3)}$.

33. It is given that on a desk are two English books, three mathematics books, one French book, and four social studies books:

E E   M M M   F   S S S S

After Theresa selects an English book, Isabelle then selects a social studies book. The desk now has these eight books on it:

~~E~~ E   M M M   F   ~~S~~ S S S

Truman then selects a book at random from the desk. You are asked to find the probability that he selects an English book. Since one of the remaining eight books is an English book:

$$P(\text{English}) = \frac{\text{Number of favorable outcomes}}{\text{Total number of possible outcomes}}$$
$$= \frac{1}{8}$$

The probability Truman selects an English book is $\mathbf{\frac{1}{8}}$.

PART III

34. In the accompanying diagram, the circumference of circle O is 16π inches, the length of $\overline{BC}$ is three-quarters of the length of diameter $\overline{AD}$, and $CE = 4$ inches.

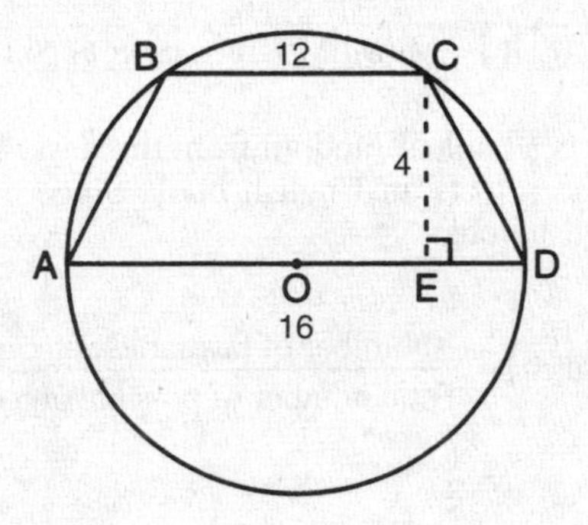

You are asked to find the area, in square inches, of trapezoid $ABCD$.

Since circumference = $\pi \times$ diameter, the diameter of circle O is 16 inches. To find the length of BC:

$$BC = \frac{3}{\cancel{4}} \times \overset{4}{\cancel{16}} \text{ inches}$$
$$= 12 \text{ inches}$$

Use the formula for the area of a trapezoid that is provided on the reference sheet:

$$\text{Area of trapezoid} = \frac{1}{2}h\left(b_1 + b_2\right)$$
$$= \frac{1}{2}CE\left(\overline{AD} + \overline{BC}\right)$$
$$= \frac{1}{2}(4)(16+12)$$
$$= (2)(28)$$
$$= 56 \text{ in}^2$$

The area of the trapezoid is **56 in^2**.

35. It is given that Robert invests $5,000 in a new savings account with a $3\frac{3}{4}\%$ interest rate compounded annually. If he makes no additional deposits or withdrawals on his account, then the amount of money, A, that Robert will have after three years grows exponentially according to the formula $A = A_0(1 + r)^n$, where A_0 is the amount initially invested at an annual rate of $r\%$ compounded for n years. Use your calculator to evaluate this formula for $A_0 = 5{,}000$, $r = 3\frac{3}{4}\% = 0.0375$, and $n = 3$:

$$\begin{aligned} A &= 5{,}000(1 + 0.0375)^3 \\ &= 5{,}000(1.0375)^3 \\ &= 5{,}583.857422 \end{aligned}$$

The amount of money he will have, to the *nearest cent*, is **$5,583.86.**

36. The accompanying table shows the number of prom tickets sold over a ten-day period.

Prom Ticket Sales

Day (x)	1	2	5	7	10
Number of Prom Tickets Sold (y)	30	35	55	60	70

- You are asked to plot these data points on the coordinate grid provided and then to draw a reasonable line of best fit. See the accompanying figure.

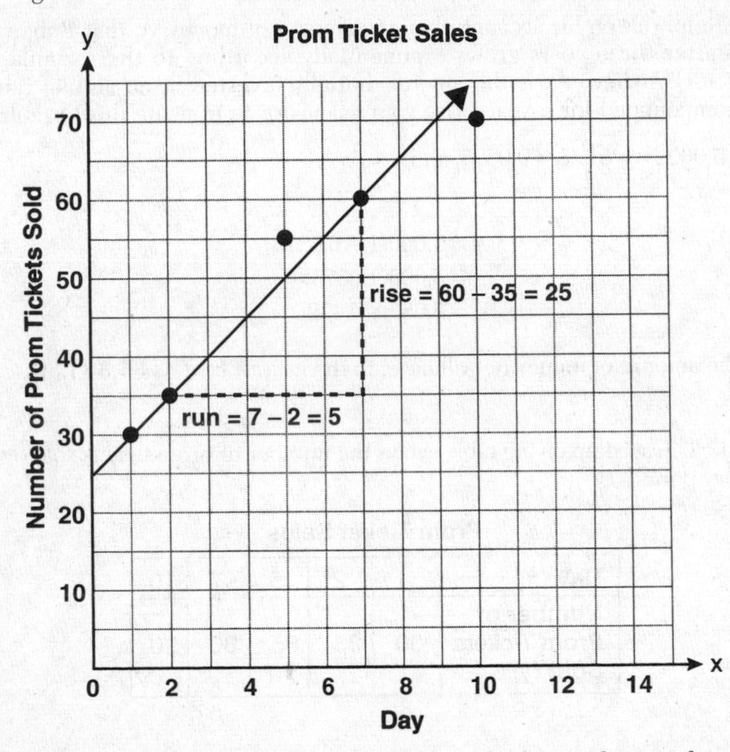

- Use a straight edge to draw a line of best fit that is close to the set of plotted points. Since you are also required to write an equation of the line of best fit, draw the line of best fit so that its y-intercept can be easily read from the graph, as shown in the accompanying figure.

- Write an equation of the line of best fit using the point-slope form of the equation of a line, $y = mx + b$. Refer to the figure to determine that the y-intercept is 25, so $b = 25$. To find the slope of the best-fit line, pick any two convenient points on the graph, such as (2,35) and (7,60). Then calculate $\frac{\text{rise}}{\text{run}}$ as shown in the accompanying figure. By using these points, slope $= m = \frac{25}{5} = 5$.

Since $m = 5$ and $b = 25$, an equation of the best-fit line drawn in the figure is $\boldsymbol{y = 5x + 25}$.

PART IV

37. A stake is to be driven into the ground away from the base of a 50-foot pole as shown in the accompanying diagram. A wire from the stake on the ground to the top of the pole is to be installed at an angle of elevation of 52°.

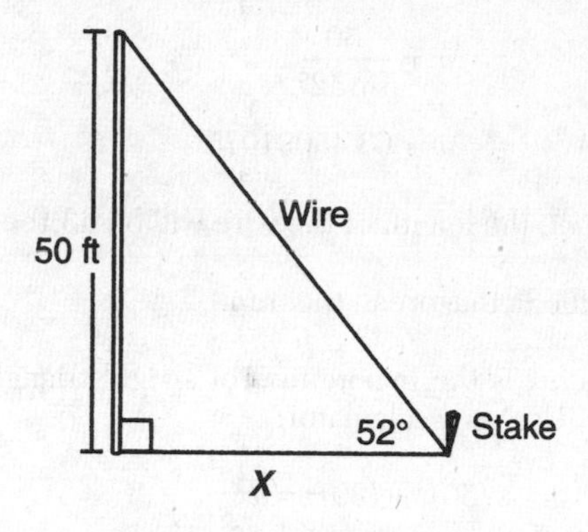

You are asked to determine how far away from the base of the pole the stake should be driven in, to the *nearest foot*. Label this distance in the figure as x. Use your calculator to find x using the tangent ratio:

$$\tan 52° = \frac{\text{Leg opposite } 52° \text{ angle}}{\text{Leg adjacent to } 52° \text{ angle}}$$

$$\tan 52° = \frac{50}{x}$$

$$x = \frac{50}{\tan 52°}$$

$$\approx 39.06428133$$

To the *nearest foot*, the stake must be driven in **39 feet** from the base of the pole.

You are also asked to find the length of the wire from the stake to the top of the pole, to the *nearest foot*. Call this length w.

Method 1: Use trigonometry.

Find w using the sine ratio and your calculator:

$$\sin 52° = \frac{\text{Leg opposite } 52° \text{ angle}}{\text{Hypotenuse}}$$

$$\sin 52° = \frac{50}{w}$$

$$w = \frac{50}{\sin 52°}$$

$$w \approx 63.45091075$$

To the *nearest foot*, the length of the wire will be **63 feet**.

Method 2: Use the Pythagorean theorem.

The wire length, w, is the hypotenuse of a right triangle whose legs measure 50 ft and 39 ft. Use your calculator:

$$(50)^2 + (39)^2 = w^2$$
$$2500 + 1521 = w^2$$
$$4021 = w^2$$
$$\sqrt{4021} = w$$
$$w \approx 63.41135545$$

To the *nearest foot*, w = **63 feet**.

38. The Fahrenheit temperature readings on 30 April mornings are given:

41°, 58°, 61°, 54°, 49°, 46°, 52°, 58°, 67°, 43°, 47°, 60°, 52°, 58°, 48°, 44°, 59°, 66°, 62°, 55°, 44°, 49°, 62°, 61°, 59°, 54°, 57°, 58°, 63°, 60°

Using the data, you are asked to complete a frequency table:

Interval	Tally	Frequency
40–44	IIII	4
45–49	~~IIII~~	5
50–54	IIII	4
55–59	~~IIII~~ III	8
60–64	~~IIII~~ II	7
65–69	II	2

Total = 30

You are also asked to construct and label a frequency histogram based on the frequency table, as shown in the accompanying figure:

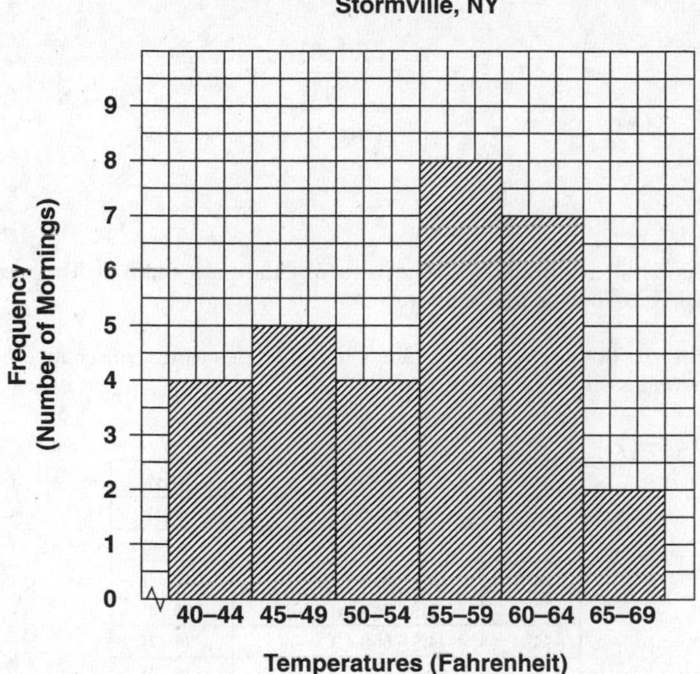

39. You are required to solve the following system of equations graphically for all values of x and y:

$$y = x^2 - 6x + 1$$
$$y + 2x = 6$$

To solve the system, find the coordinates of the points at which their graphs intersect.

First graph the quadratic equation. The graph of the quadratic equation $y = x^2 - 6x + 1$ is a parabola. An equation of the axis of symmetry of this parabola has the form $x = -\frac{b}{2a}$, where $a = 1$ and $b = -6$:

$$\begin{aligned} x &= -\frac{(-6)}{2(1)} \\ &= \frac{6}{2} \\ &= 3 \end{aligned}$$

Hence, the x-coordinate of the vertex of the parabola is 3. To graph $y = x^2 - 6x + 1$, follow the following steps.

- Step 1: Prepare a table of values that includes three consecutive integer x-values on either side of $x = 3$.

x	$x^2 - 6x + 1 = y$	(x,y)
0	$(0)^2 - 6(0) + 1 = 1$	(0,1)
1	$(1)^2 - 6(1) + 1 = -4$	(1,–4)
2	$(2)^2 - 6(2) + 1 = -7$	(2,–7)
3	$(3)^2 - 6(3) + 1 = -8$	(3,–8)
4	$(4)^2 - 6(4) + 1 = -7$	(4,–7)
5	$(5)^2 - 6(5) + 1 = -4$	(5,–4)
6	$(6)^2 - 6(6) + 1 = 1$	(6,1)

TIP: Use the table feature of your graphing calculator to create a table of values.

- Step 2: Check for symmetry in the y-values listed in the table.

Corresponding pairs of points on either side of $x = 3$ have matching y-coordinates. This confirms that (3,–8) is the vertex of the parabola.

- Step 3: Plot (0,1), (1,–4), (2,–7), (3,–8), (4,–7), (5,–4), and (6,1) on the grid provided. Then connect these points with a smooth, U-shaped curve, as shown in the accompanying figure. Label the graph with its equation.

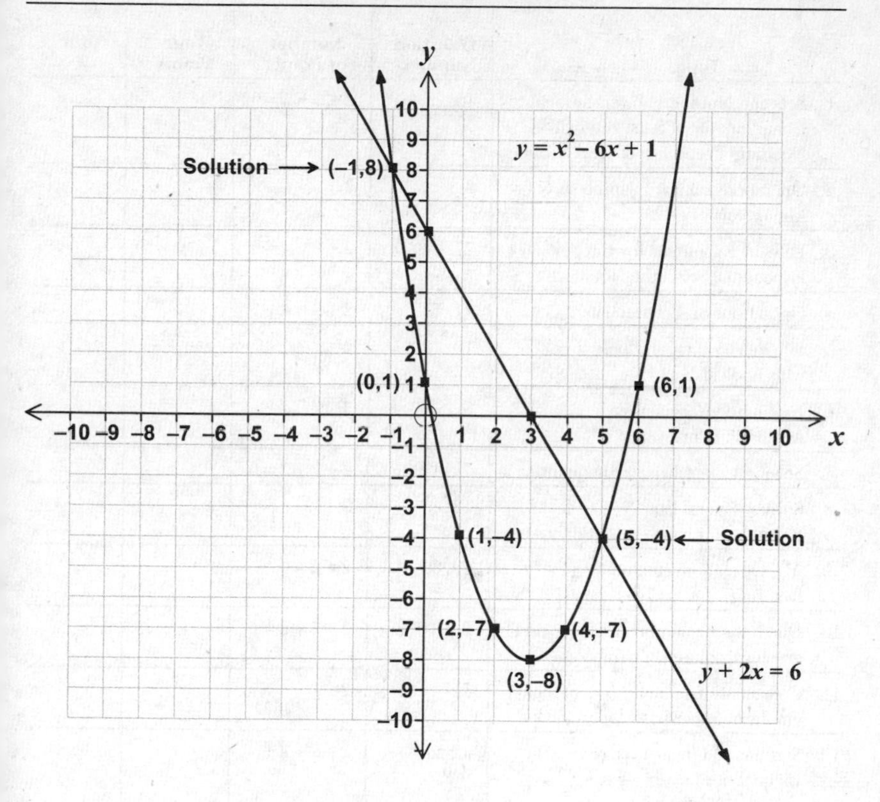

Now graph the linear equation on the same set of axes. To graph $y + 2x = 6$, find two convenient points on the line. If $x = 0$, $y + 2(0) = 6$, so $y = 6$. Hence, the line contains the point (0,6). If $y = 0$, $2x = 6$, so $x = \frac{6}{2} = 3$. The line also contains the point (3,0). Plot these two points, and draw a line through them as shown in the accompanying diagram. Label the line with its equation.

Finally, state the solution. The graphs intersect at **(−1,8)** and **(5,−4)**, which represent the solution to the given system of equations.

Topic	Question Numbers	Number of Points	Your Points	Your %
1. Sets and Numbers; Intersection and Complements of Sets; Interval Notation; Properties of Real Numbers	16, 26, 30	2 + 2 + 2 = 6		
2. Operations on Rat'l. Numbers & Monomials	3	2		
3. Laws of Exponents for Integer Exponents; Scientific Notation	27	2		
4. Operations on Polynomials	21, 23	2 + 2 = 4		
5. Square Root; Operations with Radicals	10	2		
6. Evaluating Formulas & Algebraic Expressions	6	2		
7. Solving Linear Eqs. & Inequalities	7, 14, 20	2 + 2 + 2 = 6		
8. Solving Literal Eqs. & Formulas for a Given Letter	13	2		
9. Alg. Operations (including factoring)	29, 32	2 + 2 = 4		
10. Quadratic Equations (incl. alg. and graphical solutions; parabolas)	2, 18, 24	2 + 2 + 2 = 6		
11. Coordinate Geometry (eq. of a line; graphs of linear eqs; slope)	22	2		
12. Systems of Linear Eqs. & Inequalities (algebraic & graphical solutions)	12, 25	2 + 2 = 4		
13. Mathematical Modeling (using: eqs.; tables; graphs)	4, 17	2 + 2 = 4		
14. Linear-Quadratic systems	39	4		
15. Perimeter; Circumference; Area of Common Figures (including trapezoids)	34	3		
16. Volume and Surface Area; Area of Overlapping Figures; Relative Error in Measurement	28	2		
17. Fractions and percent	—	—		
18. Ratio & Proportion (incl. similar polygons, scale drawings, & rates)	1, 11	2 + 2 = 4		
19. Pythagorean Theorem	9	2		
20. Right Triangle Trigonometry	37	4		
21. Functions (def.; domain and range; vertical line test; absolute value)	19	2		

Topic	Question Numbers	Number of Points	Your Points	Your %
22. Exponential Functions (properties; growth and decay)	35	3		
23. Probability (incl. tree diagrams & sample spaces)	8, 33	2 + 2 = 4		
24. Permutations and Counting Methods (incl. Venn diagrams)	31	2		
25. Statistics (mean, median, percentiles, quartiles; freq. dist., histograms; box-and-whisker plots; causality; bivariate data; qualitative vs. quantitative data; unbiased vs. biased samples; circle graphs)	5, 15, 38	2 + 2 + 4 = 8		
26. Line of Best Fit (including linear regression, scatter plots, and linear correlation)	36	3		
27. Nonroutine Word Problems Requiring Arith. or Alg. Reasoning	—	—		

MAP TO LEARNING STANDARDS

Content Strand	Item Numbers
Number Sense and Operations	10, 26, 27, 31
Algebra	2, 3, 4, 6, 7, 9, 12, 13, 14, 16, 17, 18, 21, 22, 23, 25, 29, 30, 32, 35, 37
Geometry	19, 20, 24, 34, 39
Measurement	1, 11, 28
Probability and Statistics	5, 8, 15, 33, 36, 38

HOW TO CONVERT YOUR RAW SCORE TO YOUR INTEGRATED ALGEBRA REGENTS EXAMINATION SCORE

Below is the conversion chart that must be used to determine your final score on the June 2009 Regents Examination in Integrated Algebra. To find your final exam score, locate in the column labeled "Raw Score" the total number of points you scored out of a possible 87 points. Since partial credit is allowed in Parts II, III, and IV of the test, you may need to approximate the credit you would receive for a solution that is not completely correct. Then locate in the adjacent column to the right the scale score that corresponds to your raw score. The scale score is your final Integrated Algebra Regents Examination score.

Regents Examination in Integrated Algebra—June 2009 Chart for Converting Total Test Raw Scores to Final Examination Scores (Scaled Scores)

Raw Score	Scaled Score	Raw Score	Scaled Score	Raw Score	Scaled Score
87	**100**	57	**81**	27	**61**
86	**99**	56	**81**	26	**60**
85	**98**	55	**81**	25	**59**
84	**96**	54	**80**	24	**57**
83	**95**	53	**80**	23	**56**
82	**94**	52	**80**	22	**54**
81	**93**	51	**79**	21	**53**
80	**92**	50	**79**	20	**51**
79	**92**	49	**78**	19	**49**
78	**91**	48	**78**	18	**48**
77	**90**	47	**78**	17	**46**
76	**89**	46	**77**	16	**44**
75	**89**	45	**77**	15	**42**
74	**88**	44	**76**	14	**40**
73	**88**	43	**75**	13	**38**
72	**87**	42	**75**	12	**36**
71	**87**	41	**74**	11	**33**
70	**86**	40	**74**	10	**31**
69	**86**	39	**73**	9	**29**
68	**86**	38	**72**	8	**26**
67	**85**	37	**71**	7	**23**
66	**84**	36	**71**	6	**20**
65	**84**	35	**70**	5	**17**
64	**84**	34	**69**	4	**14**
63	**84**	33	**68**	3	**11**
62	**83**	32	**67**	2	**7**
61	**83**	31	**66**	1	**4**
60	**82**	30	**65**	0	**0**
59	**82**	29	**64**		
58	**82**	28	**62**		

Examination August 2009
Integrated Algebra

FORMULAS

Trigonometric ratio $\sin A = \frac{\text{opposite}}{\text{hypotenuse}}$

$\cos A = \frac{\text{adjacent}}{\text{hypotenuse}}$

$\tan A = \frac{\text{opposite}}{\text{adjacent}}$

Area Trapezoid $A = \frac{1}{2}h(b_1 + b_2)$

Volume Cylinder $V = \pi r^2 h$

Surface area Rectangular prism $SA = 2lw + 2hw + 2lh$

Cylinder $SA = 2\pi r^2 + 2\pi rh$

Coordinate geometry $m = \frac{\Delta y}{\Delta x} = \frac{y_2 - y_1}{x_2 - x_1}$

PART I

Answer all 30 questions in this part. Each correct answer will receive 2 credits. No partial credit will be allowed. For each question, write in the space provided the numeral preceding the word or expression that best completes the statement or answers the question. [60]

1 If h represents a number, which equation is a correct translation of "Sixty more than 9 times a number is 375"?

(1) $9h = 375$ (3) $9h - 60 = 375$
(2) $9h + 60 = 375$ (4) $60h + 9 = 375$

1 _____

2 Which expression is equivalent to $9x^2 - 16$?

(1) $(3x + 4)(3x - 4)$ (3) $(3x + 8)(3x - 8)$
(2) $(3x - 4)(3x - 4)$ (4) $(3x - 8)(3x - 8)$

2 _____

3 Which expression represents $(3x^2y^4)(4xy^2)$ in simplest form?

(1) $12x^2y^8$ (3) $12x^3y^8$
(2) $12x^2y^6$ (4) $12x^3y^6$

3 _____

4 An online music club has a one-time registration fee of \$13.95 and charges \$0.49 to buy each song. If Emma has \$50.00 to join the club and buy songs, what is the maximum number of songs she can buy?

(1) 73 (3) 130
(2) 74 (4) 131

4 _____

5 The local ice cream stand offers three flavors of soft-serve ice cream: vanilla, chocolate, and strawberry; two types of cone: sugar and wafer; and three toppings: sprinkles, nuts, and cookie crumbs. If Dawn does not order vanilla ice cream, how many different choices can she make that have one flavor of ice cream, one type of cone, and one topping?

(1) 7
(2) 8
(3) 12
(4) 18

5 ____

6 Nancy's rectangular garden is represented in the diagram below.

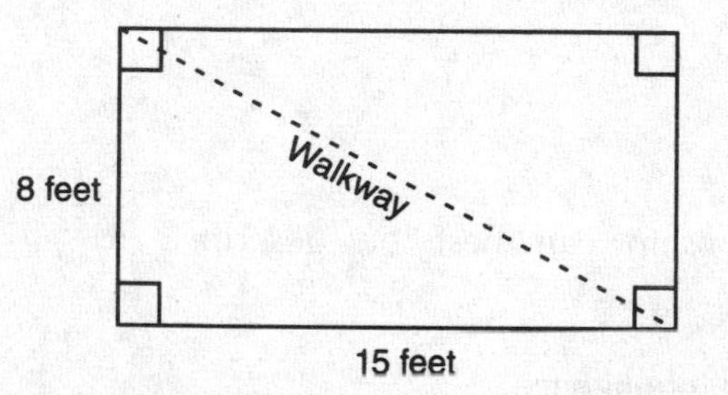

If a diagonal walkway crosses her garden, what is its length, in feet?

(1) 17
(2) 22
(3) $\sqrt{161}$
(4) $\sqrt{529}$

6 ____

7 The spinner below is divided into eight equal regions and is spun once. What is the probability of *not* getting red?

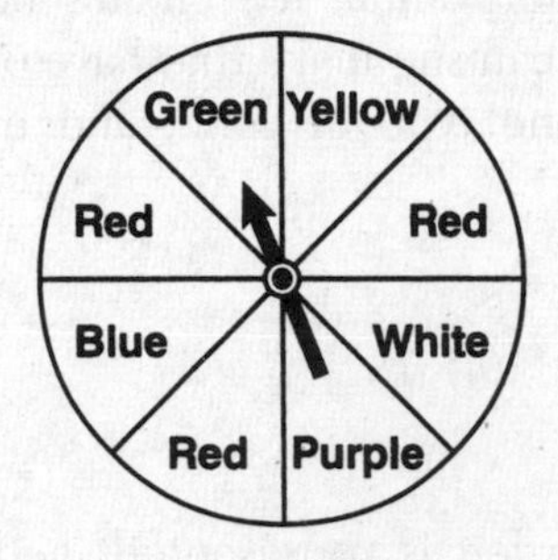

(1) $\frac{3}{5}$ (3) $\frac{5}{8}$

(2) $\frac{3}{8}$ (4) $\frac{7}{8}$ 7 ____

8 Which relationship can best be described as causal?

(1) height and intelligence
(2) shoe size and running speed
(3) number of correct answers on a test and test score
(4) number of students in a class and number of students with brown hair 8 ____

9 Solve for x: $\frac{3}{5}(x+2)=x-4$

(1) 8 (3) 15
(2) 13 (4) 23 9 ____

10 Erica is conducting a survey about the proposed increase in the sports budget in the Hometown School District. Which survey method would likely contain the *most* bias?

(1) Erica asks every third person entering the Hometown Grocery Store.
(2) Erica asks every third person leaving the Hometown Shopping Mall this weekend.
(3) Erica asks every fifth student entering Hometown High School on Monday morning.
(4) Erica asks every fifth person leaving Saturday's Hometown High School football game.

10 _____

11 Which equation represents a line parallel to the x-axis?

(1) $y = -5$ (3) $x = 3$
(2) $y = -5x$ (4) $x = 3y$

11 _____

12 Given:

$A = \{$All even integers from 2 to 20, inclusive$\}$
$B = \{10, 12, 14, 16, 18\}$

What is the complement of set B within the universe of set A?

(1) $\{4, 6, 8\}$ (3) $\{4, 6, 8, 20\}$
(2) $\{2, 4, 6, 8\}$ (4) $\{2, 4, 6, 8, 20\}$

12 _____

13 Which value of x is in the solution set of the inequality $-2(x - 5) < 4$?

(1) 0 (3) 3
(2) 2 (4) 5 13 ____

14 A tree casts a 25-foot shadow on a sunny day, as shown in the diagram below.

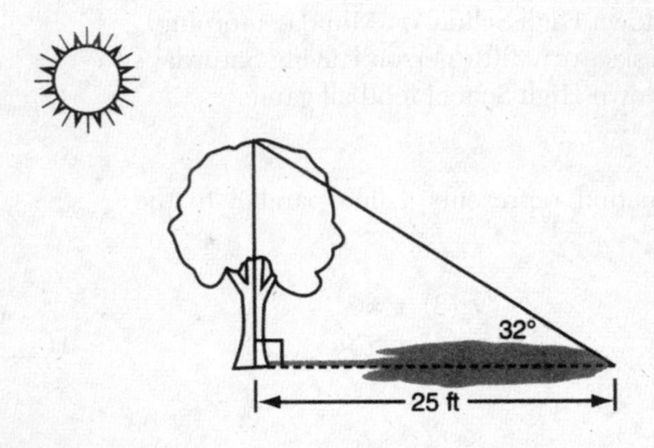

If the angle of elevation from the tip of the shadow to the top of the tree is 32°, what is the height of the tree to the *nearest tenth of a foot*?

(1) 13.2 (3) 21.2
(2) 15.6 (4) 40.0 14 ____

15 What is the slope of the line that passes through the points (–5,4) and (15,–4)?

(1) $-\frac{2}{5}$ (3) $-\frac{5}{2}$
(2) 0 (4) undefined 15 ____

16 The equation $y = -x^2 - 2x + 8$ is graphed on the set of axes below.

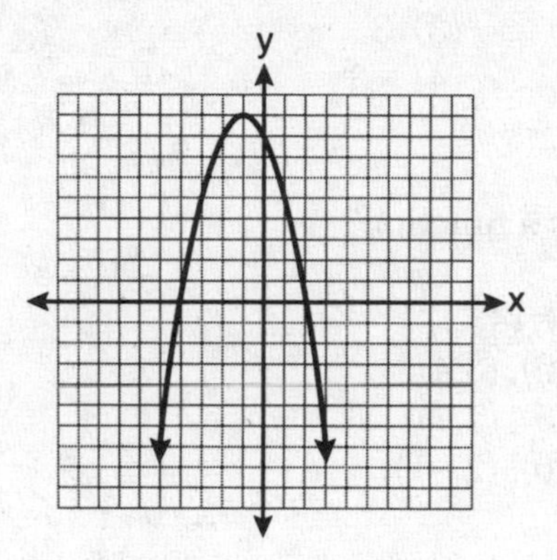

Based on this graph, what are the roots of the equation $-x^2 - 2x + 8 = 0$?

(1) 8 and 0
(2) 2 and –4
(3) 9 and –1
(4) 4 and –2

16 ____

17 What is the sum of $\frac{3}{2x}$ and $\frac{4}{3x}$ expressed in simplest form?

(1) $\frac{12}{6x^2}$
(2) $\frac{17}{6x}$
(3) $\frac{7}{5x}$
(4) $\frac{17}{12x}$

17 ____

18 Which value of x makes the expression $\frac{x^2-9}{x^2+7x+10}$ undefined?

(1) –5
(2) 2
(3) 3
(4) –3

18 _____

19 Which relation is *not* a function?

(1) {(1,5), (2,6), (3,6), (4,7)}
(2) {(4,7), (2,1), (–3,6), (3,4)}
(3) {(–1,6), (1,3), (2,5), (1,7)}
(4) {(–1,2), (0,5), (5,0), (2,–1)}

19 _____

20 What is the value of the y-coordinate of the solution to the system of equations $x - 2y = 1$ and $x + 4y = 7$?

(1) 1
(2) –1
(3) 3
(4) 4

20 _____

21 The solution to the equation $x^2 - 6x = 0$ is

(1) 0, only
(2) 6, only
(3) 0 and 6
(4) $\pm\sqrt{6}$

21 _____

22 When $5\sqrt{20}$ is written in simplest radical form, the result is $k\sqrt{5}$. What is the value of k?

(1) 20
(2) 10
(3) 7
(4) 4

22 _____

23 What is the value of the expression $|-5x + 12|$ when $x = 5$?

(1) -37 (3) 13
(2) -13 (4) 37

23 _____

24 A playground in a local community consists of a rectangle and two semicircles, as shown in the diagram below.

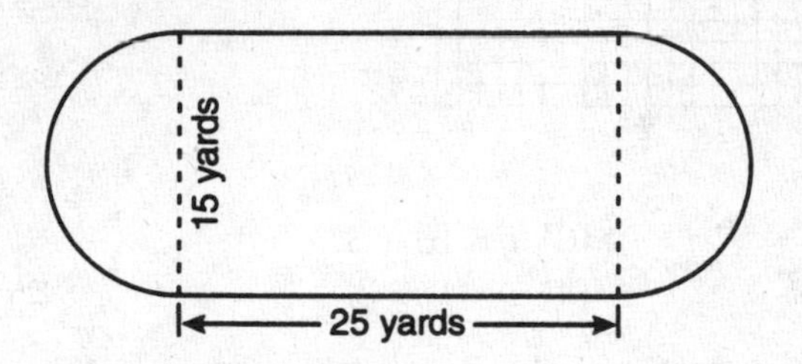

Which expression represents the amount of fencing, in yards, that would be needed to completely enclose the playground?

(1) $15\pi + 50$ (3) $30\pi + 50$
(2) $15\pi + 80$ (4) $30\pi + 80$

24 _____

25 Which equation is represented by the graph below?

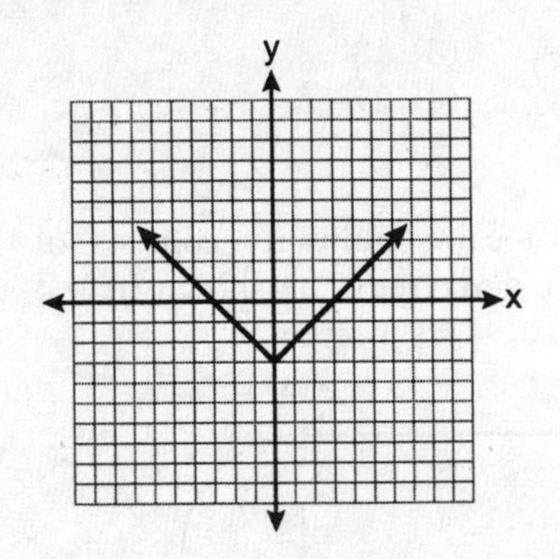

(1) $y = x^2 - 3$
(2) $y = (x - 3)^2$
(3) $y = |x| - 3$
(4) $y = |x - 3|$

25 _____

26 Carrie bought new carpet for her living room. She calculated the area of the living room to be 174.2 square feet. The actual area was 149.6 square feet. What is the relative error of the area to the *nearest ten-thousandth*?

(1) 0.1412
(2) 0.1644
(3) 1.8588
(4) 2.1644

26 _____

27 What is an equation of the line that passes through the point (3,–1) and has a slope of 2?

(1) $y = 2x + 5$
(2) $y = 2x - 1$
(3) $y = 2x - 4$
(4) $y = 2x - 7$

27 _____

28 The ages of three brothers are consecutive even integers. Three times the age of the youngest brother exceeds the oldest brother's age by 48 years. What is the age of the *youngest* brother?

(1) 14
(2) 18
(3) 22
(4) 26

28 _____

29 Cassandra bought an antique dresser for $500. If the value of her dresser increases 6% annually, what will be the value of Cassandra's dresser at the end of 3 years to the *nearest dollar*?

(1) $415
(2) $590
(3) $596
(4) $770

29 _____

30 The number of hours spent on math homework each week and the final exam grades for twelve students in Mr. Dylan's algebra class are plotted below.

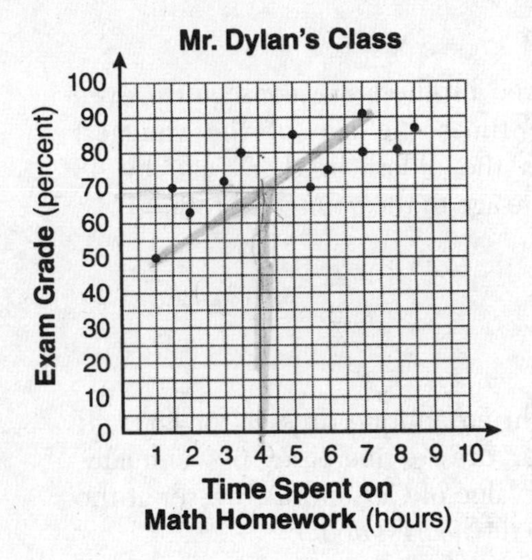

Based on a line of best fit, which exam grade is the best prediction for a student who spends about 4 hours on math homework each week?

(1) 62
(2) 72
(3) 82
(4) 92

30 _____

PART II

Answer all 3 questions in this part. Each correct answer will receive 2 credits. Clearly indicate the necessary steps, including appropriate formula substitutions, diagrams, graphs, charts, etc. For all questions in this part, a correct numerical answer with no work shown will receive only 1 credit. [6]

31 Chad complained to his friend that he had five equations to solve for homework. Are all of the homework problems equations? Justify your answer.

Math Homework

1. $3x^2 \cdot 2x^4$
2. $5 - 2x = 3x$
3. $3(2x + 7)$
4. $7x^2 + 2x - 3x^2 - 9$
5. $\frac{2}{3} = \frac{x+2}{6}$

Name *Chad*

32 The diagram below represents Joe's two fish tanks.

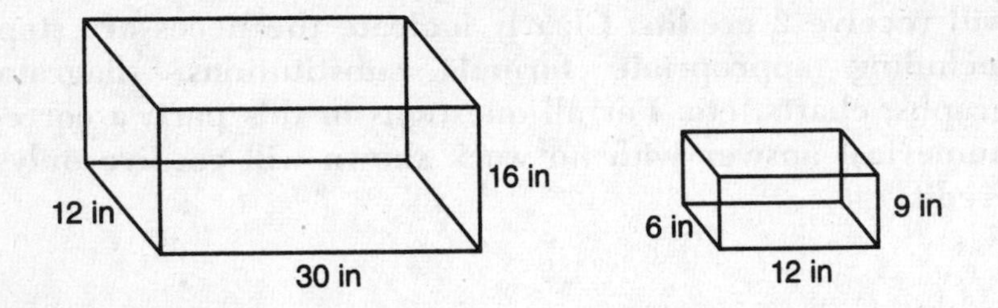

Joe's larger tank is completely filled with water. He takes water from it to completely fill the small tank. Determine how many cubic inches of water will remain in the larger tank.

33 Clayton has three fair coins. Find the probability that he gets two tails and one head when he flips the three coins.

PART III

Answer all 3 questions in this part. Each correct answer will receive 3 credits. Clearly indicate the necessary steps, including appropriate formula substitutions, diagrams, graphs, charts, etc. For all questions in this part, a correct numerical answer with no work shown will receive only 1 credit. [9]

34 Find algebraically the equation of the axis of symmetry and the coordinates of the vertex of the parabola whose equation is $y = -2x^2 - 8x + 3$.

35 At the end of week one, a stock had increased in value from \$5.75 a share to \$7.50 a share. Find the percent of increase at the end of week one to the *nearest tenth of a percent*.

At the end of week two, the same stock had decreased in value from \$7.50 to \$5.75. Is the percent of decrease at the end of week two the same as the percent of increase at the end of week one? Justify your answer.

36 The chart below compares two runners.

Runner	Distance, in miles	Time, in hours
Greg	11	2
Dave	16	3

Based on the information in this chart, state which runner has the faster rate. Justify your answer.

PART IV

Answer all 3 questions in this part. Each correct answer will receive 4 credits. Clearly indicate the necessary steps, including appropriate formula substitutions, diagrams, graphs, charts, etc. For all questions in this part, a correct numerical answer with no work shown will receive only 1 credit. [2]

37 Express in simplest form: $\dfrac{2x^2-8x-42}{6x^2} \div \dfrac{x^2-9}{x^2-3x}$

38 On the grid below, solve the system of equations graphically for x and y.

$$4x - 2y = 10$$
$$y = -2x - 1$$

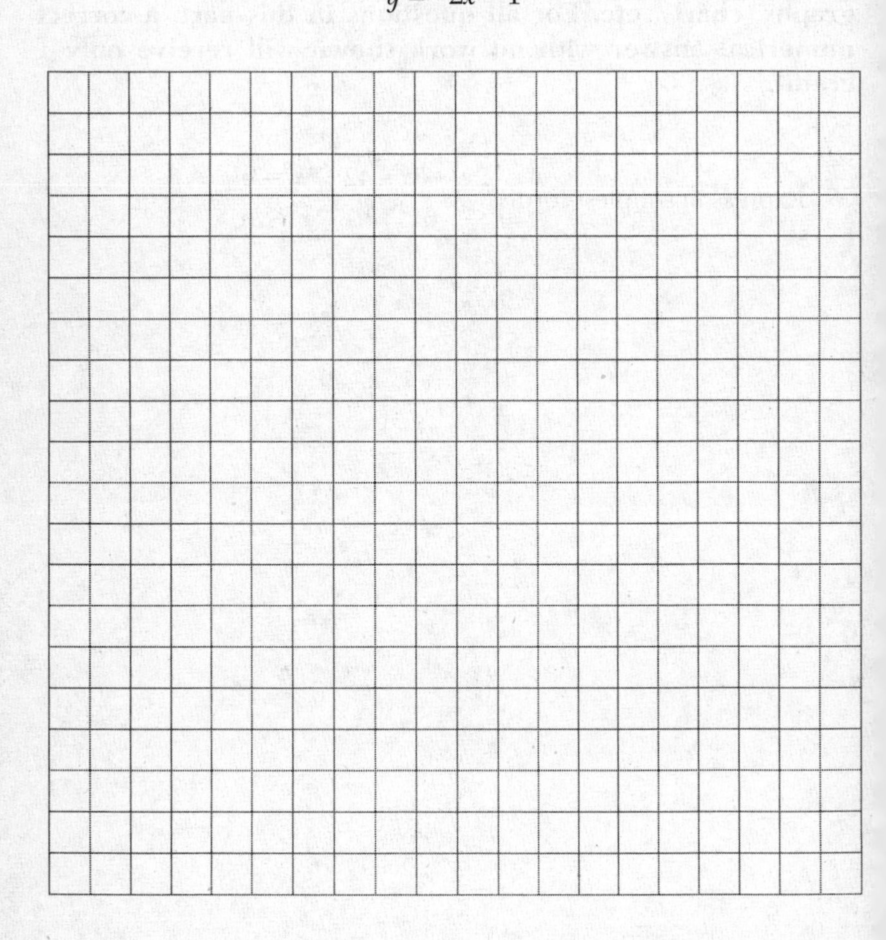

39 The test scores from Mrs. Gray's math class are shown below.

72, 73, 66, 71, 82, 85, 95, 85, 86, 89, 91, 92

Construct a box-and-whisker plot to display these data.

65 70 75 80 85 90 95 100

Answers August 2009

Integrated Algebra

Answer Key

PART I

1. (2)	**7.** (3)	**13.** (4)	**19.** (3)	**25.** (3)
2. (1)	**8.** (3)	**14.** (2)	**20.** (1)	**26.** (2)
3. (4)	**9.** (2)	**15.** (1)	**21.** (3)	**27.** (4)
4. (1)	**10.** (4)	**16.** (2)	**22.** (2)	**28.** (4)
5. (3)	**11.** (1)	**17.** (2)	**23.** (3)	**29.** (3)
6. (1)	**12.** (4)	**18.** (1)	**24.** (1)	**30.** (2)

PART II

31. No; See *Answers Explained*

32. 5,112

33. $\frac{3}{8}$

PART III

34. $x = -2$; $(-2, 11)$

35. 30.4%; No

36. Greg

PART IV

37. $\frac{x-7}{3x}$

38. $(1, -3)$; See *Answers Explained*

39. See *Answers Explained*

In **PARTS II–IV** you are required to show how you arrived at your answers. For sample methods of solutions, see the *Answers Explained* section.

Answers Explained

PART I

1. If h represents a number, then the correct translation of

Sixty more than	9 times a number	is 375
$+60$	$9h$	$= 375$

is $9h + 60 = 375$.

The correct choice is **(2)**.

2. Since the given expression, $9x^2 - 16$, represents the difference of two perfect squares, it can be factored as the product of the sum and the difference of the terms that are being squared:

$$\begin{aligned} 9x^2 - 16 &= (3x)^2 - (4)^2 \\ &= (3x + 4)(3x - 4) \end{aligned}$$

The correct choice is **(1)**.

3. Given: $(3x^2y^4)(4xy^2)$

Group like factors together: $(3 \cdot 4)(x^2 \cdot x)(y^4 \cdot y^2)$

Multiply powers of like bases by adding their exponents: $12x^3y^6$

The correct choice is **(4)**.

4. It is given that an online music club has a one-time registration fee of \$13.95 and charges \$0.49 to buy each song. A person who joins the club and buys x songs must spend $13.95 + 0.49x$ dollars. If Emma has \$50.00 to join the club and buy songs, then the maximum number of songs she can buy is the largest integer value of x for which

$$13.95 + 0.49x \leq 50.00$$

Solve for x:

$$0.49x \leq 50.00 - 13.95$$
$$0.49x \leq 36.05$$
$$\frac{0.49x}{0.49} \leq \frac{36.05}{0.49}$$
$$x \leq 73.57142857$$

Since the greatest integer value of x that satisfies the inequality is 73, the maximum number of songs that Emma can buy is 73.

The correct choice is **(1)**.

5. It is given that the local ice cream stand offers three flavors of ice cream (vanilla, chocolate, and strawberry), two types of cones, and three different toppings. If Dawn does not order vanilla ice cream, she can choose from the two remaining flavors of ice cream. The total number of different choices she can make that have one flavor of ice cream, one type of cone, and one topping can be determined using the multiplication principle of counting:

Number of flavors other than vanilla		Number of types of cones		Number of toppings		Total number of choices
$\boxed{2}$	$\times$	$\boxed{2}$	$\times$	$\boxed{3}$	$=$	12

The correct choice is **(3)**.

6. The accompanying diagram represents Nancy's rectangular garden.

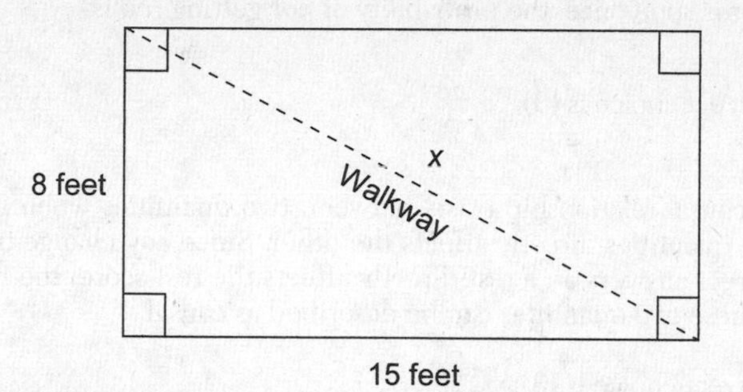

The number of feet in the length x of the diagonal walkway can be determined using the Pythagorean theorem:

$$\begin{aligned} x^2 &= 8^2 + 15^2 \\ &= 64 + 225 \\ &= 289 \\ x &= \sqrt{289} \\ &= 7 \text{ feet} \end{aligned}$$

The correct choice is **(1)**.

7. The spinner in the accompanying diagram is divided into eight equal regions distributed by color as follows:

1 yellow region
3 red regions
1 white region
1 purple region
1 blue region
1 green region

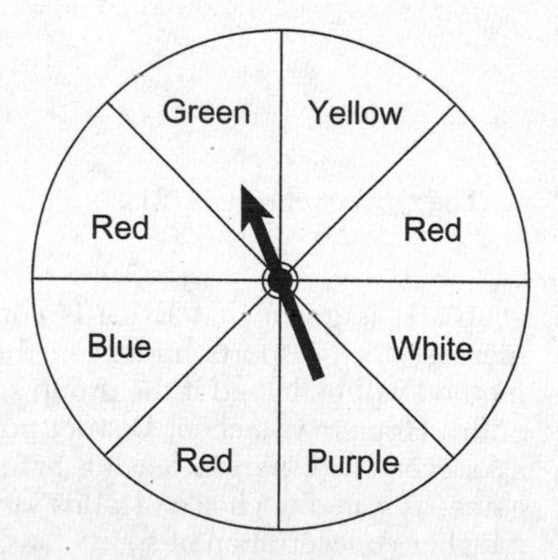

Since 3 regions are red, the remaining 5 of the 8 regions are *not* red. When the spinner is spun once, the probability of *not* getting red is $\frac{5}{8}$.

The correct choice is **(3)**.

8. A causal relationship exists between two quantities when a change in one of the quantities directly affects the other. Since any change in the number of correct answers on a test directly affects the test score, the relationship between these two quantities can be described as causal.

The correct choice is **(3)**.

9. The given equation is: $\frac{3}{5}(x+2) = x - 4$

Eliminate the fractional coefficient by multiplying both sides of the equation by 5: $\not{5}\left[\frac{3}{\not{5}_1}(x+2)\right] = 5(x-4)$

Remove the parentheses by multiplying each term inside the parentheses by the number in front of the parentheses: $3x + 6 = 5x - 20$

Collect like terms on the same side of the equation:

$$6 + 20 = 5x - 3x$$
$$26 = 2x$$
$$\frac{26}{2} = \frac{2x}{2}$$
$$13 = x$$

The correct choice is **(2)**.

10. It is given that Erica is conducting a survey about the proposed increase in the sports budget in the Hometown School District. A survey method will be biased if the group surveyed is not truly representative of the entire Hometown School District population or cannot be objective. If Erica asks every fifth person leaving Saturday's Hometown High School football game, as stated in choice (4), this sample of people would most likely include a higher concentration of sports fans. They would probably be biased in favor of spending money on sports. In contrast, people surveyed at nonathletic events or places, as described in the other answer choices, may or may not have little or no interest in sports.

The correct choice is **(4)**.

11. Any equation of the form $y = b$ describes a horizontal line that has a y-intercept of b and is parallel to the x-axis. The equation $y = -5$ describes a line parallel to the x-axis.

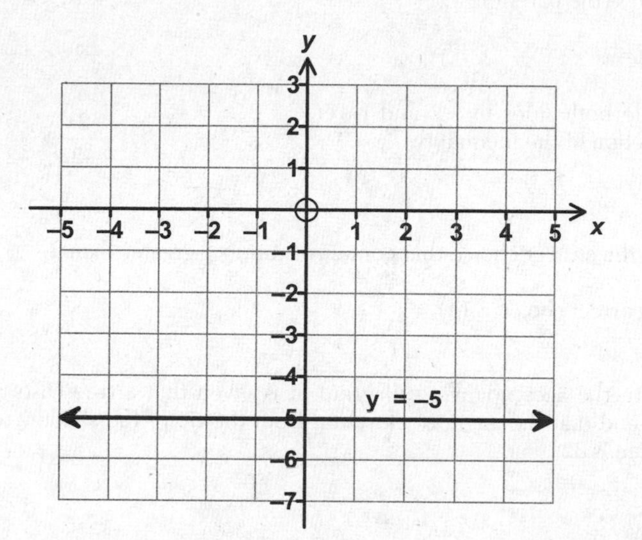

The correct choice is **(1)**.

12. It is given that:

$$A = \{\text{All even integers from 2 to 20, inclusive}\}$$
$$B = \{10, 12, 14, 16, 18\}$$

To determine the complement of set B, denoted by B', within the universe of set A, find all members of set A that are not members of set B.

$$A = \left\{ 2, 4, 6, 8, \overbrace{10, 12, 14, 16, 18}^{\text{members of set } B}, 20 \right\}$$

$$B' = \{2, 4, 6, 8, 20\}$$

The correct choice is **(4)**.

13. The given inequality is: $-2(x - 5) < 4$

Remove the parentheses: $-2x + 10 < 4$

Isolate x: $-2x < 4 - 10$

Divide both sides by -2 and reverse the direction of the inequality:

$$\frac{-2x}{-2} > \frac{-6}{-2}$$

$$x > 3$$

Find the answer choice that contains a number greater than 3.

The correct choice is **(4)**.

14. In the accompanying diagram, it is given that a tree casts a 25-foot shadow and that the angle of elevation from the tip of the shadow to the top of the tree is 32°.

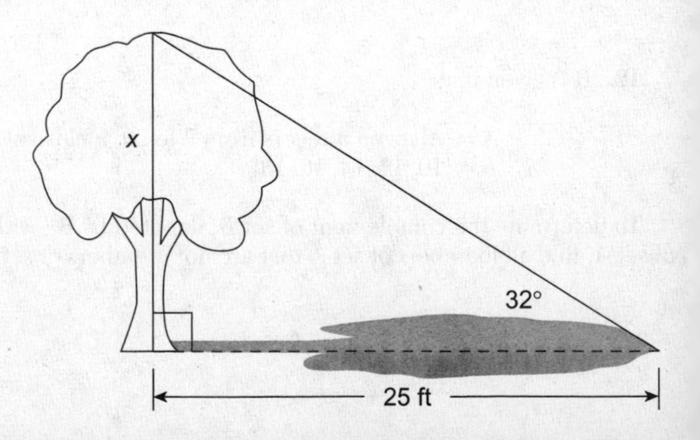

If x represents the vertical height of the tree, then x is opposite the given angle. The side of the right triangle that measures 25 feet is adjacent to the given angle. Find x using the tangent ratio and your calculator:

$$\tan 32° = \frac{\text{leg opposite } 32° \text{ angle}}{\text{leg adjacent to } 32° \text{ angle}}$$

$$\tan 32° = \frac{x}{25}$$

$$x = 25\,(\tan 32°)$$

$$= 15.6217338$$

The height of the tree, to the *nearest tenth of a foot*, is 15.6 feet.

The correct choice is **(2)**.

15. The slope, m, of a line that passes through the points (x_A, y_A) and (x_B, y_B) is given by the formula

$$m = \frac{y_B - y_A}{x_B - x_A}$$

To find the slope of a line that passes through (–5, 4) and (15, –4), use the slope formula where $(x_A, y_A) = (-5, 4)$ and $(x_B, y_B) = (15, -4)$:

$$m = \frac{-4-4}{15-(-5)}$$

$$= \frac{-8}{15+5}$$

$$= \frac{-8}{20}$$

$$= \frac{-8 \div 4}{20 \div 4}$$

$$= -\frac{2}{5}$$

The correct choice is **(1)**.

16. The real roots of a quadratic equation of the form $ax^2 + bx + c = 0$ correspond to the x-coordinates of the points at which the parabola $y = ax^2 + bx + c$ crosses the x-axis.

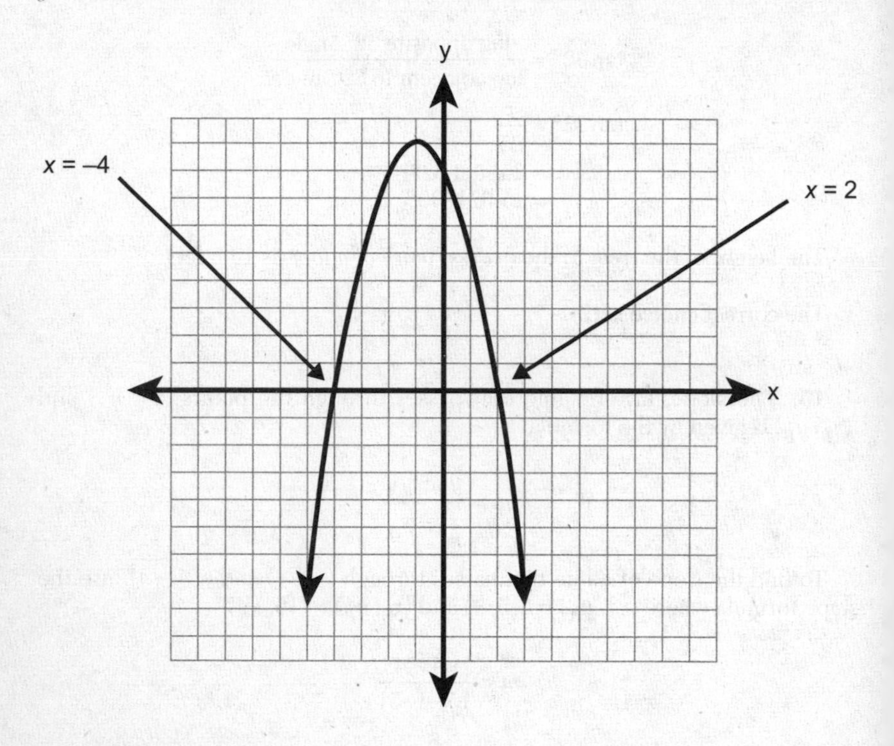

It is given that the equation $y = -x^2 - 2x + 8$ is graphed on the accompanying set of axes. Based on this graph, you are asked to find the roots of $-x^2 - 2x + 8 = 0$. Since the x-intercepts of the graph are at $x = 2$ and $x = -4$, these are the roots of the quadratic equation.

The correct choice is **(2)**.

17. The given sum is:

$$\frac{3}{2x}+\frac{4}{3x}$$

The LCD (lowest common denominator) is $6x$ since $6x$ is the smallest expression into which the denominators $2x$ and $3x$ divide evenly. Rewrite each fraction with the LCD as its denominator by multiplying the first fraction by 1 in the form of $\frac{3}{3}$ and multiplying the second fraction by 1 in the form of $\frac{2}{2}$:

$$\frac{3}{2x}\cdot\frac{3}{3}+\frac{4}{3x}\cdot\frac{2}{2}$$

Write the sum of the numerators over the common denominator:

$$\frac{9+8}{6x}$$

$$\frac{17}{6x}$$

The correct choice is **(2)**.

18. Since division by 0 is not defined, a fraction with a variable in the denominator is undefined for any value of the variable that makes the denominator equal 0. To find the values of x that make the fraction $\frac{x^2-9}{x^2+7x+10}$ undefined, set the denominator equal to 0 and solve for x:

$$x^2 + 7x + 10 = 0$$

Factor the left side of the equation as the product of two binomials:

$$(x + ?)(x + ?) = 0$$

The missing terms of the binomial factors are the two integers whose sum is +7, the coefficient of the middle term in $x^2 + 7x + 10$, and whose product is +10, the constant term in $x^2 + 7x + 10$. Since $(+5) + (+2) = +7$ and $(+5)(+10) = +10$, the missing terms are 5 and 2:

$$(x + 5)(x + 2) = 0$$

Set each factor equal to 0: $x + 5 = 0$ or $x + 2 = 0$

$$x = -5 \qquad x = -2$$

The set of answer choices includes –5.

The correct choice is **(1)**.

19. A relation is a function if for each x-value, one and only one y-value is paired with it. Answer choice (3) gives a set of ordered pairs in which the x-value of 1 is paired with two different y-values, so it does not represent a function:

$$\{(-1, 6), \mathbf{(1, 3)}, (2, 5), \mathbf{(1, 7)}\}$$

The correct choice is **(3)**.

20. You are required to find the value of the y-coordinate of the solution to a system of equations:

$$\begin{array}{r} x - 2y = 1 \\ \underline{x + 4y = 7} \end{array}$$

Subtract corresponding sides of the two equations: $0 - 6y = -6$

Solve for y: $\frac{-6y}{-6} = \frac{-6}{-6}$

$$y = 1$$

The correct choice is **(1)**.

21. The given equation is: $x^2 - 6x = 0$

Factor out x: $x(x - 6) = 0$

Set each factor equal to 0: $x = 0$ or $x - 6 = 0$

$$x = 6$$

The solution is 0 and 6.

The correct choice is **(3)**.

22. It is given that when $5\sqrt{20}$ is written in simplest radical form, the result is $k\sqrt{5}$.

Factor the number underneath the radical sign as the product of two positive integers, one of which is the greatest perfect square factor of 20: $5\sqrt{20} = 5\sqrt{4 \cdot 5}$

Write the radical over each factor of the radicand: $= 5\sqrt{4} \cdot \sqrt{5}$

Substitute 2 for $\sqrt{4}$: $= 5 \cdot 2 \cdot \sqrt{5}$

$= 10\sqrt{5}$

Since $5\sqrt{20} = 10\sqrt{5} = k\sqrt{5}$, $k = 10$.

The correct choice is **(2)**.

23. To find the value of the expression $|-5x + 12|$ when $x = 5$, replace x with 5 and then take the absolute value of the result:

$$\begin{aligned} |-5(5) + 12| &= |-25 + 12| \\ &= |-13| \\ &= 13 \end{aligned}$$

The correct choice is **(3)**.

24. It is given that the accompanying figure represents a playground that consists of a rectangle and two semicircles at opposite ends.

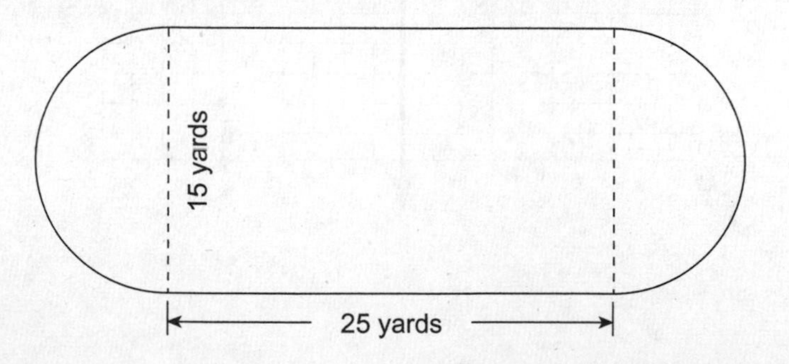

In order to find an expression that represents the amount of fencing, in yards, that would be needed to completely enclose the playground, find the total distance around the playground.

- The sum of the lengths of the two straight sides is $25 + 25 = 50$.
- The two semicircles are equivalent to one whole circle with a diameter of 15. The circumference of the whole circle is $\pi \times \text{diameter} = 15\pi$.
- The total distance around the playground is $15\pi + 50$.

The correct choice is **(1)**.

25. You are asked to identify the equation represented by the accompanying graph.

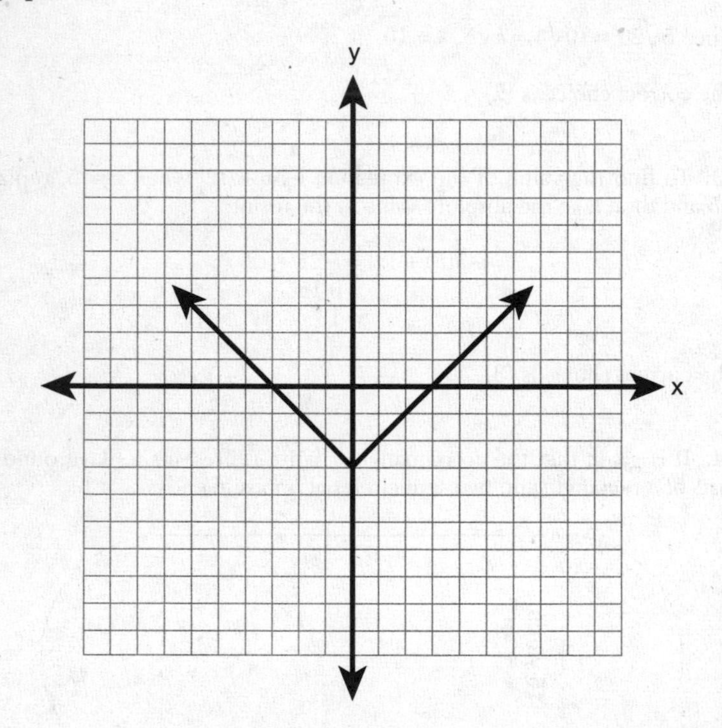

The graph shows the absolute value function $y = |x|$ shifted 3 units down. Hence, the equation of the graph is $y = |x| - 3$.

The correct choice is **(3)**.

26. To determine the relative error in a measurement, evaluate the following ratio:

$$\frac{\text{Actual value} - \text{Measured value}}{\text{Actual value}}$$

- You are given that Carrie measured the area of the living room to be 174.2 square feet.
- You are also given that the actual area is 149.6 square feet.
- The relative error in the measurement of the area of the living room is

$$\frac{174.2 - 149.6}{149.6} = \frac{24.6}{149.6}$$
$$\approx 0.1644385027$$

The relative error of the area to the *nearest ten-thousandth* is 0.1644.

The correct choice is **(2)**.

27. To find an equation of the line that passes through the point (–3, 1) and has a slope of 2, use the point-slope form of the equation of line $y - y_1 = m(x - x_1)$ where $(x_1, y_1) = (3, -1)$ and $m = 2$:

$$y - (-1) = 2(x - 3)$$
$$y + 1 = 2x - 6$$
$$y = 2x - 6 - 1$$
$$y = 2x - 7$$

The correct choice is **(4)**.

28. It is given that the ages of three brothers are consecutive even integers such that three times the age of the youngest brother exceeds the oldest brother's age by 48 years. If x represents the age of the youngest brother, then $x + 2$ and $x + 4$ represent the ages of the other brothers in increasing age order:

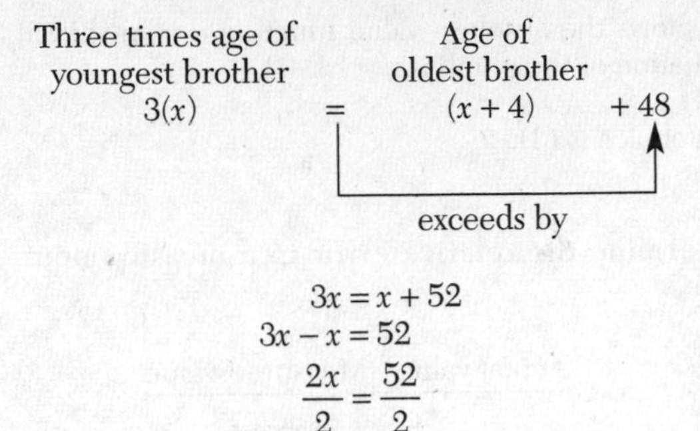

$$
\begin{aligned}
3x &= x + 52 \\
3x - x &= 52 \\
\frac{2x}{2} &= \frac{52}{2} \\
x &= 26
\end{aligned}
$$

The correct choice is **(4)**.

29. You are told that Cassandra bought an antique dresser for \$500 and that the value of her dresser increases 6% annually. To find the value of Cassandra's dresser at the end of 3 years, use the formula for exponential growth $y = A(1 + r)^n$ where A is the initial value of the dresser, r is the annual percent of increase, and y is the value of the dresser after n years. Use your calculator to evaluate the formula using $A = 500$, $r = 6\% = 0.06$, and $n = 3$:

$$
\begin{aligned}
y &= 500(1 + 0.06)^3 \\
&= 500(1.06)^3 \\
&= 595.508
\end{aligned}
$$

The value of the dresser at the end of three years to the *nearest dollar* is \$596.

The correct choice is **(3)**.

30. The number of hours spent on math homework each week and the final exam grades for twelve students in Mr. Dylan's algebra class are plotted in the accompanying graph.

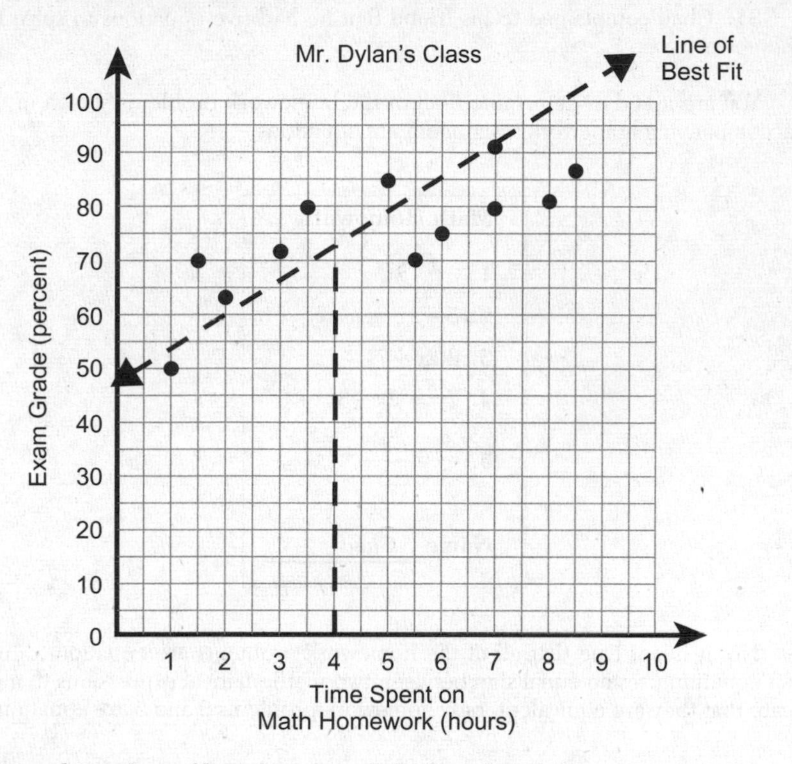

Based on a line of best fit, you are asked to identify the exam grade that is the best prediction for a student who spends about 4 hours on math homework each week.

- Estimate the line of best fit by drawing a line approximately midway between the cluster of data points as shown by the broken line in the accompanying figure.
- Locate on the line of best fit the y-value that corresponds to $x = 4$. This value appears to be somewhere between 70 and 75. Look among the answer choices for a value in this range.
- Based on the line of best fit, the best prediction is an exam grade of 72 for a student who spends about 4 hours on math homework each week.

The correct choice is **(2)**.

PART II

31. Chad complained to his friend that he had five equations to solve for homework.

You are asked to determine if all of the homework problems shown in the accompanying homework assignment are equations.

Math Homework

1. $3x^2 \cdot 2x^4$

2. $5 - 2x = 3x$

3. $3(2x + 7)$

4. $7x^2 + 2x - 3x^2 - 9$

5. $\frac{2}{3} = \frac{x+2}{6}$

Name *Chad*

No, it is not true that all of the homework problems are equations. Since an equation uses an equal sign between two mathematical expressions to indicate that they are equivalent, only homework problems 2 and 5 are equations.

32. You are told that the accompanying diagram represents Joe's two fish tanks. You are also told that Joe's larger tank is completely filled with water and that he takes water from it to completely fill the small tank.

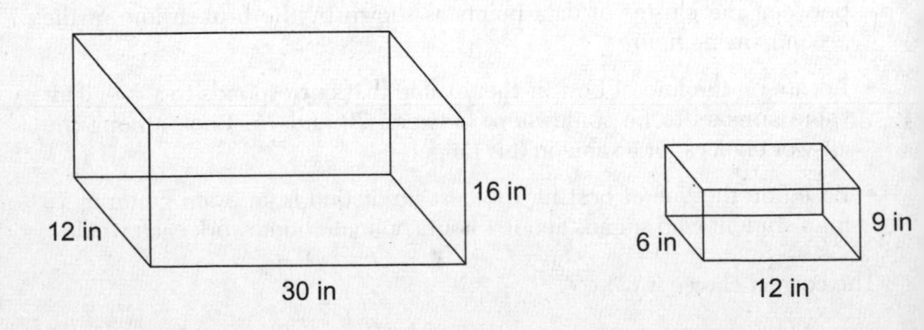

To determine how many cubic inches of water will remain in the larger tank, find the difference in the volumes of the two tanks.

- Volume of larger tank = length × width × height

$$= 30 \times 12 \times 16$$

$$= 5{,}760 \text{ in}^3$$

- Volume of smaller tank = length × width × height

$$= 12 \times 6 \times 9$$

$$= 648 \text{ in}^3$$

- Difference in volumes = $5{,}760 - 648 = 5{,}112 \text{ in}^3$

Thus, **5,112** cubic inches of water will remain in the larger tank.

33. You are required to find the probability that when Clayton flips three fair coins, he gets two tails and one head. Organize the set of all possible outcomes as a set of ordered triples where H represents getting a head and T represents getting a tail.

All Heads or All Tails	2 Heads, 1 Tail	2 Tails, 1 Head
(H, H, H) (T, T, T)	(H, H, T) (H, T, H) (T, H, H)	**(T, T, H)** **(T, H, T)** **(H, T, T)**

Since 3 of the possible 8 outcomes include two tails and one head, the probability that Clayton flips the three coins and gets two tails and one head is $\mathbf{\frac{3}{8}}$.

PART III

34. You are required to find algebraically the equation of the axis of symmetry and the coordinates of the vertex of the parabola whose equation is $y = -2x^2 - 8x + 3$.

For a parabola whose equation is $y = ax^2 + bx + c$, an equation of the axis of symmetry is given by the formula $x = -\dfrac{b}{2a}$.

- If $y = -2x^2 - 8x + 3$, then $a = -2$ and $b = -8$:

$$\begin{aligned} x &= -\frac{b}{2a} \\ &= \frac{-(-8)}{2(-2)} \\ &= \frac{8}{-4} \\ &= -2 \end{aligned}$$

See the accompanying figure.

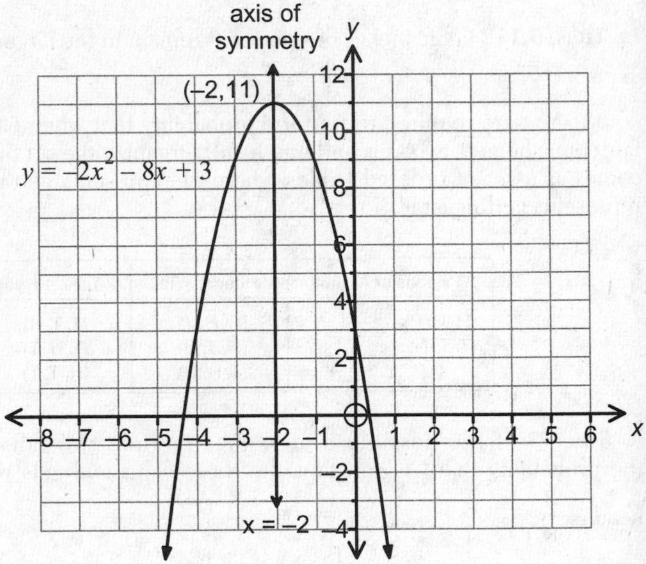

- Since the axis of symmetry is a vertical line that contains the vertex, the x-coordinate of the vertex is -2.
- Find the y-coordinate of the vertex by substituting -2 for x in the equation of the parabola:

$$\begin{aligned} y &= -2(-2)^2 - 8(-2) + 3 \\ &= -2(4) + 16 + 3 \\ &= -8 + 19 \\ &= 11 \end{aligned}$$

Hence, an equation of the axis of symmetry is $x = -2$ and the coordinates of the vertex are **(−2, 11)**.

35. At the end of week one, a stock had increased in value from \$5.75 a share to \$7.50 a share. You are asked to find the percent of increase at the end of week one to the *nearest tenth of a percent.*

$$\begin{aligned}\text{Percent increase} &= \frac{\text{Amount of increase}}{\text{Original amount}} \times 100\% \\ &= \frac{7.50 - 5.75}{5.75} \times 100\% \\ &= 0.3043478261 \times 100\% \\ &= 30.43478261\%\end{aligned}$$

The percent of increase at the end of the week, to the *nearest tenth of a percent*, was **30.4%**.

It is also given that at the end of week two, the same stock had decreased in value from \$7.50 to \$5.75. You are asked whether the percent of decrease at the end of week two is the same as the percent of increase at the end of week one.

$$\begin{aligned}\text{Percent decrease} &= \frac{\text{Amount of decrease}}{\text{Original amount}} \times 100\% \\ &= \frac{7.50 - 5.75}{7.50} \times 100\% \\ &= 0.2333333333 \times 100\% \\ &\approx 23.3\%\end{aligned}$$

No, the percent of decrease at the end of week two is not the same as the percent of increase at the end of week one. Although the amount of change for the two weeks is the same, the original amounts of the stock are different, leading to different answers.

36. It is given that the accompanying chart compares two runners. Based on the information in this chart, you are asked to state which runner has the faster rate.

Runner	Distance, in Miles	Time, in Hours
Greg	11	2
Dave	16	3

Calculate each runner's average rate by using the relationship that rate is equal to distance divided by time:

Runner	Distance, in Miles	Time, in Hours	Rate, in Miles per Hour
Greg	11	2	$\frac{11}{2} = 5.5$
Dave	16	3	$\frac{16}{3} \approx 5.33$

Greg runs at the faster rate since 5.5 > 5.33.

PART IV

37. The given division problem is:

$$\frac{2x^2-8x-42}{6x^2}\div\frac{x^2-9}{x^2-3x}$$

Change to multiplication by inverting the second fraction:

$$\frac{2x^2-8x-42}{6x^2}\cdot\frac{x^2-3x}{x^2-9}$$

Factor out the greatest common factor of each of the numerators:

$$\frac{2\left(x^2-4x-21\right)}{6x^2}\cdot\frac{x(x-3)}{x^2-9}$$

Factor the quadratic trinomial in the first numerator as the product of two binomials. Since the second denominator represents the difference of two perfect squares, factor it as product of the sum and difference of the terms that are being squared:

$$\frac{2(x-7)(x+3)}{6x^2}\cdot\frac{x(x-3)}{(x+3)(x-3)}$$

Divide out any factor that is common to both a numerator and a denominator since their quotient is 1:

$$\frac{\cancel{2}(x-7)\overset{1}{\cancel{(x+3)}}}{\underset{3x}{\cancel{6x^2}}}\cdot\frac{\cancel{x}\overset{1}{\cancel{(x-3)}}}{\cancel{(x+3)}\cancel{(x-3)}}$$

Multiply the remaining factors in the numerator, and multiply the remaining factors in the denominator:

$$\frac{x-7}{3x}$$

The quotient in simplest form is $\mathbf{\frac{x-7}{3x}}$.

38. You are required to solve the following system of equations graphically for x and y:

$$4x - 2y = 10$$
$$y = -2x - 1$$

Graph the two linear equations on the same set of axes.

- To graph $4x - 2y = 10$, first solve for y. Since $2y = 4x - 10$, $y = 2x - 5$. Find two convenient points on the line. If $x = 0$, $y = 2(0) - 5$ so $y = -5$. Hence, the line contains the point $(0, -5)$. If $x = 5$, $y = 2(5) - 5 = 5$. The line also contains the point $(5, 5)$. Plot these two points, and draw a line through them as shown in the accompanying diagram. Label the line with its equation.

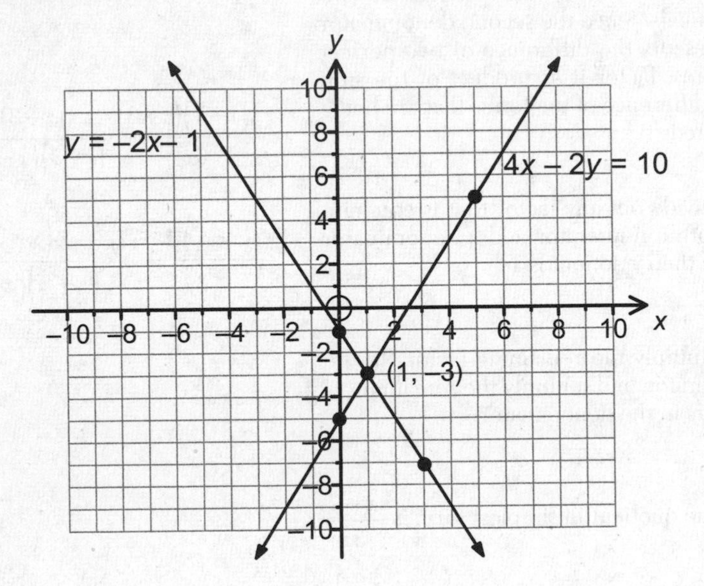

- To graph $y = -2x - 1$, find two convenient points on the line. If $x = 0$, $y = -2(0) - 1$ so $y = -1$. Hence, the line contains the point $(0, -1)$. If $x = 3$, $y = -2(3) - 1 = -7$. The line also contains the point $(3, -7)$. Plot these two points, and draw a line through them as shown in the accompanying diagram. Label the line with its equation.

The graphs intersect at **(1, –3)**, which represents the solution to the given system of equations.

39. The test scores from Mrs. Gray's math class are shown below.

72, 73, 66, 71, 82, 85, 95, 85, 86, 89, 91, 92

To construct a box-and-whisker plot to display these data, first arrange the 12 data values in ascending numerical order:

66, 71, 72, 73, 82, 85, 85, 86, 89, 91, 92, 95

Identify the 5 key values needed to construct a box-and-whisker plot: the minimum score, the maximum score, the median, and the first and third quartiles. The minimum score is 66, and the maximum score is 95. The median or second quartile (Q_2) is the score that divides the set of scores into two groups, each having the same number of scores. Since there are 12 scores, the median is midway between the 6th and 7th scores, so $Q_2 = 85$. The first quartile is the score that separates the set of scores below the median into two equal groups, so $Q_1 = 72.5$. Similarly, the third quartile is the score that separates the set of scores above the median into two equal groups, so $Q_3 = 90$.

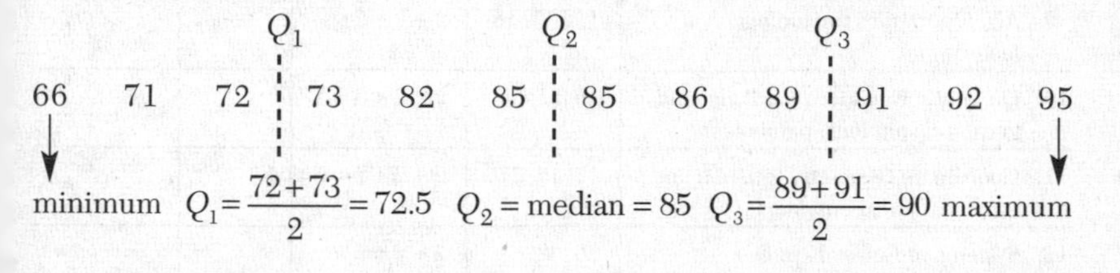

Draw a rectangular box whose vertical sides are aligned with the scores that represent the first and third quartiles. Then draw a vertical bar inside the rectangle that is aligned with the median score of 85 on the number line, as shown in the accompanying figure.

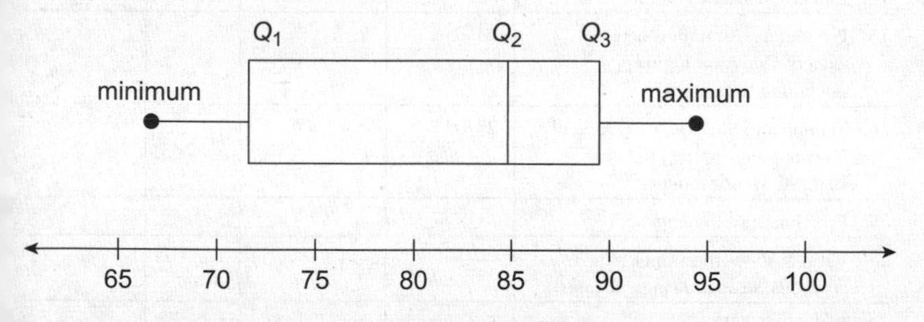

Draw a horizontal whisker from the left vertical side of the box to the minimum score of 66. Draw another horizontal whisker from the right vertical side of the box to the maximum score of 95, as shown in the accompanying figure.

Topic	Question Numbers	Number of Points	Your Points	Your Percentage
1. Sets and Numbers; Intersection and Complements of Sets; Interval Notation; Properties of Real Numbers	12	2		
2. Operations on Rat'l. Numbers & Monomials	—	—		
3. Laws of Exponents for Integer Exponents; Scientific Notation	—	—		
4. Operations on Polynomials	3, 37	$2 + 4 = 6$		
5. Square Root; Operations with Radicals	22	2		
6. Evaluating Formulas & Algebraic Expressions	23	2		
7. Solving Linear Eqs. & Inequalities	4, 9, 13	$2 + 2 + 2 = 6$		
8. Solving Literal Eqs. & Formulas for a Given Letter	—	—		
9. Alg. Operations (including factoring)	1, 2,17, 18	$2 + 2 + 2 + 2 = 8$		
10. Quadratic Equations (incl. alg. and graphical solutions; parabolas)	16, 21, 34	$2 + 2 + 3 = 7$		
11. Coordinate Geometry (eq. of a line; graphs of linear eqs; slope)	11, 15, 27	$2 + 2 + 2 = 6$		
12. Systems of Linear Eqs. & Inequalities (algebraic & graphical solutions)	20, 38	$2 + 4 = 6$		
13. Mathematical Modeling (using: eqs.; tables; graphs)	—	—		
14. Linear-Quadratic systems	—	—		
15. Perimeter; Circumference; Area of Common Figures (including trapezoids)	24	2		
16. Volume and Surface Area; Area of Overlapping Figures; Relative Error in Measurement	26, 32	$2 + 2 = 4$		
17. Fractions and Percent	—	—		
18. Ratio & Proportion (incl. similar polygons, scale drawings, & rates)	—	—		
19. Pythagorean Theorem	6	2		
20. Right Triangle Trigonometry	14	2		
21. Functions (def.; domain and range; vertical line test; absolute value)	19, 25	$2 + 2 = 4$		

Topic	Question Numbers	Number of Points	Your Points	Your Percentage
22. Exponential Functions (properties; growth and decay)	29	2		
23. Probability (incl. tree diagrams & sample spaces)	7, 33	2 + 2 = 4		
24. Permutations and Counting Methods (incl. Venn diagrams)	5	2		
25. Statistics (mean, median, percentiles, quartiles; freq. dist., histograms; box-and-whisker plots; causality; bivariate data; qualitative vs. quantitative data; unbiased vs. biased samples; circle graphs)	8, 10, 39	2 + 2 + 4 = 8		
26. Line of Best Fit (including linear regression, scatter plots, and linear correlation)	30	2		
27. Nonroutine Word Problems Requiring Arith. or Alg. Reasoning	28, 31, 35, 36	2 + 2 + 3 + 3 = 10		

MAP TO LEARNING STANDARDS

Key Ideas	Item Numbers
Number Sense and Operations	22, 23, 35
Algebra	1, 2, 3, 4, 6, 9, 11, 12, 13, 14, 15, 17, 18, 20, 21, 27, 28, 29, 31, 34, 37
Geometry	16, 19, 24, 25, 32, 38
Measurement	26, 36
Statistics and Probability	5, 7, 8, 10, 30, 33, 39

HOW TO CONVERT YOUR RAW SCORE TO YOUR INTEGRATED ALGEBRA REGENTS EXAMINATION SCORE

Below is the conversion chart that must be used to determine your final score on the August 2009 Regents Examination in Integrated Algebra. To find your final exam score, locate in the column labeled "Raw Score" the total number of points you scored out of a possible 87 points. Since partial credit is allowed in Parts II, III, and IV of the test, you may need to approximate the credit you would receive for a solution that is not completely correct. Then locate in the adjacent column to the right the scale score that corresponds to your raw score. The scale score is your final Integrated Algebra Regents Examination score.

Regents Examination in Integrated Algebra—August 2009 Chart for Converting Total Test Raw Scores to Final Examination Scores (Scaled Scores)

Raw Score	Scale Score	Raw Score	Scale Score	Raw Score	Scale Score
87	**100**	57	**81**	27	**61**
86	**98**	56	**81**	26	**60**
85	**97**	55	**81**	25	**58**
84	**96**	54	**80**	24	**57**
83	**95**	53	**80**	23	**55**
82	**94**	52	**79**	22	**54**
81	**93**	51	**79**	21	**52**
80	**92**	50	**79**	20	**51**
79	**91**	49	**78**	19	**49**
78	**90**	48	**78**	18	**47**
77	**90**	47	**77**	17	**45**
76	**89**	46	**77**	16	**43**
75	**88**	45	**76**	15	**41**
74	**88**	44	**76**	14	**39**
73	**87**	43	**75**	13	**36**
72	**87**	42	**75**	12	**34**
71	**86**	41	**74**	11	**32**
70	**86**	40	**73**	10	**29**
69	**86**	39	**73**	9	**27**
68	**85**	38	**72**	8	**24**
67	**84**	37	**71**	7	**21**
66	**84**	36	**71**	6	**18**
65	**84**	35	**70**	5	**15**
64	**84**	34	**69**	4	**12**
63	**83**	33	**68**	3	**9**
62	**83**	32	**67**	2	**6**
61	**83**	31	**66**	1	**3**
60	**82**	30	**65**	0	**0**
59	**82**	29	**64**		
58	**82**	28	**62**